RECENT ADVANCES IN REGRESSION METHODS

STATISTICS: Textbooks and Monographs

A SERIES EDITED BY

Volume 1: The Generalized Jackknife Statistic, *H. L. Gray and W. R. Schucany*

Volume 2: Multivariate Analysis, *Anant M. Kshirsagar*

Volume 3: Statistics and Society, *Walter T. Federer*

Volume 4: Multivariate Analysis: A Selected and Abstracted Bibliography, 1957-1972, *Kocherlakota Subrahmaniam and Kathleen Subrahmaniam* (out of print)

Volume 5: Design of Experiments: A Realistic Approach, *Virgil L. Anderson and Robert A. McLean*

Volume 6: Statistical and Mathematical Aspects of Pollution Problems, *John W. Pratt*

Volume 7: Introduction to Probability and Statistics (in two parts) Part I: Probability; Part II: Statistics, *Narayan C. Giri*

Volume 8: Statistical Theory of the Analysis of Experimental Designs, *J. Ogawa*

Volume 9: Statistical Techniques in Simulation (in two parts), *Jack P. C. Kleijnen*

Volume 10: Data Quality Control and Editing, *Joseph I. Naus* (out of print)

Volume 11: Cost of Living Index Numbers: Practice, Precision, and Theory, *Kali S. Banerjee*

Volume 12: Weighing Designs: For Chemistry, Medicine, Economics, Operations Research, Statistics, *Kali S. Banerjee*

Volume 13: The Search for Oil: Some Statistical Methods and Techniques, *edited by D. B. Owen*

Volume 14: Sample Size Choice: Charts for Experiments with Linear Models, *Robert E. Odeh and Martin Fox*

Volume 15: Statistical Methods for Engineers and Scientists, *Robert M. Bethea, Benjamin S. Duran, and Thomas L. Boullion*

Volume 16: Statistical Quality Control Methods, *Irving W. Burr*

Volume 17: On the History of Statistics and Probability, *edited by D. B. Owen*

Volume 18: Econometrics, *Peter Schmidt*

Volume 19: Sufficient Statistics: Selected Contributions, *Vasant S. Huzurbazar (edited by Anant M. Kshirsagar)*

Volume 20: Handbook of Statistical Distributions, *Jagdish K. Patel, C. H. Kapadia, and D. B. Owen*

Volume 21: Case Studies in Sample Design, *A. C. Rosander*

Volume 22: Pocket Book of Statistical Tables, *compiled by R. E. Odeh, D. B. Owen, Z. W. Birnbaum, and L. Fisher*

Volume 23: The Information in Contingency Tables, *D. V. Gokhale and Solomon Kullback*

Volume 24: Statistical Analysis of Reliability and Life-Testing Models: Theory and Methods, *Lee J. Bain*

Volume 25: Elementary Statistical Quality Control, *Irving W. Burr*

Volume 26: An Introduction to Probability and Statistics Using BASIC, *Richard A. Groeneveld*

Volume 27: Basic Applied Statistics, *B. L. Raktoe and J. J. Hubert*

Volume 28: A Primer in Probability, *Kathleen Subrahmaniam*

Volume 29: Random Processes: A First Look, *R. Syski*

Volume 30: Regression Methods: A Tool for Data Analysis, *Rudolf J. Freund and Paul D. Minton*

Volume 31: Randomization Tests, *Eugene S. Edgington*

Volume 32: Tables for Normal Tolerance Limits, Sampling Plans, and Screening, *Robert E. Odeh and D. B. Owen*

Volume 33: Statistical Computing, *William J. Kennedy, Jr. and James E. Gentle*

Volume 34: Regression Analysis and Its Application: A Data-Oriented Approach, *Richard F. Gunst and Robert L. Mason*

Volume 35: Scientific Strategies to Save Your Life, *I. D. J. Bross*

Volume 36: Statistics in the Pharmaceutical Industry, *edited by C. Ralph Buncher and Jia-Yeong Tsay*

Volume 37: Sampling from a Finite Population, *J. Hájek*

Volume 38: Statistical Modeling Techniques, *S. S. Shapiro*

Volume 39: Statistical Theory and Inference in Research, *T. A. Bancroft and C.-P. Han*

Volume 40: Handbook of the Normal Distribution, *Jagdish K. Patel and Campbell B. Read* (in press)

Volume 41: Recent Advances in Regression Methods, *Hrishikesh D. Vinod and Aman Ullah*

OTHER VOLUMES IN PREPARATION

RECENT ADVANCES IN REGRESSION METHODS

Hrishikesh D. Vinod

Economic Analysis Section
American Telephone and Telegraph Company
Piscataway, New Jersey

Aman Ullah

Department of Economics
University of Western Ontario
London, Ontario, Canada

MARCEL DEKKER, INC. New York and Basel

Library of Congress Cataloging in Publication Data

Vinod, Hrishikesh D., [date]
Recent advances in regression methods.

(Statistics, textbooks and monographs ; v. 41)
Includes index.
1. Regression analysis. I. Ullah, Aman. II. Title.
III. Series.
QA278.2.V56 519.5'36 81-15079
ISBN 0-8247-1574-8 AACR2

MARCEL DEKKER, INC.

270 Madison Avenue, New York, New York 10016

Current printing (last digit):

10 9 8 7 6 5 4 3 2

PRINTED IN THE UNITED STATES OF AMERICA

To my parents, Appa and Vahini,
and my wife, Arundhati (H.D.V.)

To my wife, Shobha (A.U.)

PREFACE

Since the exposition of the Gauss Markov theorem, and especially during the past fifty years or so, practically all fields in the natural and social sciences have widely and sometimes blindly used the Ordinary Least Squares (OLS) method. It gives us the Best Linear Unbiased Estimator (BLUE), which is also equivalent to the maximum likelihood (ML) estimator for estimation of the underlying normal regression relationship. It was in the late 1950's that Charles Stein brought to the attention of statisticians the fact that there is a better alternative to OLS under certain conditions. For a decade, his surprising result went largely unnoticed. In the late 1960's a series of papers was published giving interpretations and variants of what began to be called the Stein-Rule estimators. All these developments were primarily aimed at suggesting a family of biased estimators with potentially smaller Mean Squared Error (MSE) than OLS. Independent development parallel to this came from Hoerl and Kennard in 1970 who suggested another biased ridge regression estimator with potentially smaller MSE as an alternative to OLS. Ridge regression is often thought to be particularly applicable when the applied researchers in any field of natural or social science face the problem of multicollinearity in their data — a serious disease. Again, a series of papers appeared on the scene soon after the seminal work on ridge regression by Hoerl and Kennard was published.

Thus, "improved estimation" of the linear regression model has been a focus of scholarly writing over the past twenty-five years, and particularly in the past decade. The number of contributions has been so numerous that almost anyone who is not an active researcher in this area is finding the pace too fast to keep up with. Graduate students are also discovering that more and more of their required readings in standard courses in Statistics, Econometrics, Biomedicine, Psychology, Engineering, and so on are covering these topics; as a rule they have to read them from various journals. Often, the local libraries do not even subscribe to the journals from otherwise remote disciplines that may contain the appropriate references. An increasing number of university departments in various sciences are offering courses involving applications of Stein-Rule, ridge regression, etc., on regression models with their own data.

This book attempts to bring together the recent developments in the area of "improved estimation" and inference in linear models in the form of a graduate level textbook or a supplementary text. We hope that this book will be an important step in bridging the above mentioned gap in the published literature.

Our intended readers are graduate students in Statistics, Economics,

Psychology, Bio-medicine, and Engineering, among others. The book can serve as a text for theoretical and applied courses in Statistics and Econometrics. We hope that this book will also serve as a handy guide to these somewhat difficult theoretical advances for the non-experts who are merely interested in an overview of the theory with an emphasis on potential applications, or as a useful starting point for a more serious research adventure. For the experts in the area it is intended to be a convenient reference book.

The book has thirteen chapters, most of which are based upon the published papers in numerous learned journals. Some of these papers were written by the authors themselves. The chapters highlight the important aspects of the main contributions, comment on the results obtained, and establish a relationship with earlier results. We include the technical aspects of the material in the simplest possible way, and discuss applications. This is one of the reasons why we think that this book is somewhat different from many other advanced books. We give a unified treatment of theory and tests by using a restricted least squares model. The shrinkage factors are used to compare various shrinkage type, ridge and Stein-Rule estimators. There are a number of sections, explanatory notes, figures, etc. included in this book which are simply unavailable in other books. For example, the canonical reduction of the original regression model using the singular value decomposition is explained, which is rarely mentioned in the available textbooks.

The discussion of multicollinearity in the usual textbooks is often misleading and incomplete. The use of eigenvectors and eigenvalues of the correlation matrix among regressors is more reliable than certain other *ad hoc* measures of multicollinearity. This book explains the related material to practitioners in simple geometrical terms.

The Stein-Rule estimator for the mean of the normal variable and its relevance to ordinary regression is not clear from the published literature. Stein's "unbiased" estimate of the Mean Squared Error of arbitrary biased estimators has considerable practical relevance, which is clarified in this book.

In later chapters the biased estimation techniques based on Stein-Rule or ridge regression is shown to lead to "improvements" over the usual estimators in the so-called distributed lag models, the models with autocorrelation and heteroscedasticity, Zellner's seemingly unrelated regression (SUR) equations and simultaneous equations models.

When there are two or more dependent variables, Hotelling's canonical correlation analysis represents a natural multivariate extension of the regression model. This is extensively used in Psychology, Education, etc. However, the coefficients of the fitted model are known to be highly unstable with respect to data perturbations. Some of these difficulties can be avoided by using ridge regression type ideas. This is also true for the so-called discriminant analysis. Our twelfth chapter shows that ridge regression type modifications can make such multivariate techniques more meaningful to the practitioner.

Our final chapter deals with non-normal errors where we have included several new minimax-type results, and a short discussion of a few aspects of robust regression methods.

This book makes extensive use of Kadane's small-sigma asymptotics to analyze the sampling properties of various newer estimators. We have included an explanatory note in Chapter 6 to give an elementary discussion of this topic. We are encouraged by the fact that the main results based on small sigma asymptotics are identical with the corresponding exact result. We recommend it as a valuable research tool. We have also made considerable use of Bayesian methods including Lindley's hyperparameter model. We offer an integration of classical and Bayesian methods without being doctrinaire.

ACKNOWLEDGMENTS

We would like to thank L. Magee, A. Hoque of University of Western Ontario for reading entire early drafts and M. Leques of AT&T for help in preparing the early chapters of the manuscript. R. L. Obenchain, W. E. Taylor, and P. S. Brandon of Bell Labs and M. Burnside and P. B. Linhart of AT&T, have each commented on at least one chapter. C. L. Mallows of Bell Laboratories commented on about half the book. V. K. Srivastava from University of Lucknow, India, has commented in detail on some sections of the manuscript. We are grateful to Chandar, Farebrother, Obenchain and others for permitting us to use some of their unpublished material. David Hendry made critical comments of Chapter 5. R. A. L. Carter commented on some early chapters. M. Gower, G. Moore, C. Patuto and E. Quinzel typed parts of the manuscript with patience, efficiency and skill.

Vinod wishes to thank Bell Laboratories and AT&T for research support. Ullah acknowledges the research support of NSERC, and a leave fellowship from Canada Council. A part of his work on the book was done during his visit to Monash University, Australia and C.O.R.E. in Belgium. The research facilities at both these places are greatly appreciated.

Although many colleagues, students and friends have contributed directly or indirectly in the preparation of this book, they bear no responsibility whatever for any shortcomings or errors.

Hrishikesh D. Vinod
Aman Ullah

TABLE OF CONTENTS

Preface v

1. Linear Regression Model 1

1.1 Introduction 1
1.2 Specification of the Model 2
1.3 Canonical Reduction Using Singular Value Decomposition 5
1.4 Standardization of the Model 8
1.5 A Simple Example of Two Regressors 10

Explanatory Notes 17
Exercises 25

2. Criteria for Good Regression Estimators: MSE, Consistency, Stability, Robustness, Minimaxity and Bayesian 'MELO'ness 29

2.1 Introduction 29
2.2 Univariate Concepts 30
2.3 Multivariate Concepts 31
2.4 Asymptotic Properties 36
2.5 Efficiency (Minimum Variance) Properties of OLS 38
2.6 Stability Properties 40
2.7 Admissibility Properties (Unbeaten Somewhere) 45
2.8 Minimax Properties 47
2.9 Acceptable Range of Shrinkage Factor for Reducing the MSE of OLS 49
2.10 Bayes Theorem 55

Exercises 58

3. Restricted Least Squares and Bayesian Regression 61

3.1 Restricted Least Squares Estimator (RLS) 61
3.2 General Linear Hypothesis Tests 63
3.3 Properties of Restricted Estimators (RLS) 65
3.4 Mixed Regression, Stochastic Restrictions Model 70

3.5 Preliminary Test Estimator (PT) 76
3.6 Bayesian Analysis of Regression 77
3.7 Inequality Restricted (IR) Least Squares Estimator 87
3.8 Need for Judgment in Empirical Research 88

Exercises 89

4. Autoregressive Moving Average (ARMA) Regression Errors and Heteroscedasticity 90

4.1 Durbin-Watson-Vinod-Wallis Tests 90
4.2 Tests for Regression Coefficients 93
4.3 Derivation of Bounds on F Tests Based on Eigenvalues of Ω 94
4.4 Eigenvalues of Ω for ARMA Processes and Tabulation of Bounds 99
4.5 Estimation Allowing for AR(1) Errors 102
4.6 "Exact" DW Test Without Tables 107
4.7 Comments on MA Errors 109
4.8 Comments on General AR and/or MA processes 109
4.9 The Heteroscedasticity Problem 111

Exercises 118

5. Multicollinearity and Stability of Regression Coefficients 120

5.1 What is Multicollinearity? 120
5.2 Multicollinearity and Inestimability 123
5.3 Effects of Multicollinearity 124
5.4 An Example of the Effect of Perturbation 129
5.5 Final Remarks 129

Exercises 131

6. Stein-Rule Shrinkage Estimator 134

6.1 Motivation for Shrinkage: Baseball Example 134
6.2 Stein-Rule in the Regression Context 142
6.3 Properties of the Stein-Rule Estimator 149
6.4 Some Extensions of the Stein-Rule Estimator 153
6.5 Final Remarks 158

Explanatory Notes 159
Exercises 167

7. Ridge Regression 169

7.1 Introduction and Historical Comments 169
7.2 Properties of the Ridge Estimator and Choice of Biasing Parameters k 173
7.3 Bayesian and Non-Bayesian Interpretation of Ridge Methods 183
7.4 Computational Shortcuts for Ordinary Ridge 189
7.5 An Illustrative Example for Ridge Solution Selection 191

Explanatory Note 200
Exercises 203

8. Further Ridge Theory and Solutions 206

8.1 Introduction 206
8.2 Operational Values of Shrinkage Factors 208
8.3 Bias and MSE of Double f-Class Generalized Ridge Estimator 210
8.4 Bias and MSE for "Operational" Ordinary Ridge Estimator 213
8.5 MSE Matrix for Generalized Ridge Involving a Combination of OLS and Restricted Least Squares Estimators 217

Explanatory Note 222
Exercises 225

9. Estimation of Polynomial Distributed Lag Models 226

9.1 Introduction 226
9.2 Estimation Under Nonstochastic Smoothness Priors 227
9.3 Estimation Under Stochastic Smoothness Priors 229
9.4 Ridge Estimation Without Smoothness Priors 237
9.5 Ridge Estimation with Nonstochastic Smoothness Priors 237
9.6 Final Remarks 238

Exercises 238

10. Multiple Sets of Regression Equations 241

10.1 Introduction 241
10.2 Seemingly Unrelated Regressions (SUR) Model 242
10.3 Temporal Cross-section Model 248
10.4 Random Coefficient Regression Model 251
10.5 Lindley's Hyperparameter Model 255

Exercises 257

11. Simultaneous Equations Model 262

11.1 Model Specification 262
11.2 Further Specifications of the System 265
11.3 Problem of Identification 267
11.4 Estimation of Structural Systems 271
11.5 Full Information Estimators 285

Exercises 289

12. Canonical Correlations, and Discriminant Analysis with Ridge-Type Modification 292

12.1 Canonical Correlations 292
12.2 Discriminant Analysis 300

Exercise 303

13. Improved Estimators Under Nonnormal Errors and Robust Regression 305

13.1 Introduction 305
13.2 A General Class of Improved Estimators 306
13.3 Properties of β Under Nonnormal Disturbances 308
13.4 A Review of Minimaxity Results Under Normal and Nonnormal Errors 313
13.5 Robust Location Estimators for Nonnormal Mean 317
13.6 Robust Regression for Nonnormal Errors 323
13.7 Iteratively Reweighted Least Squares 325
13.8 Nonnormal Errors in Multivariate Models 326

Explanatory Notes 326
Exercises 331

References 333

Index 359

RECENT ADVANCES IN REGRESSION METHODS

1

Linear Regression Model

1.1 INTRODUCTION

The least squares regression model is used for studying various types of relations including technical, behavioral, static, and dynamic ones, both at the individual (micro) and collective (macro) levels. Adrian Marie Legendre [1805] was the first to publish results concerning the method of least squares, although others may have used it before him; e.g. Galileo Galilei [1632] came very close to proposing a theory of errors related to the least squares method. Gauss [1806] postulates that when any number of equally good data regarding an unknown quantity are available, their arithmetic mean is the "most probable" value. From this, Gauss derives his normal law of error or what we call the Gaussian or the normal distribution. Later Gauss [1823] and Markoff [1900] developed the theory of least squares for estimation of parameters in a general linear regression model which has proved to be useful in a great many fields of application.

In this chapter we study the general linear regression between a dependent variable and a set of explanatory variables (regressors). Various canonical forms of this relationship are given. Further, several concepts related to canonical reductions are explained both algebraically and geometrically by using a simple example of two regressors. Since this topic is covered extensively in various textbooks, we have concentrated on items that may not be generally available.

1.2 SPECIFICATION OF THE MODEL

Consider a linear regression model as

$$y_t = \beta_1 x_{1t} + \beta_2 x_{2t} + \cdots + \beta_p x_{pt} + u_t \tag{1}$$

where $\beta_1, \ldots, \beta_p$ are unknown regression coefficients; y_t are observations on the dependent variable; x_{it} are observations on the p regressors; u_t are true unknown errors (shocks or disturbances); $t = 1, \ldots, T$ is the index for numbering observations, and $i = 1, \ldots, p$ is the index for numbering the variables and their coefficients. We may rewrite (1) in matrix notation as

time

parameters

$$y = X\beta + u \tag{2}$$

where y is a $T \times 1$ vector of observable random variables; X is a $T \times p$ matrix of known constants (non-stochastic regressors); β is a $p \times 1$ vector of unknown regression coefficients; u is a $T \times 1$ vector of disturbances.

We often use the following conventional assumptions:

Assumption 1 Non-stochastic X

X is a non-stochastic matrix of regressors.

Assumption 2 Full rank of $X'X$

The rank of X is p.

Assumption 3 Asymptotically full rank of $X'X/T$

$$\lim_{T\to\infty} \frac{X'X}{T} = Q$$

where Q is a finite and non-singular matrix of rank p. This assumption is required primarily for consistency and other asymptotic results in Econometrics mentioned in later chapters.

Assumption 4 Normality of the disturbances

$$u \sim N(0, \sigma^2 I).$$

The $T \times 1$ vector u of errors has a multivariate normal distribution.

This assumption is required primarily for various tests of significance. It also implies that the disturbances are homoscedastic, i.e. $V(u_t) = \sigma^2$ for all $t = 1, \ldots, T$, and that they are serially independent, i.e. $cov(u_t, u_{t'}) = 0$ for all $t \neq t' = 1, \ldots, T$.

Most of the estimators discussed in the book have been given non-Bayesian interpretations and do not need the normality assumption for studying their properties. Chapter 13 deals with the implications of non-normality. This assumption of normality is generally essential for maximum likelihood estimation. Explanatory Note 1.1 at the end of this chapter clarifies the concept of maximum likelihood estimation for readers who might not be familiar with it.

Under an appropriate subset of these assumptions the conditional expectation of y given the regressors X, and the conditional variance covariance matrix of y are:

$$E(y|X) = X\beta \; ; \; V(y|X) = E[y - E(y|X)][y - E(y|X)]' = \sigma^2 I \qquad (3)$$

Thus we have $p + 1$ "true unknown" parameters β and σ^2 in our model (2).

A condition that will be used occasionally is that one of the regressors is a constant term; that is, $x_{it} = 1$ for all t and for some $1 \le i \le p$. Alternatively, we could add an intercept in (2) and write $y = \beta_0 + X\beta + u$. This has been taken up in a later section.

The ordinary least squares (OLS) estimator of $\beta_1, \ldots, \beta_p$ which minimizes

$$S = \sum_{t=1}^{T} (y_t - \beta_1 x_{1t} - \cdots - \beta_p x_{pt})^2 , \qquad (4)$$

or in matrix notation: $S = (y - X\beta)'(y - X\beta)$, is obtained by taking the derivative of S with respect to β and equating it to zero. This gives for $\beta = b$,

$$X'Xb = X'y \qquad (5)$$

a set of equations known as the least squares normal equations. Now, if X is of rank p, $(X'X)^{-1}$ exists and we have

$$b = (X'X)^{-1}X'y \qquad (6)$$

which is called the OLS estimator. This value of β corresponds to a minimum of S because the second-order partial derivative of S with respect to β is a positive definite matrix $2X'X$.

Once β has been estimated by b, we can write

$$\hat{u} = y - Xb \qquad (7)$$

as the estimator for the error (residual) vector u. Further, the sum of squares of estimated residuals divided by $T - p$, viz.,

$$s^2 = \frac{\hat{u}'\hat{u}}{T - p} \qquad (8)$$

can be shown to be a consistent and unbiased estimator of σ^2. We note that the sum of the estimated residuals (unless one of the regressors is a constant) is not zero. However, $X'\hat{u} = 0$. Thus, on the basis of the estimated regression $y = Xb + \hat{u}$,

$$y'y = b'X'Xb + \hat{u}'\hat{u} \qquad (9)$$

where $y'y$ is the total sum of squares (SST), $b'X'Xb$ is the sum of squares due to regression (SSR), and $\hat{u}'\hat{u}$ is the sum of squares due to errors (SSE). The multiple correlation coefficient (a measure of goodness-of-fit) is then defined as

$$R^2 = \frac{b'X'Xb}{y'y} = \frac{SSR}{SST} = 1 - \frac{\hat{u}'\hat{u}}{y'y} \tag{10}$$

This is interpreted as a measure of the proportion of variation due to regression to the total variation in y around the value zero. If one of the x's is constant then R^2 is usually defined as $1 - \dfrac{\hat{u}'\hat{u}}{(y - \bar{y})'(y - \bar{y})}$ where $\bar{y}$ is the mean value of the observations on y.

Regarding the sampling properties of b we note that

$$E(b) = E(X'X)^{-1}X'y = (X'X)^{-1}X'E(y) = \beta \tag{11}$$

and the variance covariance matrix

$$\begin{aligned} V(b) &= E(b - \beta)(b - \beta)' = (X'X)^{-1}X'E(uu')X(X'X)^{-1} \\ &= \sigma^2(X'X)^{-1} . \end{aligned} \tag{12}$$

The OLS estimator b is the best linear unbiased estimator in the sense that in the class of linear and unbiased estimators this has the smallest variance. The proof for this is given in Chapter 2.

Under assumption 4 the OLS estimator is the same as the maximum likelihood (ML) estimator. To show this we first note that the vector y follows a multivariate normal distribution with the mean vector $X\beta$ and variance covariance matrix $\sigma^2 I$, i.e., $y \sim N(X\beta, \sigma^2 I)$. Thus we can write the likelihood function as

$$\begin{aligned} L &= L(\beta, \sigma^2|y) = f(y|\beta, \sigma^2) = f(y_1, \ldots, y_T|\beta, \sigma^2) \\ &= \frac{1}{(2\pi)^{T/2}(\sigma^2)^{T/2}} \exp\left[-\frac{1}{2\sigma^2}(y - X\beta)'(y - X\beta)\right] \\ &= \frac{1}{(2\pi)^{T/2}(\sigma^2)^{T/2}} \exp\left[-\frac{1}{2\sigma^2}\sum_{t=1}^{T}(y_t - \beta_1 x_{1t} \cdots - \beta_p x_{pt})^2\right] \end{aligned}$$

and the log of likelihood as

$$\log L = \log L(\beta, \sigma^2|y) = -\frac{T}{2}\log(2\pi\sigma^2) - \frac{1}{2\sigma^2}(y - X\beta)'(y - X\beta) .$$

It is then clear that for any given σ^2 the maximization of $\log L$ with respect to β is the same as the minimization of $S = (y - X\beta)'(y - X\beta)$. Thus the ML estimator of β will be identical to its OLS estimator, viz., b. One can also verify that $\partial \log L/\partial\beta = 0$ and $\partial \log L/\partial\sigma^2 = 0$ provide the ML estimators of β and σ^2, respectively, as b and

$$\hat{\sigma}^2 = \frac{1}{T}(y - Xb)'(y - Xb) . \tag{13}$$

We note that in the expression for $\hat{\sigma}^2$ the divisor is sample size T whereas in s^2 it is $T - p$, the degrees of freedom. Further, although $E(s^2) = \sigma^2$, in the case of $\hat{\sigma}^2$, $E(\hat{\sigma}^2) \neq \sigma^2$.

1.3 CANONICAL REDUCTION USING SINGULAR VALUE DECOMPOSITION

It is always possible to write the $T \times p$ matrix X of regressors as

$$X = H \Lambda^{1/2} G' , \tag{14}$$

where the columns of the $T \times p$ matrix H are the sample principal coordinates of X standardized in the sense that $H'H = I$; the $p \times p$ diagonal matrix $\Lambda^{1/2}$ gives the ordered singular values of X, $\lambda_1^{1/2} \geq \cdots \geq \lambda_p^{1/2}$; the columns of the $p \times p$ matrix G (rows of G'), say g_i, give the direction cosine vectors (see Explanatory Note 1) which orients the ith principal axes of X with respect to the given axes, and G is orthogonal. Writing $X'X = G \Lambda G'$, it is clear that the squared singular values λ_i are the eigenvalues of $X'X$, and the columns of G are eigenvectors of $X'X$.

The above decomposition is called the singular value decomposition (SVD) of X (Belsley and Klema [1974], Rao [1973, p. 42]). It has assumed prominence recently because it allows considerable conceptual flexibility. Furthermore, it is possible to implement this decomposition using modern computers by a simple call to a subroutine (Businger and Golub, 1965). The novelty here is in the matrix H, not in G, since the latter is the well-known matrix of eigenvectors noted below, in Explanatory Note 1.2.

The SVD of (14) implies the following for the OLS estimator. We can rewrite (6) as

$$b = (X'X)^{-1}X'y = G \Lambda^{-1/2} H'y . \tag{15}$$

Similarly, (12) can be written as

$$\begin{aligned} V(b) &= \sigma^2 (X'X)^{-1} = \sigma^2 \, G \Lambda^{-1} G' , \\ &= \sigma^2 \sum_{i=1}^{p} \lambda_i^{-1} g_i g_i' . \end{aligned} \tag{16}$$

1.3.1 Parameterization Giving Uncorrelated Components of b

Upon substitution of the SVD of (14) into $y = X\beta + u$ of (2) we have

$$y = H \Lambda^{1/2} G' \beta + u . \tag{17}$$

Instead of treating β as the unknowns, we may consider a different parameterization where we define

$$\gamma = G'\beta \tag{18}$$

as the unknown $p \times 1$ vector of elements $\gamma_1, \gamma_2, \ldots, \gamma_p$. Now (17) becomes

$$y = X^* \gamma + u \tag{19}$$

where

$$X^* = H\Lambda^{1/2} \tag{20}$$

such that

$$X^{*\prime}X^* = \Lambda^{1/2}H'H\Lambda^{1/2} = \Lambda \tag{21}$$

is a diagonal matrix. This parameterization is popular in ridge regression literature (see Chapter 6) although its relation to the original parameterization is not clearly expressed when one does not exploit the SVD. This parameterization has regressors $x_1^*, x_2^*, ..., x_p^*$ comprising the columns of X^* which are uncorrelated (but not orthonormal).

The OLS estimator of the transformed parameter γ is given by

$$c = G'b = G'G\Lambda^{-1/2}H'y = \Lambda^{-1/2}H'y \tag{22}$$

upon substituting b for β in (18). Alternatively, using the SVD we can write the estimator of γ directly from (19) as

$$c = (X^{*\prime}X^*)^{-1}X^{*\prime}y = \Lambda^{-1/2}H'y \tag{23}$$

where we use (20) and (21). Similarly, the variance-covariance matrix for c is

$$V(c) = \sigma^2(X^{*\prime}X^*)^{-1} = \sigma^2\Lambda^{-1} \tag{24}$$

which is diagonal. This shows the important property of this parameterization, that the elements $c_1, c_2, ..., c_p$ of c are uncorrelated.

Now, note that $c = G'b$ implies that

$$b = Gc = \sum_{i=1}^{p} g_i c_i = G\Lambda^{-1/2}H'y \tag{25}$$

where we premultiply by G and use the fact that $G'G = I$. Thus, c_i may be regarded as uncorrelated components of b. The jth element b_j is a weighted sum of $c_1, c_2, ..., c_p$ with the weights given by the jth row of G. In other words,

$$b_j = \sum_{i=1}^{p} g_{ji} c_i \ . \tag{26}$$

If one is worried about the signs or relative magnitude of the elements b_i, it is instructive to study the matrix of weights G, i.e, the eigenvectors of $X'X$. This will be explained later with the help of an example. Also, we note that because (25) is the same as (15) one can estimate the model (19) first and then get back to (2) through (25).

1.3.2 Mean of a Multinormal Variable Parameterization

Another parametrization often considered in the literature is in terms of the mean of a multivariate normal variable. Denote

$$\alpha = \Lambda^{1/2}G'\beta \tag{27}$$

to write $y = X\beta + u$ as $y = H\alpha + u$. For this parameterization note that $H'H = I$ implies that the regressors $h_1, h_2, ..., h_p$ comprising the columns of H are orthogonal. However, reducing the original problem $y = X\beta + u$ to $y = H\alpha + u$, which has orthogonal regressors, does not mean that we have eliminated the multicollinearity problem implicit in the eigenvalue spectrum $\lambda_1, \lambda_2, ..., \lambda_p$ of these data.

For example, let λ_p be very small. Now consider the OLS estimator of α:

$$a = \Lambda^{1/2}G'b = \Lambda^{1/2}G'G\ \Lambda^{-1/2}H'y = H'y\ , \tag{28}$$

where we substitute b for β in (27). Clearly, a is indirectly affected by the high variance arising from smallness of λ_p because b is affected by it. The simplicity of $a = (H'H)^{-1}\ H'y = H'y$ with orthogonal regressors is deceptive, and the multicollinearity problem has been simply swept under the rug.

Next, we note that the variance-covariance matrix for a is

$$V(a) = \sigma^2 I\ . \tag{29}$$

Thus, we have

$$a \sim N(\alpha, \sigma^2 I) \tag{30}$$

and the problem of estimating α is similar to that of estimating the (simple) mean of a p-variate normal variable a with the common variance σ^2 for all p variables. This also implies that we are dealing with the estimation of α in the model.

$$H'y = H'H\alpha + H'u$$

or

$$a = \alpha + H'u$$

where $H'u \sim N(0,\ \sigma^2 I)$.

1.3.3 Decomposition of R^2

It is instructive to note the following relations. Denote

$$R_{yi} = h_i'\,y\,(y'y)^{-1/2} = h_i'\,y/||y|| \tag{31}$$

as the sample correlation coefficient between y and the ith principal coordinates of X contained in the ith column of H; $||y|| = (y'y)^{1/2}$ is the Euclidean length of y. For the OLS estimator, the regression sum of squares from (9) is

$$SSR = b'X'Xb = y'X(X'X)^{-1}X'y\ . \tag{32}$$

where $b = (X'X)^{-1}X'y$. Using $X = H\Lambda^{1/2}G'$, $G'G = I$ and $H'H = I$, we can write (32) as

$$SSR = y'HH'y \; . \tag{33}$$

where HH' is a $T \times T$ matrix (not identity) which should not be confused with the $p \times p$ matrix $H'H = I$.

Thus, the squared multiple correlation from (10) is given by $R^2 = \frac{SSR}{SST} = \frac{y'HH'y}{y'y} = \sum_{i=1}^{p} R_{yi}^2$. This shows that SVD decomposes R^2 into p components measuring the contribution of each principle direction. Clearly, the relatively large values of R_{yi}^2 are important for the basic model.

1.4 STANDARDIZATION OF THE MODEL

By standardization we mean here changing the origin and sometimes also the scale of data on y and on the x's. The change of origin is normally done by taking the deviations from their respective means, and the scale is changed by dividing by the respective standard deviation. We must note, however, that one should ensure before doing standardization that one of the x's is constant or alternatively that there is an intercept in (2). If it is not so, the estimation of the β's from standardized data will not be the same as that from unstandardized data. Thus let us write first the model (1) as

$$y_t = \beta_0 + x_{1t}\beta_1 + \cdots + x_{pt}\beta_p + u_t \tag{34}$$

where β_0 is a scalar intercept.

It would be convenient to standardize (34) as

$$y_t^s = y_t - \bar{y}; \quad x_{it}^s = (x_{it} - \bar{x}_i) \tag{35}$$

where the bar denotes the sample mean, s represents standardized and $i = 1, \ldots, p$ as before.

Now the standardized model in the matrix notation is

$$y^s = X^s\beta + u^s \tag{36}$$

where y^s is $T \times 1$ vector with the elements y_t^s, X^s is $T \times p$ with the elements x_{it}^s, β is as before, and u^s is a $T \times 1$ vector with the elements $u_t^s = u_t - \bar{u}$. Alternatively, since $y^s = Ay$, $X^s = AX$ and $u^s = Au$, one can write (36) as

$$Ay = AX\beta + Au \tag{37}$$

where

$$A = [I - \frac{\iota\iota'}{T}] \tag{38}$$

is an $T \times T$ idempotent matrix such that $A = A^2$; ι is an $T \times 1$ column vector of T unit elements such that $\iota'y^s = 0$ and $\iota'X^s = 0$. For readers who may not be familiar with the projection matrix A we have included explanatory Note 1.5 and Exercise 1.7 at the end of the chapter. We note that

$$E(Au) = 0 \quad \text{and} \quad V(Au) = \sigma^2 A \ . \tag{39}$$

The OLS estimator of β is

$$b = (X^{s\prime} X^s)^{-1} X^{s\prime} y^s = (X' A X)^{-1} X' A y \ . \tag{40}$$

The estimator for β_0 is simply

$$b_0 = \bar{y} - \sum_{i=1}^{p} b_i \bar{x}_i = \frac{1}{T} \iota'(y - Xb) \tag{41}$$

because β_0 equals $\bar{y} - \sum_{i=1}^{p} \beta_i \bar{x}_i$.

If we do not standardize the model (34), it can be written as

$$y = X^* \delta + u; \quad X^* = [\iota \ X], \quad \delta = \begin{bmatrix} \beta_0 \\ \beta \end{bmatrix} . \tag{42}$$

In this case the OLS estimator of δ is

$$\begin{aligned} d = \begin{bmatrix} b_0 \\ b \end{bmatrix} &= (X^{*\prime} X^*)^{-1} X^{*\prime} y \\ &= \begin{bmatrix} \iota'\iota & \iota' X \\ X'\iota & X'X \end{bmatrix}^{-1} \begin{bmatrix} \iota' y \\ X' y \end{bmatrix} . \end{aligned} \tag{43}$$

Using matrix inversion, one can obtain from above the expressions for b_0 and b. These will be identical with (41) and (40), respectively.

An alternative standardization, which converts the matrix $X'X$ into a correlation matrix and is often used in applied work, is as follows.

Let y_t^s denote the standardized y values as before, and let

$$z_{it} = \frac{x_{it} - \bar{x}_i}{\sqrt{T-1}(SD)_i} \ ; \quad (SD)_i^2 = \frac{1}{T-1} \sum_{t=1}^{T} (x_{it} - \bar{x}_i)^2 \ . \tag{44}$$

Then we can write the model (34) as

$$y_t^s = z_{1t} \beta_1^s + \cdots + z_{pt} \beta_p^s + u_t^s \tag{45}$$

where

$$\beta_i^s = \sqrt{T-1} \ (SD)_i \beta_i \tag{46}$$

are the standardized parameters. In the matrix notation

$$y^s = Z \beta^s + u^s \tag{47}$$

where $y^s = Ay$, $Z = AXD^{-1}$, $\beta^s = D\beta$, $u^s = Au$ and

$$D = \sqrt{T-1} \begin{bmatrix} (SD)_1 & \cdot & 0 \\ \vdots & \cdot & \\ 0 & 0 & (SD)_p \end{bmatrix} . \tag{48}$$

is a $p \times p$ diagonal matrix. In this case the standardized matrix $Z'Z$ is the correlation matrix. Note that if we had not used $\sqrt{T-1}$ in the denominator of z_{it} (i.e. had only considered $(SD)_i$), then the resulting matrix for $X'X$ would not have led to a correlation matrix.

The OLS estimator of β^s is

$$\begin{aligned} b^s &= (Z'Z)^{-1}Z'y^s \\ &= (D^{-1}X'AXD^{-1})^{-1}\, D^{-1}X'Ay \\ &= D(X'AX)^{-1}\, X'Ay \end{aligned} \tag{49}$$

and that of the original parameter β is

$$\begin{aligned} b &= D^{-1}b^s \\ &= (X'AX)^{-1}\, X'Ay \ . \end{aligned} \tag{50}$$

This is identical to (40). Thus the standardization of data by either method ultimately gives equivalent OLS results, and they both are identical with the OLS results derived directly from (34) using unstandardized data.

1.5 A SIMPLE EXAMPLE OF TWO REGRESSORS

When $p = 2$, the (2×2) matrix $Z'Z$ has unity along the diagonal, and r_{12}, the correlation coefficient between z_1 and z_2, is the off-diagonal element. When multicollinearity is present r_{12} becomes large. As an example of multicollinear data consider the United States data for 1921-1929 ($t = 1, ..., 9$) which is a subset of L. R. Klein's [1950] famous data in his first econometric model (see Theil [1971], p. 456).

Let y_t = (41.9, 45, 49.2, 50.6, 52.6, 55.1, 56.2, 57.3, 57.8) denote aggregate consumption, x_{1t}= (12.4, 16.9, 18.4, 19.4, 20.1, 19.6, 19.8, 21.1, 21.7) denote aggregate income from profits, and x_{2t}= (28.2, 32.2, 37, 37, 38.6, 40.7, 41.5, 42.9, 45.3) denote aggregate income from wages. The means and standard deviations are: $\bar{y} = 51.744$, $\bar{x}_1 = 18.822$, $\bar{x}_2 = 38.156$, $SD(y) = 5.5853$, $SD(x_1) = 2.7865$, $SD(x_2) = 5.3463$. Also, $r_{12} = 0.943031$ which is large, but is not the limiting value (unity). The Euclidean length of the vector y is $(y'y)^{1/2} = 15.7976$. Denote standardized variables by $z_{it} = (x_{it} - \bar{x}_i)/\sqrt{T-1}\,(SD)_i$ for $i = 1, 2$, as before. This standardization serves the purpose of ensuring that $Z'Z$ is a correlation matrix. The numerical values are exhibited in the Explanatory Note 1.3.

Figure 1.1 is a 3-dimension plot of unstandardized y_t and standardized z_{1t} and z_{2t} where one pillar is plotted for each of the nine observations, with some smoothing when two pillars are close to each other to give a proper perspective. Figure 1.2 plots z_{1t} against z_{2t} and shows the two principal axes which involves a counterclockwise rotation of the original axes about the origin through an angle $\theta = \pi/4 = 45$ degrees preserving their orthogonality. This rotation is explained in Explanatory Note 1.4. The coordinates in terms of the rotated axes are $w(z_{1t}, z_{2t})$ and $w(-z_{1t}, z_{2t})$, respectively, where w denotes

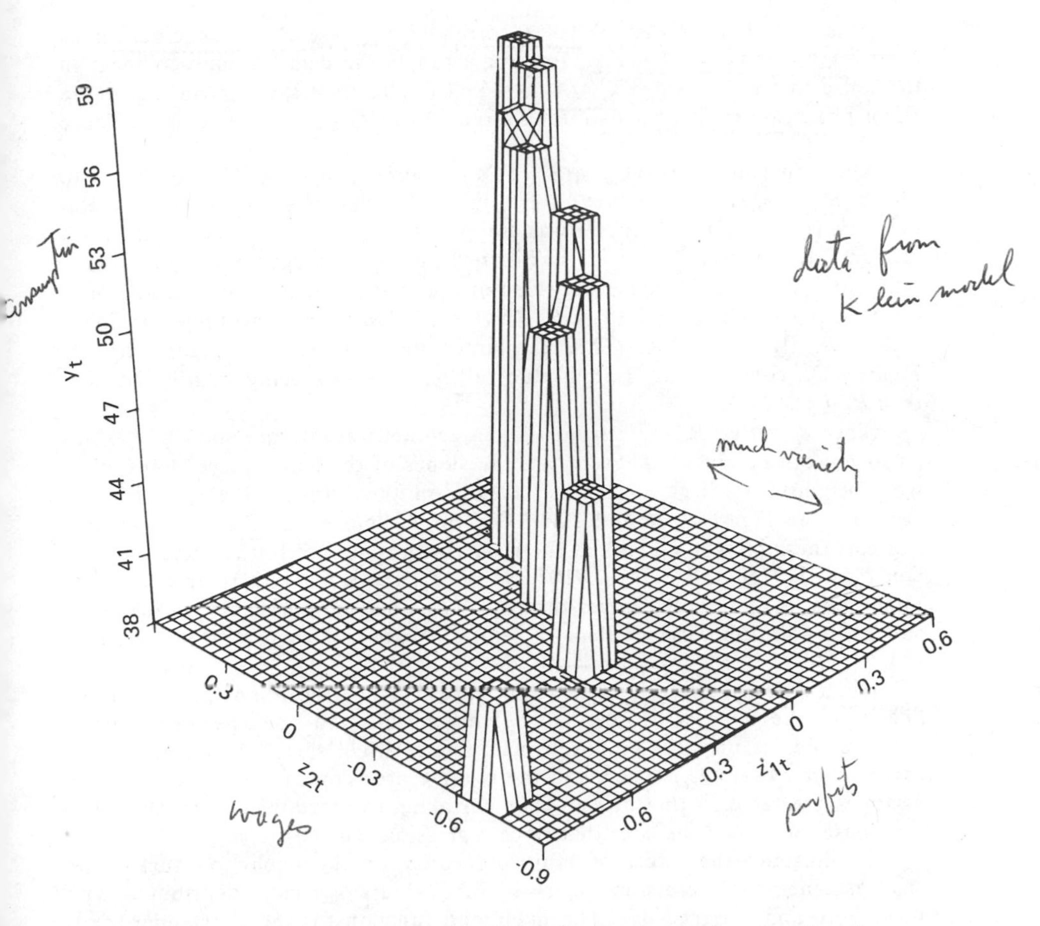

Figure 1.1 Pillar Plot or Ray y and Standardized z_1 and z_2 for $T = 9$

$\cos\theta = \sin\theta = 1/\sqrt{2}$, since $\theta = \pi/4$. Algebraically, these coordinates are a transformation from Z to ZG where $G = [g_1, g_2]$ is the matrix of eigenvectors. For $p = 2$, $g_1' = (\cos\theta, \sin\theta) = w(1,1)$ and $g_2' = (-\sin\theta, \cos\theta) = w(-1, +1)$. The SVD of (14) implies that $ZG = H\Lambda^{1/2}$, therefore the principal axes are marked $h_i\lambda_i^{1/2}$, $(i = 1, 2)$ in Figure 1.2. The numerical values of h_i are found in the Explanatory Note 1.3. The co-ordinates of the nine data points given in the two columns of H are scaled to give equal "spread" along the two axes. Note that "spread" is a geometrical concept suggesting the width of the region wherein the data points lie. It is seen to be proportional to $\lambda_i^{1/2}$ in

Figure 1.2.

It is well known that a 2×2 correlation matrix $Z'Z$ has eigenvalues $\lambda_1 = 1 + r_{12}$ and $\lambda_2 = 1 - r_{12}$. For our example the data are multicollinear in the sense that r_{12} is large; $\lambda_1^{1/2}/\lambda_2^{1/2} = 5.84$ implies that the "spread" along the major principal axis is over five times larger than along the minor axis in Figure 1.2.

Recall that the (relatively large) $T \times p$ matrix H need not be computed to obtain $R_{yi} = h_i' y / ||y||$. However, for this example, $H = [h_1, h_2]$, with elements h_{it} ($i = 1, 2; t = 1, ..., 9$) was computed to be able to plot y_t against h_{1t} in Figure 1.3a and y_t against h_{2t} in Figure 1.3b. Now the geometrical representation of R_{yi} is clear, because the slope of the least squares regression line of y_t on h_{1t} drawn in Figure 1.3a is $R_{y1} ||y|| = 15.298$. A similar slope in Figure 1.3b is $R_{y2} ||y|| = 3.0816$. Thus the larger slope of the least squares line in Figure 1.3a compared to 1.3b is due to $R_{y1}(= .9684)$ being relatively larger than $R_{y2}(= .1951)$.

Since $c_i = ||y|| R_{yi}/\lambda_i^{1/2}$ from (22), a geometrical interpretation of c_i similar to R_{yi} can be given. These c_i are the slopes of the least squares lines when we plot unstandardized y_t against $h_{it}\lambda_i^{1/2}$ on the horizontal axis instead of against h_{it} in Figures 1.3a and 1.3b. Our three dimensional Figure 1.1 already suggests these plots if we look at the y data with reference to the principal axes. Our $c_1 = 10.9748$ and $c_2 = 12.9107$. Now, c_2, the slope with reference to the minor axis, is numerically larger than c_1, but c_2 may be unreliable because c_2 is subject to a greater variance than c_1. Note that $var(c_2)/var(c_1) = 34.1068$ (see (24)).

Geometrically, there is inadequate "spread" along the minor principal axis, and hence the slope c_2 is "seen" to be very sensitive to small perturbations of the data, i.e., is unstable or unreliable. The relative "spread" along the major axis is over five times greater, and the relative precision of the slope c_1 is over 34 times greater than that of c_2. Thus shrinking c_2 more than c_1 makes intuitive sense. We shall see later that ridge regression achieves this.

To illustrate the effect of multicollinearity on the likelihood surface we shall assume that regression errors u_t of (34) are normally distributed with mean zero and variance σ^2. The likelihood function for the (unstandardized) model (34) in terms of the original units of measurement is

$$(2\pi\sigma^2)^{-T/2} \exp\left[\frac{-1}{2\sigma^2} \sum_{t=1}^{T} (y_i - \beta_0 - \beta_1 x_{1t} - \beta_2 x_{2t})^2\right].$$

Using the OLS estimate 1.0056 for σ^2, setting $\beta_0 = \bar{y} - \beta_1\bar{x}_1 - \beta_2\bar{x}_2$, taking the natural log, and ignoring the term $(-T/2)ln\ 2\pi$, we have a slightly modified log likelihood function.

Now, truncating the depth of the plot at -10 we have a three-dimensional plot of the slightly modified log likelihood in Figure 1.4a. The scale on the base is chosen in such a way that, roughly speaking, the center corresponds to the OLS estimates: $b_1 = -.17$ and $b_2 = 1.12$. The standard errors of the OLS

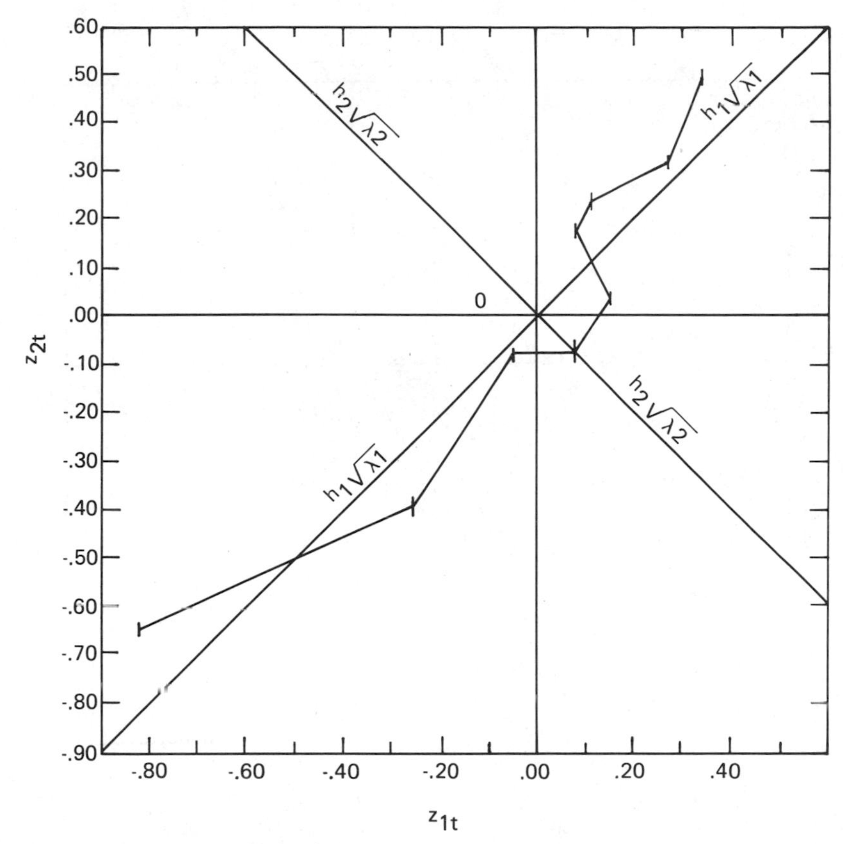

Figure 1.2 Plot of Standardized Regressor Data: Greater Spread Along First Principal Axis

estimates are: $SE_1 = 0.0382$ and $SE_2 = 0.1993$, and the range of β_i in the plot is approximately $\pm 3\ SE_i$.

The OLS solution b_1 is at the maximum of the log likelihood surface. We show below that from the viewpoint of economic theory b_i can be naively misinterpreted. In Keynesian terminology β_1 is the marginal propensity to consume from profit income. The estimate $b_1 = -.17$ may be naively interpreted to mean that when the income from profits increases by one dollar, consumption declines by 17 cents. Similarly, $b_2 = 1.12$ may suggest that an increase of one dollar in wage earner's income leads to an increase in their consumption of one dollar and 12 cents. Since this implies imminent bankruptcy of our wage

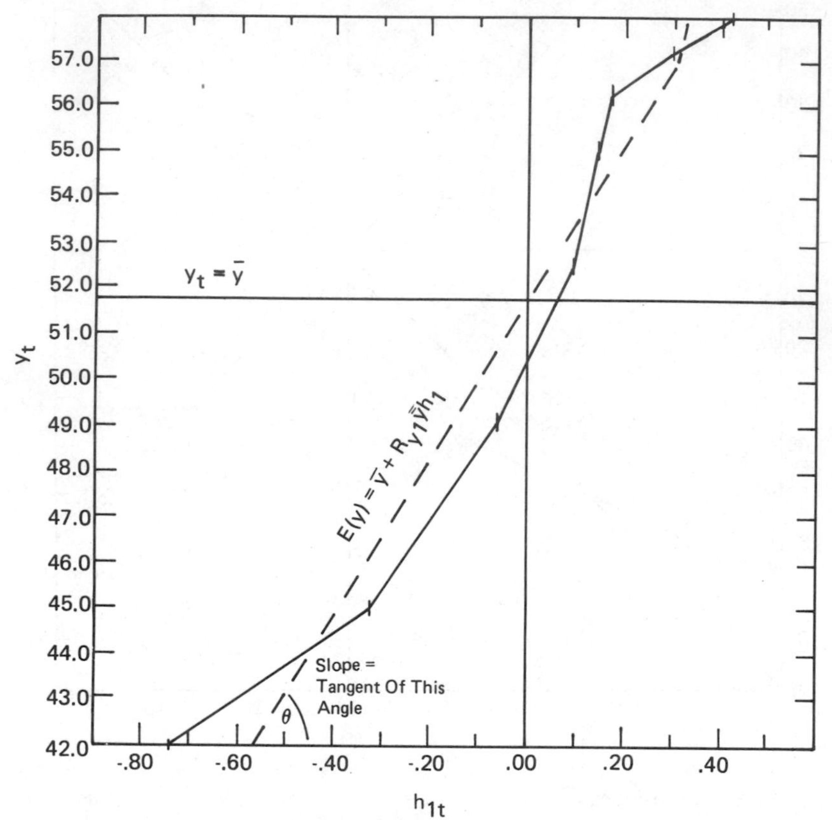

Note: This Slope is larger than in Figure 1.3b

$$\bar{\bar{y}} = \left(\sum_{1}^{T} y_t^2\right)^{1/2}$$

Figure 1.3a Relation Between y_t and h_{1t}

earner, the OLS solution is unacceptable.

The purpose in plotting the likelihood surface is to reveal the presence of obliquely oriented ridges, and possible tradeoffs among estimates of β_1 and β_2. For merely obtaining the maximum of the likelihood the OLS solution is adequate, and no plotting is needed. Rao [1973] mentions the limitations of the maximum likelihood approach. Our plot reveals that a family of solutions along the ridge in Figure 1.4a is worthy of consideration. For example, there is

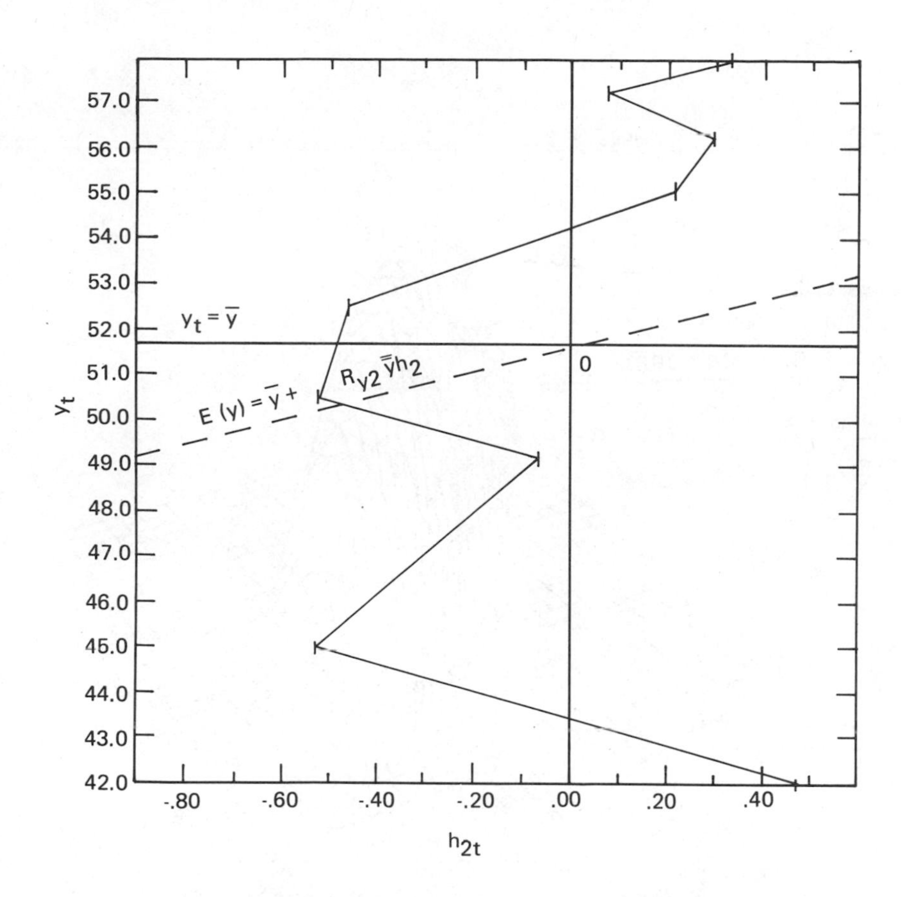

Figure 1.3b Relation Between y_t and h_{2t}

(visually) only a modest reduction in the log likelihood from about −3.03 at OLS to about −4.20 at the place marked GRR, PCR* and ORR. The solution is approximately $\beta_1 = 0.4$ and $\beta_2 = 0.8$, which is less likely to be naively misinterpreted. We shall discuss in the following chapters the GRR, PCR*, ORR and PCR estimators as special cases of generalized ridge regression solutions.

Figure 1.4b is a rotation and rescaling of Figure 1.4a where we have γ_1 and γ_2 on the base axes, and where the OLS solution is marked where γ_i equals c_i defined from $b = Gc$, $b_1 = w(c_1 - c_2)$ and $b_2 = w(c_1 + c_2)$, where $w = (1/2)^{1/2}$. It is clearly possible to write the likelihood for the model (34) in standardized units as:

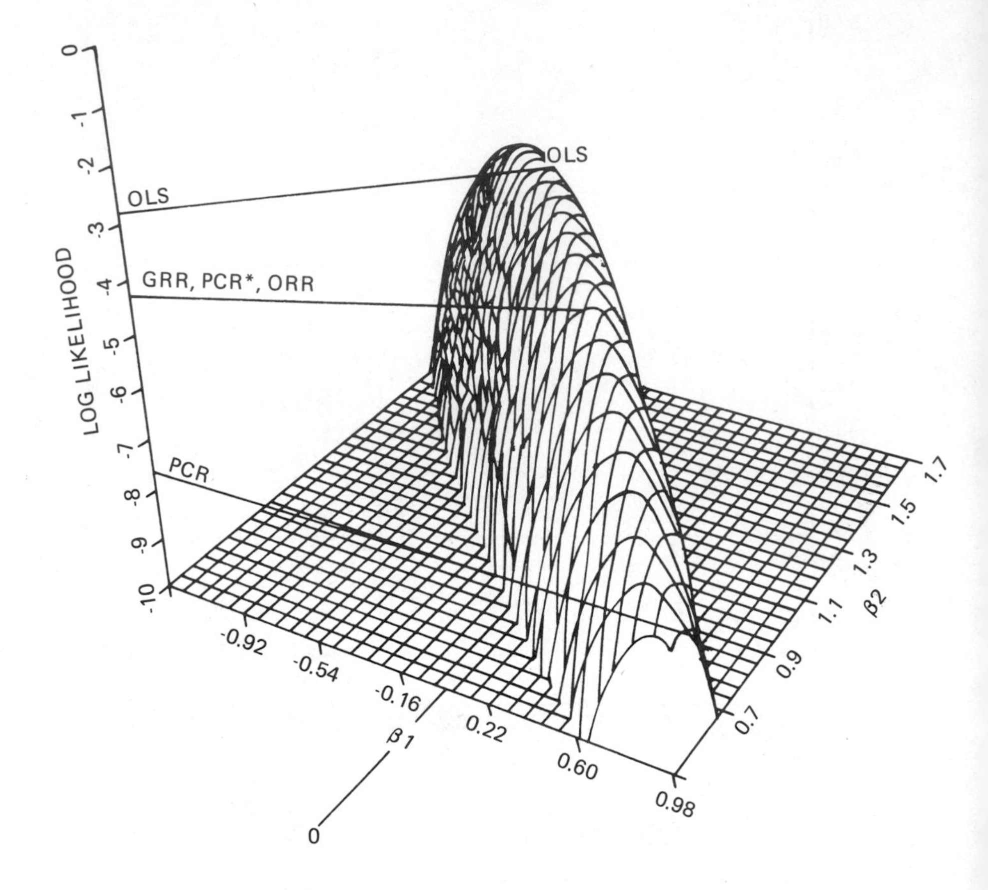

Figure 1.4a Modified Log Likelihood Surface

$$(2\pi\sigma^2)^{-T/2} \exp\left[\frac{-1}{2\sigma^2} \sum_{t=1}^{T} [y_t^s - w(\gamma_1 - \gamma_2)z_{1t} - w(\gamma_1 + \gamma_2)z_{2t}]^2\right]$$

In Figure 1.4b, if one wishes to consider other solutions in the neighborhood of the OLS solution, those which shrink the relatively imprecise component c_2 relatively more are "seen" to be most appropriate. The purpose of shrinking c_2 is to attack the source of the inappropriate signs and magnitudes of b_i; namely the presence of imprecise c_2. A drastic solution is to choose $c_2 = 0$ and it is marked as PCR in Figures 1.4a and 1.4b, which will be seen to be a Principal Component Regression (PCR) solution in the following chapters.

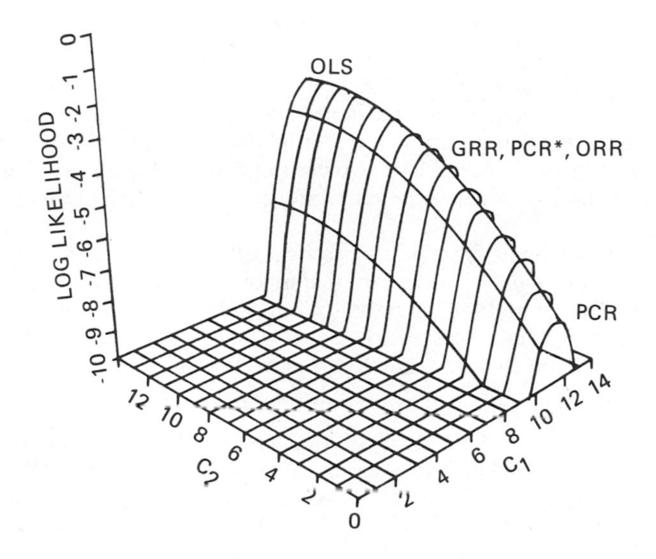

Figure 1.4b Modified Log Likelihood in Terms of Uncorrelated Components

We have shown the effect of multicollinearity on the likelihood surface and have given an intuitive explanation of the shrinking of coefficients with ridge regression type methods. Note that it is the signs and magnitudes of b_1 and b_2, not so much the forecast of y, that may be naively misinterpreted.

EXPLANATORY NOTES

Explanatory Note 1.1

Maximum Likelihood Estimation

Consider a random sample $y_1, \ldots, y_T$ of T observations represented by a T-dimensional random variable y. Suppose a known probability distribution of y is $f(y) = f(y \mid \theta)$ which depends on some unknown parameters θ. For example, θ may refer to $p + 1$ parameters β and σ^2 in a regression model.

Before the data is available, $f(y \mid \theta) = f(y_1, \ldots, y_T \mid \theta)$ represents the joint density of T random outcomes for fixed θ. However, after the data is obtained we look for the possible values of of θ which might have generated the fixed set of T elements of y in the data. The appropriate function to look at for this purpose is the *likelihood function* $L(\theta|y)$, an idea originated by Sir R. A. Fisher. This function is of the same form as $f(y|\theta)$ but y is now fixed and θ is variable. Thus, $L(\theta|y) = f(y|\theta)$, and in the case when $y_1, \ldots, y_T$ are independent

$$L(\theta|y) = f(y_1, \ldots, y_T|\theta) = \prod_{t=1}^{T} f(y_t|\theta)$$

where $f(y_t|\theta)$ is the density of the random variable y_t. If θ is a scalar parameter, the likelihood function can be plotted with θ on the horizontal axis and $L(\theta|y)$ on the vertical axis. In practice, it is often convenient to work with the log likelihood function, $\log L(\theta|y)$. According to the "likelihood principle" the likelihood function contains all the data information about the unknown parameters.

Given the likelihood function, we can maximize it or equivalently its log to determine the "most likely" value of the unknown parameter θ for our data. For this, the calculus methods may be used to differentiate $L(\theta|y)$ with respect to θ and setting the derivative equal to zero. Thus, the first order (necessary) condition for the likelihood function to be at its maximum is given by $\frac{\partial}{\partial\theta} L(\theta|y) = 0$. The sufficient condition requires, in addition, that the second derivative be negative. The value of the parameter θ which maximizes the likelihood function is called the maximum likelihood estimate. We note that this procedure would give sensible results if the likelihood function is well behaved in the sense that it is unimodal and can be approximated by a quadratic function over an extensive region near the maximum.

Explanatory Note 1.2

Eigenvectors as Direction Cosine Vectors

The cosine of an angle θ, $\cos\theta$, is the length of the side "adjacent" to the angle θ divided by the length of the hypotenuse in a right angled triangle OPR. Consider an example having two regressors (e.g. see Section 1.5). The cosine of a 45° angle line with respect to the axes for X_1 and X_2 is given by $1/\sqrt{2}$ in Figure 1.5. Verify by the Pythagorean theorem that the hypotenuse is $\sqrt{2}$. For brevity we denote the scalar $1/\sqrt{2}$ by w.

The (direction) cosine of the angle between the line $0Z$ and the line $0X_1$ is $w(= 1/\sqrt{2})$. Now, the X_2 axis makes an angle $\phi = 135°$ with the line $0Z$. Here, we use the relation $\cos\chi = -\cos(180^0 - \chi)$ to note that $\cos\phi = \cos 135° = -\cos(180° - 45°) = -\cos 45° = -1/\sqrt{2}$. This shows that the (direction) cosines of the angle between lines $0Z$ and $0X_2$ is $-1/\sqrt{2}$ $(= -w)$. It is customary to express the orientation of a line like $0Z$ by a direction cosine vector $(w, -w)$ where $w = 1/\sqrt{2}$. Note that the length of the

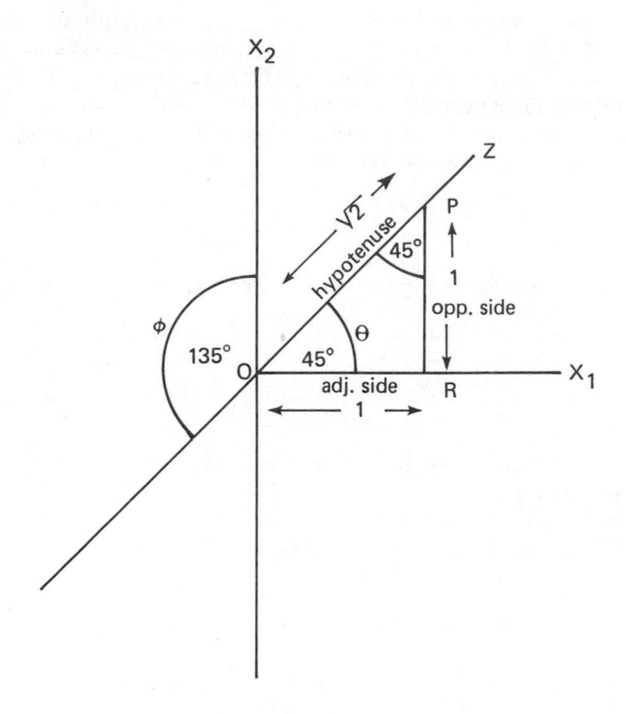

Figure 1.5 Direction Cosines of Vector *OZ*

"adjacent side" was arbitrarily chosen to be 1 in the above definition of the two cosines. The Euclidian length of any vector (a,b) is simply the square root of the sum of squares of its elements, that is, $(a^2+b^2)^{1/2}$. The geometrical representation of any vector is an arrow through the origin. The representation of our direction cosine vector $(w,-w)$ is the arrow $0P$ where the length of the vector is $(w^2+w^2)^{1/2}=1^{1/2}=1$.

The coordinates of the point P are (w,w) with respect to the X_1 and X_2 axes. This vector of coordinates should not be confused with the vector of direction cosines $(w,-w)$. When the elements of a direction cosine vector are normalized so that its Euclidian length is 1, it is sometimes referred to as a "normalized direction cosine vector". In the absence of normalization, $(cw,-cw)$, where c is any arbitrary constant defining the length of the "adjacent side", gives us the orientation of the line $0Z$. Since the constant c can even be negative, the direction cosine vector is equivalently given by $(-w,w)$. The length of $(-w,w)$ remains 1, suggesting that a general definition of our normalized direction cosine vector is $\pm(w,-w)$.

For the standardized data, the correlation matrix is $(X'X)=\begin{bmatrix}1 & r\\ r & 1\end{bmatrix}$. The eigenvalues may be numbered in decreasing order of magnitude (for $r > 0$) by $\lambda_1 = 1 + r$ and $\lambda_2 = 1 - r$, and the corresponding eigenvectors are $g_1 = (w, w)$ and $g_2 = (-w, w)$ respectively. We verify these assertions from the usual definitions of the eigenvectors g_i of the 2×2 matrix Λ with eigenvalues λ_i. We may now verify the basic relation of eigenvalues and eigenvectors of a matrix: $X'Xg_i = \lambda_i g_i$. Note that for $i = 1$

$$\begin{pmatrix} 1 & r \\ r & 1 \end{pmatrix} \begin{pmatrix} w \\ w \end{pmatrix} = (1 + r) \begin{pmatrix} w \\ w \end{pmatrix}$$

implies the two relations $w + rw = w + rw$, and $rw + w = w + rw$, which are clearly true.
Similarly, for $i = 2$

$$\begin{pmatrix} 1 & r \\ r & 1 \end{pmatrix} \begin{pmatrix} -w \\ w \end{pmatrix} = (1 - r) \begin{pmatrix} -w \\ w \end{pmatrix}$$

implies the two relations $-w + rw = w + rw$, and $-rw + w = w - rw$, which are also clearly true.

The usual eigenvalue-eigenvector decomposition of $X'X$ is $X'X = G\Lambda G'$, which may be verified by explicit multiplication as follows:

$$\begin{pmatrix} 1 & r \\ r & 1 \end{pmatrix} = \begin{pmatrix} w & -w \\ w & w \end{pmatrix} \begin{pmatrix} 1+r & 0 \\ 0 & 1-r \end{pmatrix} \begin{pmatrix} w & w \\ -w & w \end{pmatrix}$$

$$= \begin{pmatrix} w+rw & -w+rw \\ w+rw & w-rw \end{pmatrix} \begin{pmatrix} w & w \\ -w & w \end{pmatrix}$$

$$= \begin{pmatrix} w^2+rw^2+w^2-rw^2 & w^2+rw^2-w^2+rw^2 \\ w^2+rw^2-w^2+rw^2 & w^2+rw^2+w^2-rw^2 \end{pmatrix}$$

$$= \begin{pmatrix} 2w^2 & 2rw^2 \\ 2rw^2 & 2w^2 \end{pmatrix}$$

$$= \begin{pmatrix} 1 & r \\ r & 1 \end{pmatrix},$$

where $w = 1/\sqrt{2}$ implies $2w^2 = 1$.

In higher dimensions than 2 it is possible to write computer programs that give the decomposition $X'X = G\Lambda G'$ where $X'X$ is a symmetrix matrix. These computer programs are written from the complicated analytical expressions for eigenvalues and eigenvectors developed by John VonNeumann. IBM's [1968] scientific subroutines contain a FORTRAN subroutine DEIGEN to implement this in double precision arithmetic. Single precision arithmetic is not recommended for this operation, because then the subroutine may give inaccurate results.

Explanatory Note 1.3

Obtaining the H Matrix of the Singular Value Decomposition

The singular value decomposition of $T \times p$ matrix X is given by $X = H\Lambda^{1/2}G'$ where H is $T \times p$, and both Λ and G are $p \times p$. Now $G'G = I$ implies that $XG = H\Lambda^{1/2}$, and $XG\Lambda^{-1/2} = H$.

This shows that knowing X we can always compute the H matrix. The first step is to obtain G and Λ by the formula $X'X = G\Lambda G'$, which involves calling IBM's subroutine DEIGEN mentioned in the Explanatory Note 1.2.

For the 2×2 example, H may be obtained by direct multiplication.

The standardized X matrix is given by

$$X = \begin{bmatrix} -0.8149 & -0.6584 \\ -0.2439 & -0.3938 \\ -0.0536 & -0.0764 \\ 0.0733 & -0.0764 \\ 0.1621 & 0.0294 \\ 0.0987 & 0.1683 \\ 0.1241 & 0.2212 \\ 0.2890 & 0.3137 \\ 0.3651 & 0.4725 \end{bmatrix}.$$

The G matrix is $\begin{bmatrix} w & -w \\ w & w \end{bmatrix}$, with $w = 2^{-1/2}$, and Λ is $\begin{bmatrix} 1+r & 0 \\ 0 & 1-r \end{bmatrix}$ with $r = 0.9430$ being the correlation coefficient between x_1 and x_2, which are the columns of the X matrix above. The H matrix is obtained by post-multiplying the standardized X by G and then by $\Lambda^{-1/2}$. We have

$$H = \begin{bmatrix} -.7474 & 0.4635 \\ -.3235 & -.4440 \\ -.0659 & -.0675 \\ -.0016 & -.4434 \\ .0971 & -.3930 \\ .1354 & .2061 \\ .1752 & .2876 \\ .3057 & .0732 \\ .4249 & .3181 \end{bmatrix}.$$

A good check on the arithmetic computations is to verify that $H'H$ is the identity matrix.

Explanatory Note 1.4

Rotation of Axes to the Principal Axes

Consider a two-dimensional Euclidian space spanned by the basis vectors e_1 and e_2 along the axes X_1 and X_2. The coordinates of e_1 are (1,0) suggesting that the vertical length is zero. The coordinates of e_2 are (0,1) where the horizontal length is zero at e_2.

We rotate the system of axes X_1 and X_2 to a new system denoted by H_1 and H_2. The unit-circle is drawn in the figure with radius 1. The basis vectors of the rotated system of axes are now given by the vectors Oh_1 and Oh_2 from the unit circle. From the definitions $\cos\theta = \textit{adjacent side/hypotenuse}$, and $\sin\theta = \textit{opposite side/hypotenuse}$, it can be verified that the X_1 coordinate of the point h_1 is $\cos\theta$ since the hypotenuse is unity on the unit circle. Similarly, other coordinates can be verified. We write the 2×1 vector h_1 as a weighted sum of two 2×1 vectors e_1 and e_2 with weights $\cos\theta$ and $\sin\theta$:

$$h_1 = \cos\theta\, e_1 + \sin\theta\, e_2$$

Similarly, $h_2 = (-\sin\theta)e_1 + \cos\theta\, e_2$. Hence the coordinates of the rotated system of axis e_1 and e_2 are related to the original coordinates by the relations

$$H_1 = \cos\theta X_1 + \sin\theta X_2\,,$$

$$H_2 = -\sin\theta X_2 + \cos\theta X_2\,,$$

$$\begin{pmatrix} H_1 \\ H_2 \end{pmatrix} = \begin{pmatrix} \cos\theta & \sin\theta \\ -\sin\theta & \cos\theta \end{pmatrix} \begin{pmatrix} X_1 \\ X_2 \end{pmatrix}.$$

Verify that the inverse of the above rotational transformation matrix R is known from the relation $RR^{-1} = I$ with

$$\begin{pmatrix} \cos\theta & \sin\theta \\ -\sin\theta & \cos\theta \end{pmatrix} \begin{pmatrix} \cos\theta & -\sin\theta \\ \sin\theta & \cos\theta \end{pmatrix} = \begin{pmatrix} 1 & 0 \\ 0 & 1 \end{pmatrix}.$$

where we use the well-known relation $\sin^2\theta + \cos^2\theta = 1$. Thus, the inverse of R is also its transpose for arbitrary θ suggesting that R is an orthogonal transformation. The word orthogonal refers to the geometrical idea of a perpendicular, i.e., the fact that the new axes H_1 and H_2 are perpendicular to each other. Using R^{-1} we can write

$$\begin{pmatrix} X_1 \\ X_2 \end{pmatrix} = \begin{pmatrix} \cos\theta & -\sin\theta \\ \sin\theta & \cos\theta \end{pmatrix} \begin{pmatrix} H_1 \\ H_2 \end{pmatrix}$$

When $\theta = 45^\circ$ this rotation for standardized regressor data will express them in terms of the principal axis. This follows from the fact that the eigenvector matrix G of the correlation matrix $X'X$ coincides with the R^{-1} matrix above. Note that when $\theta = 45^\circ = \pi/4$, $\sin\theta = \cos\theta = 2^{-1/2}$.

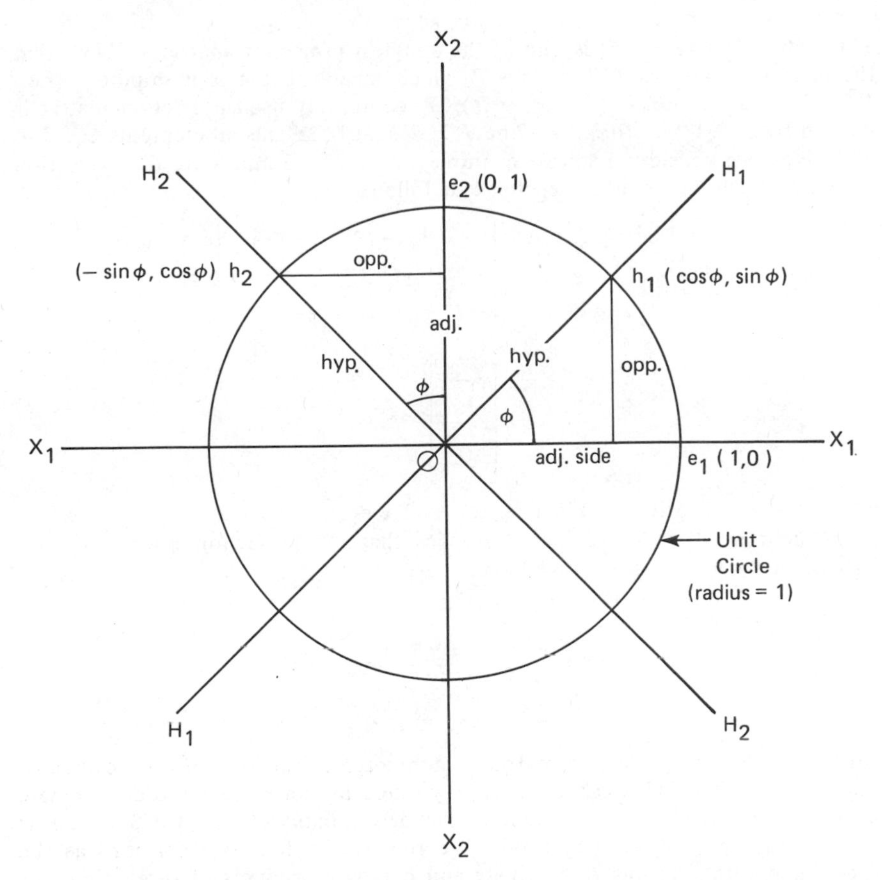

Figure 1.6 Rotation of Co-ordinate Axes: Relationship between the Old Basis Vectors e_1, e_2 and the New Basis Vectors h_1 and h_2

These ideas extend readily to higher dimensions, although we do not discuss these extensions.

Explanatory Note 1.5

Measurement from Means, Projection Matrix

The word projection refers to causing a shadow to fall upon a surface. In a parallel projection the image figure is obtained from the original figure by drawing from the point of the original figure certain parallel straight lines.

In algebraic terms let z denote a $T \times 1$ vector such that $z'z = \sum_{t=1}^{T} z_t^2 = 1$.

The matrix $P = (I - zz')$ is one of the simplest projection matrices. Note that the mátrix P satisfies $P'P = P^2 = P$ which means that it is idempotent (i.e., $P^2 = P$) and symmetric (i.e., $P' = P$). A commonly used projection matrix is defined by $z_t = 1/\sqrt{T}$ for all t. The $T \times T$ matrix zz' has all elements equal to T^{-1}. Now we consider a simple example where $T = 3$ and write the projection matrix P operating on a 3×1 vector x as follows:

$$Px = (I - zz')x = \begin{bmatrix} 1-T^{-1} & -T^{-1} & -T^{-1} \\ -T^{-1} & 1-T^{-1} & -T^{-1} \\ -T^{-1} & -T^{-1} & 1-T^{-1} \end{bmatrix} \begin{bmatrix} x_1 \\ x_2 \\ x_3 \end{bmatrix}$$

$$= \begin{bmatrix} x_1 - \bar{x} \\ x_2 - \bar{x} \\ x_3 - \bar{x} \end{bmatrix}$$

where $x = [x_1\ x_2\ x_3]'$ is a 3×1 vector, and $\bar{x} = x_1 + x_2 + x_3/3$

Note that $P' = (I - zz') = P$ implies that the projection matrix is symmetric. Also, using $z'z = 1$, we have

$$P^2 = (I - zz')(I - zz')$$

$$= I - 2zz' + zz'zz'$$

$$= I - zz' = P$$

suggesting that the projection matrix is idempotent. Furthermore, we can show that P is singular. This can be directly verified by noting that the determinate of $I - zz'$ is zero. For our example the determinant of our 3×3 matrix is obtained by the following method. We rewrite the first two columns as the fourth and fifth columns respectively and compute products of three elements along the three North-West to South-East diagonals with a positive sign added to the products of three elements along three diagonals in the North-East and South-West direction with a negative sign. For brevity, we write $a = -T^{-1}$, and

$$\begin{matrix} + & + & + & & \\ \uparrow & \uparrow & \uparrow & & \\ 1+a & a & a & 1+a & a \\ a & 1+a & a & a & 1+a \\ a & a & 1+a & a & a \\ \downarrow & \downarrow & \downarrow & & \\ - & - & - & & \end{matrix}$$

The determinant is given by

$$\det = (1 + a)^3 + a^3 + a^3 - 3(1 + a)a^2 ,$$

where $a = -T^{-1}$, and $T = 3$ $(3a = -1)$. Now

$$\det = 1 + 3a + 3a^2 + a^3 + 2a^3 - 3a^3 - 3a^2$$

$$= 1 + 3a$$

$$= 0$$

Since this determinant is zero, we also know that one of the eigenvalues of the projection matrix is zero. In particular, the eigenvalue associated with the eigenvector z can be seen to be zero. This follows from the relation $Pz = (I - zz')z = 0$. Verify that the matrix zz' is also symmetric idempotent and singular and projects a vector x onto a vector of "fitted" values $\bar{x}$.

Recentering is a Matrix Translation

The geometrical interpretation of this projection is that it centers the data at the mean values. For the general case, verify that an unstandardized matrix of regressors X may be centered at the means by the projection operator. This projection amounts to a so-called "translation" of the coordinate axes, where the original coordinates x_{1t}, x_{2t} for $t = 1, \ldots, T$ are replaced by $x_{1t} - \bar{x}_1$ and $x_{2t} - \bar{x}_2$ respectively.

A more complicated projection matrix to be discussed later is given by the idempotent matrix

$$M = (I - X(X'X)^{-1}X')$$

which is used to project the vector of observations on the dependent variable y onto the space spanned by regressors. Note that My gives the vector of deviations from the fitted ordinary least squares regression plane. This is analogous to the projection matrix P which gives a vector of deviations from the arithmetic mean (see Exercise 1.7).

EXERCISES

1.1 Let A and C be the $p \times p$ nonsingular matrices and let B be a $p \times r$ matrix. Then show that

$$\begin{bmatrix} A & B \\ B' & C \end{bmatrix}^{-1} = \begin{bmatrix} A^{-1}+A^{-1}BD^{-1}B'A^{-1} & -A^{-1}BD^{-1} \\ -D^{-1}B'A^{-1} & D^{-1} \end{bmatrix},$$

where $D = C - B'A^{-1}B$ (reference Theil 1971). Use this result to derive the ordinary least squares estimators b_0 and b in the equation (43) of the text.

1.2 Consider the following simple linear regressions

$$(1)\ y_t = \beta_0 + \beta_1 x_{1t} + \beta_2 x_{2t} + u_t$$

$$(2)\ y_t^s = \beta_0^s + \beta_1^s x_{1t}^s + \beta_2^s x_{2t}^s + u_t^s \quad t = 1,\dots,T$$

where $x_{it}^s = (x_{it} - \bar{x}_i)/\sqrt{T-1}(SD)_i$, $\beta_i^s = \sqrt{T-1}(SD)_i\beta_i$ and $(SD)_i = \sum_{t=1}^{T} (x_{it}-\bar{x}_i)^2/(T-1)$ for i=0,1,2. Further $y_t^s = y_t - \bar{y}$ and $u_t^s = u_t - \bar{u}$. Show that for the model (2)

$$(a)\ X'X = \begin{bmatrix} 1 & r_{12} \\ r_{21} & 1 \end{bmatrix}$$

$$(b)\ X'y = \begin{bmatrix} r_{1y} \\ r_{2y} \end{bmatrix} (\sqrt{T-1})(SD)_y$$

$$(c)\ (X'X)^{-1} = \frac{1}{1-r_{12}^2} \begin{bmatrix} 1 & -r_{12} \\ -r_{21} & 1 \end{bmatrix}$$

1.3 Let X be the $T \times p$ matrice of p regressors. Show that if we express the regressors as deviations from sample means then the matrix representation is $[I - \frac{\iota\iota'}{T}]X$ where ι is a $T \times 1$ vector of unit elements. Show also that $[I - \frac{\iota\iota'}{T}]$ is an idempotent matrix.

1.4 Show that the least squares estimator of β in the linear regression $y = \beta_0 + X\beta + u$ is identical with the least squares estimator of β in the standardized model $y^s = X^s\beta + u$ given in (36). What can you say about this result when $\beta_0 = 0$?

1.5 (a) Show that if there is no intercept term in the model $y = X\beta + u$, and we estimate β by the least squares method, then the sum of the elements of the estimated residual vector is not equal to zero. Prove that $X'\hat{u} = 0$.

(b) For the model in (a) prove that $0 \leq R^2 = 1 - \frac{\hat{u}'\hat{u}}{y'y} \leq 1$

(c) For the model in (a) show that $R^2 = 1 - \frac{\hat{u}'\hat{u}}{(y-\bar{y})'(y-\bar{y})}$ can become negative. For which model *this* R^2 will be assured to satisfy $0 \leq R^2 \leq 1$?

What suggestions would you give to a user of regression model on the basis of your findings in (a) to (c)?

1.6 The attached Table 1.1 gives Bell System data on natural logs of Bell System output, capital stock, labor input and percent direct distance dialing from L. H. Mantell [1974] for 1947 to 1971.

Consider a production function defined by regressing ln$Y$ on ln$K$, ln$L$ and %DDD. Analyze the singular value decomposition of the matrix of regressors. Note that there will be slight differences in numerical values of H matrix

TABLE 1.1

Year	ln$Y$	ln$K$	ln$L$	%DDD
1	8.00092	9.18878	13.4066	.0
2	8.03427	9.28464	13.5667	.0
3	8.13558	9.41229	13.6173	.0
4	8.18849	9.48801	13.6407	.0
5	8.24698	9.54717	13.6211	.0
6	8.36117	9.62905	13.6508	.0
7	8.40838	9.69390	13.7083	.0
8	8.46605	9.74783	13.7453	.0
9	8.51579	9.80995	13.7520	5.80000
10	8.58582	9.86590	13.7713	9.50000
11	8.66553	9.93576	13.8210	13.50000
12	8.72724	10.0281	13.8164	17.7000
13	8.78094	10.0996	13.7922	22.0000
14	8.84376	10.1626	13.7422	26.1000
15	8.88746	10.2288	13.7291	31.4000
16	8.94073	10.2954	13.7191	36.8000
17	9.0089	10.3630	13.7024	41.7000
18	9.05924	10.4268	13.7068	45.8000
19	9.12649	10.4967	13.7323	49.7000
20	9.20504	10.5675	13.7676	53.2000
21	9.30229	10.6397	13.8178	56.4000
22	9.37683	10.7048	13.8617	59.6000
23	9.45875	10.7679	13.8838	62.7000
24	9.53405	10.8361	13.8763	65.3000
25	9.61677	10.9415	14.0042	68.5000

obtained by Businger and Golub [1965] methods and the indirect method $H = XG\Lambda^{-1/2}$.

1.7 Plot the sample mean as a perpendicular projection in the regression model: $y = \beta\iota + u$, where ι denotes a $T \times 1$ vector of ones. The plot should also show $\hat{u}$ the residual. Also plot the bivariate model with 2 regressors and show its residual sum of squares. Can you show "restricted" residual sum of squares and F values? (Reference Margolis [1979]).

2

Criteria for Good Regression Estimators: MSE, Consistency, Stability, Robustness, Admissibility, Minimaxity, and Bayesian "melo"ness

2.1 INTRODUCTION

Since an estimator of a parameter is usually regarded as a random variable with a certain distribution, many properties of the estimator can be discussed in terms of the usual theory of probability distributions.

Consider the familiar regression model: $y = X\beta + u$, $Eu = 0$, $E(u\ u') = \sigma^2 I$. For the familiar significance tests we usually assume that y is a multi-variate normal variable with the mean

$$E(y) = X\beta \tag{1}$$

and the variance covariance matrix

$$V(y) = \sigma^2 I\ . \tag{2}$$

Note that the OLS estimator b is clearly a linear transformation of a p-variate normal variable which follows from its definition: $b = (X'X)^{-1}X'y$ and the well known fact that linear transformations of normal variables are also normal. Hence the ith element b_i of b is normally distributed with mean β_i and variance

$$V(b_i) = \sigma^2(X'X)_i^{-1}\ , \tag{3}$$

where the subscript i denotes the ith diagonal element of the matrix. The measured b_i is only one of several probable realizations of a random variable, which is centered near the true unknown parameter. The vector b is multivariate normal with mean vector β and variance covariance matrix $\sigma^2(X'X)^{-1}$.

2.2 UNIVARIATE CONCEPTS

2.2.1 Definition: Univariate Unbiasedness

An estimator b_i of β_i is said to be unbiased if its expected value coincides with the true unknown parameter β_i, that is $E(b_i) = \beta_i$.

If $E(b_i) \neq \beta_i$, then b_i is a biased estimator with the magnitude of bias as $E(b_i) - \beta_i$.

2.2.2 Definition: Univariate Efficiency

One unbiased estimator is said to be more efficient than another if its variance is smaller than the variance of the other.

Consider two unbiased estimators b_{i1} and b_{i2} of β_i satisfying $E(b_{i1}) = E(b_{i2}) = \beta_i$. Let

$$\begin{aligned} V(b_{i1}) &= E(b_{i1} - Eb_{i1})^2 = E(b_{i1} - \beta_i)^2 \\ V(b_{i2}) &= E(b_{i2} - \beta_i)^2 . \end{aligned} \tag{4}$$

Then the estimator b_{i1} is more efficient than b_{i2} if $V(b_{i1}) < V(b_{i2})$. If $V(b_{i1}) = V(b_{i2})$ then both are equally efficient.

Note that we restrict our attention to unbiased estimators, while comparing the variances. Once we admit "biased" estimators it is almost ridiculous to compare variances alone. To understand this point clearly, consider a "silly" estimator which says that

$$b_i = 4 , \tag{5}$$

where 4 is a specified constant. Since 4 is a fixed constant (not a random variable) its variance is zero. Since variance cannot be negative, $b_i = 4$ is the most efficient of all the available estimators. The trouble is that $b_i = 4$ is biased unless β_i happens to be 4 also.

2.2.3 Definition: Univariate Mean Squared Error (MSE)

If b_i is a biased estimator of β_i, then its mean squared error is defined by

$$\begin{aligned} MSE(b_i) &= E(b_i - \beta_i)^2 = E(b_i - Eb_i + Eb_i - \beta_i)^2 \\ &= E(b_i - Eb_i)^2 + [E(b_i) - \beta_i]^2 + 2E(b_i - Eb_i)(Eb_i - \beta_i) . \end{aligned} \tag{6}$$

Now since the third term is zero we get

$$MSE(b_i) = V(b_i) + [Bias(b_i)]^2 , \tag{7}$$

where

$$Bias(b_i) = Eb_i - \beta_i . \tag{8}$$

2.2.4 Univariate MSE Criterion

If b_{i1} and b_{i2} are two biased (or unbiased) estimators of β_i, then b_{i1} is preferred to b_{i2} if $MSE(b_{i1}) < MSE(b_{i2})$. Thus the MSE criterion may be used to obtain a preference ordering among the various estimators.

2.3 MULTIVARIATE CONCEPTS

2.3.1 Definition: Multivariate Unbiasedness

A vector estimator b of a $p \times 1$ parameter vector β is said to be unbiased if $E(b) = \beta$.

2.3.2 Definition: Multivariate Efficiency

If the two unbiased estimators b_1 and b_2 of β have variance-covariance matrices v^1 and v^2 respectively, b_1 is more efficient than b_2 if $v^2 - v^1$ is a positive semi-definite matrix.

Now, we show that the OLS estimator b is more efficient than any other linear estimator which is also unbiased. In other words, b is "best" in the class of "linear unbiased estimators" (BLUE).

2.3.3 Theorem: (Gauss-Markov) b is BLUE of β

Proof: Consider any estimator linear in y, say $\tilde{\beta} = Cy$. Let $C = (X'X)^{-1}X' + Z$ where Z is a non-stochastic matrix. Then using the fact that $E(y) = X\beta$ we write

$$E(\tilde{\beta}) = E[(X'X)^{-1}X' + Z][X\beta]$$

$$= \beta + ZX\beta$$

so that for $\tilde{\beta}$ to be unbiased we require ZX = 0. Now, the covariance matrix of $\tilde{\beta}$ is

$$E(\tilde{\beta}-\beta)(\tilde{\beta}-\beta)' = E[(X'X)^{-1}X' + Z]uu'[(X'X)^{-1}X' + Z]' \tag{9}$$

$$= \sigma^2[(X'X)^{-1}X' IX(X'X)^{-1} + ZIX(X'X)^{-1}$$

$$+ (X'X)^{-1}X' IZ' + ZIZ']$$

$$= \sigma^2(X'X)^{-1} + \sigma^2 ZZ', \quad \textit{since } ZX = 0 .$$

But ZZ' is a positive semidefinite matrix, which shows that the variance covariance matrix of $\tilde{\beta}$ equals the variance covariance matrix of b plus a positive semi-definite matrix. Hence b is efficient relative to any other linear unbiased estimator of β.

The Gauss-Markov theorem also implies that any linear combinations $L'b$ are BLUE for $L'\beta$. This is an important implication.

2.3.4 Multivariate MSE

Denote the MSE of a biased estimator b of β by MSE(b). We have

$$MSE(b) = E(b-\beta)'(b-\beta) = E(b-Eb+Eb-\beta)'(b-Eb+Eb-\beta)$$

$$= E(b-Eb)'(b-Eb) + (Eb-\beta)'(Eb-\beta) \tag{10}$$

$$= Tr[V(b)] + [Bias(b)]'[Bias(b)] ,$$

where *Tr* represents trace (i.e. sum of diagonals). In statistical decision theory $E(b-\beta)'(b-\beta)$ is called the expected squared loss function or *risk function* with respect to squared loss function.

If we define the Euclidian length of a vector v by $||v|| = (v'v)^{1/2}$, then MSE(b) in (10) measures the average of the squared Euclidian distance between b and β. Thus an estimator with low MSE will be close to the true parameter. This is shown in two figures, 2.1 and 2.2. In Figure 2.1 the Euclid^ian distance between b and β is larger than the corresponding distance with respect to a biased estimator.

2.3.5 Multivariate WMSE

The weighted MSE with the positive semidefinite weight matrix W is defined by

$$WMSE(b) = E(b - \beta)'W(b - \beta) \tag{11}$$

2.3.6 Predictive MSE

The predictive MSE is defined by

$$PMSE(b) = E(y - Xb)'(y - Xb) = E(b - \beta)'X'X(b - \beta) \tag{12}$$

According to the model $y = X\beta + u$, note that the "true unknown" forecast of y corresponding to r "future" values of the regressors given by a matrix X_f of dimension $r \times p$ is simply $X_f \beta$. If the future values are simply a repeat of the available X values, the true forecast of y is $X\beta$. When β is replaced by any estimator b the "predictive" MSE for arbitrary X_f is defined by

$$PMSE(b, X_f) = E(y - X_f b)'(y - X_f b)$$

$$= E(b - \beta)'X_f'X_f(b - \beta) . \tag{12a}$$

The important point is that PMSE is similar to a weighted MSE, where the weights are known if X_f is specified.

It should be noted that using the eigenvalue decomposition $X'X = G\Lambda G'$ we can write (12) as follows

$$PMSE(b) = E(G'b - G'\beta)'\Lambda(G'b - G'\beta)$$

$$= E(c - \gamma)'\Lambda(c - \gamma') , \tag{12b}$$

where c is a p-vector estimating the uncorrelated component vector γ of the canonical model defined in equations (22) and (18) of Chapter 1 respectively. In general

$$PMSE(b, X_f) = E(c - r)'G'X_f'X_f G(c - r) . \tag{12c}$$

2.3.7 Matrix MSE

The matrix MSE is defined by a $p \times p$ matrix

$$MtxMSE(b) = E(b - \beta)(b - \beta)' \tag{13}$$

We can also write (13) as

$$MtxMSE(b) = E(b - Eb + Eb - \beta)(b - Eb + Eb - \beta)' \quad (13a)$$

$$= E(b - Eb)(b - Eb)' + (Eb - \beta)(Eb - \beta)'$$

$$= V(b) + [Bias(b)][Bias(b)]' .$$

If we take trace on both sides of (13a) we get (10), that is

$$Tr[MtxMSE(b)] = MSE(b) = E(b - \beta)'(b - \beta) .$$

2.3.8 Multivariate MSE Criteria

Consider two competing estimators b_1 and b_2. If the difference $\Delta = MtxMSE(b_2) - MtxMSE(b_1)$ is positive definite, then b_1 is to be preferred to b_2. To check up positive definiteness of the difference Δ it is easier in practice to check whether the scalar $\eta'\Delta\eta > 0$ for all $\eta \neq 0$, where η is any nonzero column vector. If $\eta'\Delta\eta > 0$ for all $\eta \neq 0$ is true, then it implies that $\mathrm{MSE}(\eta' b_2) > MSE(\eta' b_1)$ for all such η. On the other hand, if $MSE(\eta' b_2) > MSE(\eta' b_1)$ for all $\eta \neq 0$, then $\eta'\Delta\eta > 0$. This is because

$$\eta' E(b_1 - \beta)(b_1 - \beta)'\eta = Tr[E(b_1 - \beta)(b_1 - \beta)'\eta\eta']$$

$$= E[Tr(b_1 - \beta)(b_1 - \beta)'\eta\eta']$$

$$= E(b_1 - \beta)'\eta\eta'(b_1 - \beta)$$

$$= E(\eta' b_1 - \eta'\beta)'(\eta' b_1 - \eta'\beta)$$

$$= MSE(\eta' b_1)$$

and similarly $\eta' E(b_2 - \beta)(b_2 - \beta)'\eta = MSE(\eta' b_2)$. Thus a strong criterion, (Wallace [1972]), for the estimator b_1 to be preferred to another estimator b_2 can be written as

$$MSE(\eta' b_2) \geq MSE(\eta' b_1) , \quad \text{for all } \eta \neq 0 .$$

This is equivalent to Δ "$\geq$" 0, where

$\Delta = MtxMSE(b_2) - MtxMSE(b_1)$ = a positive semi-definite matrix. For a strict inequality we need Δ to be positive definite. The quotes around the inequality remind us that it refers to matrices.

A weaker criterion for b_1 to be preferred to b_2 is that $MSE(b_2) \geq MSE(b_1)$.

Chengand Iglarsh [1976] consider an alternative definition of "closeness to truth" in terms of a confidence interval criterion. They compare the minimum

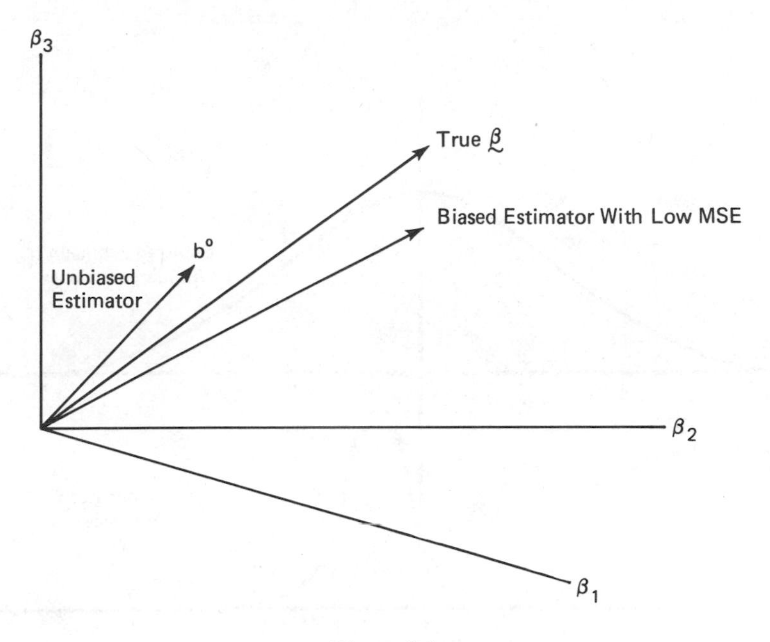

Figure 2.1

value of $\tilde{\epsilon}$ for which the following inequality is satisfied for all i:

$$Pr[|b_{i1} - \beta_i| < \tilde{\epsilon}] < Pr[|b_{i2} - \beta_i| < \tilde{\epsilon}] , \tag{14}$$

where Pr denotes probability. The results with this criterion are harder to prove, and are rarely shown to be different from the MSE criterion. In fact, this is similar to Pitman's [1937] criterion known to yield *intransitive ranking* (e.g. estimator a is preferred to b which in turn is preferred to c, and yet a may not be preferred to c) of estimators, Johnson [1950]. Recently, Bibby and Toutenburg [1977] have given a good discussion of these non-standard "closeness" criteria applied to the prediction problem. This criterion may be helpful in situations where MSE is not defined due to nonexistence of the first two moments. Efron's [1975] "A shaggy statistician story" is relevant here. According to this story a shaggy statistician suggested a shrinkage estimator which is "closer" by the criterion (14) more than 50% of times, compared to the usual maximum likelihood estimator suggested by another candidate in a job interview. The shaggy statistician got the job. Our private simulation of

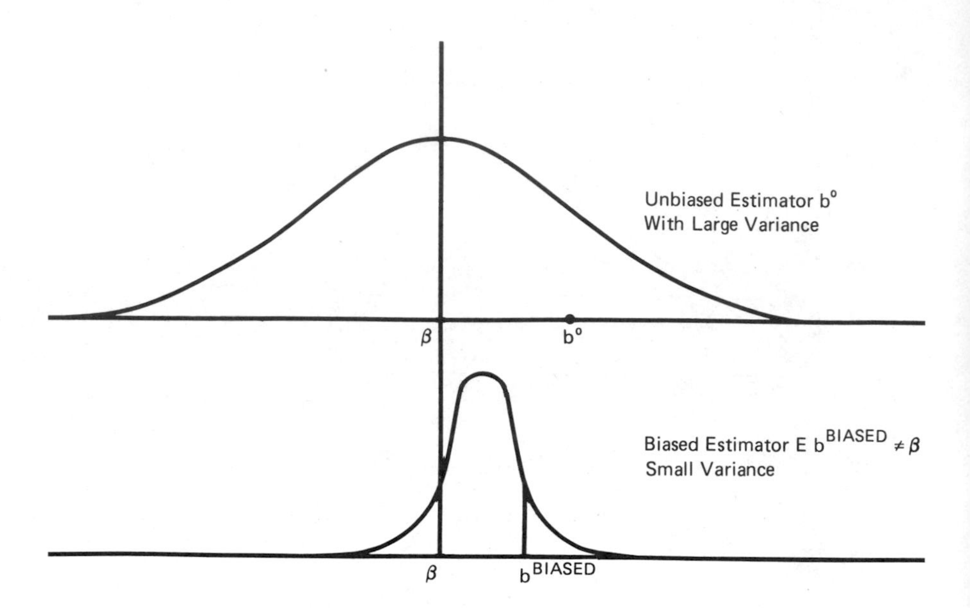

Figure 2.2

Efron's estimator suggests that this particular shrinkage estimator can have much larger MSE than sample mean, and does not produce meaningful practical improvement. The difficulty is that when it is better than the mean it is so by a small quantity, whereas when it is worse it is worse by a large quantity.

2.4 ASYMPTOTIC PROPERTIES

Let $b_i^{(T)}$ denote an estimate of β_i based on T observations with mean $Eb_i^{(T)}$ and variance $V(b_i^{(T)}) = E(b_i^{(T)} - Eb_i^{(T)})^2$. As T changes from T to $T+1$, $T+2$, $T+3$, ... , etc. we have sequences $b_i^{(T)}$, $Eb_i^{(T)}$ and $V(b_i^{(T)})$. The notation of this section does not necessarily refer to the regression model.

2.4.1 Definition: Asymptotic Unbiasedness

If for any ϵ ($\epsilon > 0$) there is a T_o such that $|Eb_i^{(T)} - \beta_i| < \epsilon$ for all $T \geq T_o$ then β_i is an asymptotic expectation of $b_i^{(T)}$. We write

$$ASY{\cdot}E[b_i^{(T)}] = \beta_i \ , \ \ or \ \ \lim_{T\to\infty} E[b_i^T] = \beta_i \tag{15}$$

Since the asymptotic expectation coincides with the true parameter, $b_i^{(T)}$ is said to be asymptotically unbiased. If the expectation does not exist, we must consider a modified definition. We require the mean of the limiting distribution of $\sqrt{T}\,(b_i^{(T)} - \beta_i)$ to be zero.

2.4.2 Definition: Asymptotic Variance

Asymptotic variance may be defined in terms of $ASY{\cdot}E$ as follows

$$\begin{aligned} ASY{\cdot}V[b_i^{(T)}] &= ASY{\cdot}E\,(b_i^{(T)} - Eb_i^{(T)})^2 \\ &= \lim_{T\to\infty} E\,(b_i^{(T)} - E\,[b_i^{(T)}])^2 \,. \end{aligned} \tag{16}$$

The ($ASY{\cdot}E$) and ($ASY{\cdot}V$) are summary statistics describing the distribution of $b_i^{(T)}$ as sample size increases. If $ASY{\cdot}E\,(b_i^{(T)}) = \beta_i$, then the bias decreases as sample size increases. If $ASY{\cdot}V(b_i^{(T)}) = 0$, then the distribution of $b_i^{(T)}$ is getting more and more concentrated around its mean as $T \to \infty$. An estimator $b_i^{(T)}$ of β_i satisfying both conditions $ASY{\cdot}E\,(b_i^{(T)}) = \beta_i$, and $ASY{\cdot}Var\,(b_i^{(T)}) = 0$ is intuitively desirable. As in the case of asymptotic unbiasedness, we should consider the variance of the limiting distribution for the awkward case when the variance does not exist.

However, these are large sample properties, which may be irrelevant in small samples.

2.4.3 Definition: Probability Limit (Univariate)

If for any arbitrary $\epsilon > 0$, $\lim_{T\to\infty} Pr\,[|b_i^{(T)} - \beta_i| \geq \epsilon] = 0$, then we write $\operatorname*{plim}_{T\to\infty} b_i^{(T)} = \beta_i$.

2.4.4 Definition: Consistency

An estimator $b_i^{(T)}$ of β_i is said to be consistent if $\operatorname*{plim}_{T\to\infty} b_i^{(T)} = \beta_i$. A multivariate extension is straightforward with vectors $b^{(T)}$ and β replacing the scalars $b_i^{(T)}$ and β_i. Consistency requires that a measure of the distribution of the estimator spread around the true value is zero in the limit.

A sufficient condition which is convenient to check the consistency of an estimator involves checking whether or not: $ASY \cdot E(b_i^{(T)}) = \beta_i$, and $ASY \cdot V(b_i^{(T)}) = 0$. For example the OLS estimator b is a consistent estimator of β because $ASY \cdot Eb = \beta$ and $ASY \cdot V(b) = 0$. To prove this, note that $X'X$ is assumed to be a positive definite (nonsingular) matrix. Now $\lim_{T \to \infty} (\frac{1}{T} X'X) = Q$ is a finite nonsingular matrix. Hence the inverse Q^{-1} is also finite. Now, note that

$$\lim_{T \to \infty} (X'X)^{-1} = \lim_{T \to \infty} \frac{T}{T}(X'X)^{-1} = \lim_{T \to \infty} \frac{1}{T}(\frac{X'X}{T})^{-1} \tag{17}$$

$$= 0 \cdot Q^{-1} = 0.$$

We emphasize that consistency does not necessarily imply unbiasedness and vice versa. It is possible to have examples where an estimator is consistent but neither unbiased nor asymptotically unbiased. A popular example is the estimator of the mean of the Cauchy population. (See Exercise 2.2, below.)

2.5 EFFICIENCY (MINIMUM VARIANCE) PROPERTIES

We can use the so-called Cramér-Rao lower bound for the unbiased estimator of β to prove that b is efficient. It is convenient to discuss the Cramér-Rao lower bound on the variance in multivariate terms.

2.5.1 Definition: (CUAN)

An estimator b is consistent uniformly asymptotically normal (CUAN) if it is consistent, if $\sqrt{T}(b - \beta)$ converges in distribution to $N(0,\Sigma)$, and if the convergence is uniform over any compact (closed and bounded) subset of the parameter space.

2.5.2 Definition: (Relatively Asympotically Efficient)

If b is a CUAN estimator with asymptotic covariance matrix Σ/T and b^1 is a CUAN estimator with asymptotic covariance matrix Ω/T, then b is asymptotically efficient relative to b^1 if the matrix $\Omega - \Sigma$ is positive semidefinite.

2.5.3 Definition: (Asymptotically Efficient)

A CUAN estimator is asymptotically efficient if it is asymptotically efficient relative to any other CUAN estimator.

2.5.4 Definition: (Regular Density)

A density $f(\cdot,\beta)$ is regular with respect to its first derivative if

$$E\left[\frac{\partial \log f(\cdot,\beta)}{\partial \beta_i}\right] = 0\,, \quad (i = 1,2,...)\,.$$

2.5.5 Definition: (Information Matrix)

Let $y = (y_1,...,y_T)$ be a random sample from a population with density $f(\cdot,\beta)$ which is regular with respect to its first two derivatives. Let the likelihood function be

$$L(y,\beta) = \prod_{t=1}^{T} f(y_t,\beta)\,.$$

Then the information matrix is defined as

$$I = -E\left[\frac{\partial^2 \log L(y,\beta)}{\partial\beta\partial\beta'}\right]$$

with the i, j^{th} element given by

$$I_{ij} = -E\left[\frac{\partial^2 \log L(y,\beta)}{\partial\beta_i\partial\beta_j}\right].$$

2.5.6 Theorem: (Cramer-Rao)

Suppose that b is any unbiased estimator, with covariance matrix Σ. Then the matrix $\Sigma - I^{-1}$ is positive semidefinite ($\Sigma - I^{-1}$ "$\geq$" 0). Thus I^{-1} provides a lower bound to the covariance matrix of an estimator. Here I represents the information matrix.

An intuitive interpretation of this Theorem is that whenever a specific estimator has the covariance matrix which equals the lower bound I^{-1}, it should be "most efficient" in some sense. We note in passing that an estimator may be efficient even though its covariance matrix does not quite attain this lower bound. For such estimates, efficiency is shown in some other manner, such as by showing that the estimator is a function of a so-called "sufficient" statistic, [Rao 1973, p. 130]. Roughly speaking, a sufficient estimator is one that contains all the information in the sample about the parameter. A precise characterization of sufficiency requires a certain factorization of the likelihood function.

2.5.7 Corollary (To Cramer-Rao Theorem)

A sufficient condition for a CUAN estimator to be asymptotically efficient is that its asymptotic covariance matrix (Σ/T) be equal to the lower bound

$$\frac{1}{T}\lim_{T\to\infty}\left(\frac{1}{T}I\right)^{-1}. \tag{18}$$

Thus, we may write the log likelihood of the sample, differentiate it with respect to the parameters to construct the information matrix I. Next, we note that the Cramer-Rao lower bounds for the variance of OLS estimator b is $\sigma^2(X'X)^{-1}$. This can be obtained by differentiating the log likelihood twice with respect to the parameters.

2.6 STABILITY PROPERTIES

Since OLS has several impressive theoretical properties, including efficiency, a careful study of b must be the first step in many kinds of empirical research. However, OLS may be unsatisfactory because it may have high MSE, and may not be usable in certain practical situations. Most practitioners are aware of continuing data revisions, data additions, etc. If an estimated regression equation is to be useful, it should not be too sensitive to small data changes when the data matrix X is ill-conditioned in the sense described below. We use certain classical concepts in perturbation theory developed by Householder, Turing, Varga, Von Neumann, Wilkinson and others. Wilkinson [1965] is regarded by numerical analysts as a modern day classic reference for these developments.

2.6.1 Stability Under Data Perturbation

A less formal criterion for choosing among estimators may be based on its "stability", "resistance" or "robustness" under certain perturbations in the available data.

We consider perturbations of the X matrix and y vector which arise due to "measurement errors", data revisions and modifications made from time to time by the data producing agencies (government or private) in addition to the rounding errors. We are mainly interested in assessing the effects of these perturbations on the OLS estimator b. We shall see that b is unstable with respect to such perturbations in the presence of multicollinearity. A geometrical discussion of this is found in Swindel [1974] and Moulaert [1976]. Chapter 5 discusses multicollinearity in detail.

For studying perturbation theory, it is helpful to consider the singular value decomposition, (Belsley and Klema [1974], Rao [1973], p. 42) of the $T \times p$ matrix X,

$$X = H \Lambda^{1/2} G' , \tag{19}$$

where the matrices Λ and G are as defined in Section 1.3 of Chapter 1 (See Eq. (14)). The H matrix is a $T \times p$ matrix of the coordinates of the observations along the principal axes of X standardized in the sense that $H'H = I$. The eigenvectors g_i are the direction cosine vectors which orient the ith principal axis of X with respect to the given axes. The $\lambda_i^{1/2}$ (i = 1,...,p) are traditionally called the singular values of X, which seems to be a misnomer (they are not all zero for a singular matrix). The notation $X^+ = (X'X)^{-1}X'$ represents the so-called Moore-Penrose generalized inverse of X (See Rao, 1973). Denote the "matrix 2-norm" of X by $\|X\|_2 = \max_i (\lambda_i^{1/2})$ which equals the largest singular value of X. For example, $\|I\|_2 = 1$, where I is the identity matrix. In matrix algebra, the norm of a matrix is a measure of its size in some sense, Wilkinson [1965, pp 56-60]. An alternative matrix norm is the Euclidian norm which is defined as the square root of the sum of squares of all elements, similar to Euclidian distances.

2.6.2 Definition: Condition Number of a Matrix

The "condition number" of matrix X is given by $K^{\#} = \|X\| \cdot \|X^+\|$, where $\| \ \|$ denotes any "consistent" matrix norm defined by the property that the norm of the identity matrix must be unity, i.e., $\|I\| = 1$. Since the eigenvalue lambdas are arranged according to our convention in a declining (non-increasing) order of magnitude, the "2-norm" leads to the condition number

$$K^{\#} = (\lambda_1/\lambda_p)^{1/2} . \tag{20}$$

(Hint: the largest singular value of X^+ is $(\lambda_p)^{-1/2}$, by arrangement)

To study the effect of perturbation in y and X on the regression coefficients, it is useful to have the $K^{\#}$ available. The precise bounds involving $K^{\#}$ are discussed in Chapter 5. Here we are merely giving a useful definition.

2.6.3 Definition: Robustness

Mallows [1979] defined robustness to be a collection of three attributes: (i) Resistance: insensitive to the presence of a small number of bad data values; (ii) smoothness: the technique should respond only gradually to the injection of a small number of gross errors, to perturbations in data and to small changes in the model; (iii) Breadth: applicability to a wide variety of situations.

2.6.4 Definition: Hat Matrix and Studentized Residuals

The fitted residuals for the model (1) are

$$\begin{aligned} \hat{u} &= y - Xb \\ &= (I - X(X'X)^{-1}X')y \\ &= (I - \hat{H})y. \end{aligned}$$

The "Hat matrix" gives the predicted (fitted) value of y, $\hat{y}$ from the observed values: $\hat{y} = Hy$. Let the elements of the "Hat matrix" be denoted by

$$\hat{h}_{tj} = x_t'(X'X)^{-1} x_j \ , \tag{21}$$

where x_t' and x_j' denote the tth and j^{th} ROWS of X, i.e. they are t^{th} and j^{th} observations on all p regressors.

Hoaglin and Welch [1978] discuss additional properties of the hat matrix including the fact that $o \leq \hat{h}_{tt} \leq 1$. Note that $\hat{h}_{tt}$ can be obtained simply by computing the regression once again with all y_t replaced by $y_t + 1$ and finding the predicted y for the second regression $\hat{y}_t^{(+1)}$. It is easy to verify that

$$\hat{h}_{tt} = \hat{y}_t^{(+1)} - \hat{y}_t \ .$$

Recall from (19) the singular value decomposition $X = H\Lambda^{1/2}G'$, where H is $T \times p$ matrix with elements H_{tj} (say) containing normalized principal component coordinates, and G contains the eigenvectors of $X'X$. Using this the hat matrix can be written as

$$\begin{aligned} \hat{H} &= X(X'X)^{-1}X' = H\Lambda^{1/2}G'G\Lambda^{-1}G'G\Lambda^{1/2}H' \\ &= HH', \end{aligned}$$

$$\hat{h}_{tt} = \sum_{j=1}^{p} H_{tj}\, H_{jt}' = \sum_{j=1}^{p} H_{tj}^2 \ , \qquad (since\ H_{jt}' = H_{tj})$$

If $\hat{h}_{tt}$ is a given constant, this is an equation of an ellipse. For t^{th} observation, $\hat{h}_{tt}$ measures its "leverage," i.e. the extent to which it stands out in the space spanned by X. It is proportional to the so-called Mahalanobis distance of x_t from the center of the data at $\bar{x}$.

Given any observation (say t^{th}) for the p regressors the fitted y is given by $\hat{y}_t = x_t'b$ by plugging in the values of the observation in the model. Now the variance

$$V(\hat{y}_t) = V(x_t'b) = x_t'\ V(b)\ x_t$$

$$= x_t'[\sigma^2(X'X)^{-1}]x_t$$

$$= \sigma^2\ \hat{h}_{tt} \qquad \text{(21a)}$$

from (21). Recall that the variance covariance matrix of the true errors u is $\sigma^2 I$. For the fitted errors $\hat{u} = (I - \hat{H})y$ the covariance matrix is given by $V(\hat{u}) = \sigma^2(I - \hat{H})$. Hence the variance for t^{th} residual is $V(\hat{u}_t) = \sigma^2(1 - \hat{h}_{tt})$, while the standard error is given by its square root. Thus the "studentized" residual is given by

$$Student\,(\hat{u}_t) = \hat{u}_t/\hat{s}\,(1 - \hat{h}_{tt})^{1/2}\ , \qquad \text{(21b)}$$

where $\hat{s}^2 = (y-Xb)'(y-Xb)(T-p-1)^{-1}$ estimates σ^2. If the notation p includes one regressor for the intercept this denominator is $(T-p)$ instead of $(T-p-1)$. If $Student\,(\hat{u}_t)$ is large, this suggests that t^{th} observation has a large residual.

2.6.5 Influence Function

Omitting the tth observation, and recomputing all regression coefficients by OLS say $b_{(-t)}$, we can assess the "influence" of the tth observation on b empirically. One measure of "empirical influence function" may be the squared Euclidian distance between b and $b_{(-t)}$, which is the sum of squared differences: $\sum_{i=1}^{p} (b_i - b_{(-t)i})^2$. More generally one can consider a *weighted* sum of squared differences. Along these lines, Cook [1977] first writes the normal theory $(1-\alpha)100\%$ confidence ellipsoid for the unknown β. This is given by the set of all possible β^* vectors which satisfy

$$\frac{(\beta^*-b)'\,X'X(\beta^*-b)}{ps^2} < F(p,T-p-1,1-\alpha) \qquad \text{(22)}$$

where $F(p,T-p-1,1-\alpha)$ is the $1-\alpha$ probability point of the central F variable with p and $T-p-1$ degrees of freedom respectively. Now the influence of tth residual may be measured in terms of these confidence ellipses provided we substitute the vector $b_{(-t)}$ in place of β^* in (22) as:

$$D_t = (b_{(-t)} - b)'X'X(b_{(-t)} - b)/ps^2 \tag{22a}$$

Cook [1977] shows that the cumbersome computation of (22) can be avoided by using the hat matrix. It can be shown that

$$D_t = \frac{[Student\,(\hat{u}_t)]^2}{p} \, \frac{V(\hat{y}_t)}{V(y - Xb)_t} \, ,$$

$$D_t = \frac{[Student\,(\hat{u}_t)]^2}{p} \, \frac{\hat{h}_{tt}}{(1 - \hat{h}_{tt})} \tag{22b}$$

An obvious use of these distance measures is to find out the so-called outliers, or "influential" or extremely unusual points. The distance function D_t is an example of an empirical influence function. It uses observed data.

A rigorous definition of the theoretical influence function given by Hampel [1974] is based on the concept of Prokhorov's distance between two probability distributions. It uses measure theoretic ideas which are beyond the scope of this book.

We may simplify matters by considering the special case of the arithmetic mean, $\bar{y}$, defined for a cumulative density function F having mean $\mu = \int y dF(y)$. We know that the estimator (i.e. statistic) $\bar{y}$ estimates μ if the underlying distribution function is F. This may be stated as: $\bar{y}\{F\} = \mu$, where the curly braces denote an operation of working on a cumulative distribution function. Now we contaminate F by another cumulative density δ_{y^0} which is a unit step function at y^0. In other words $\delta_{y^0} = 1$ for all $y \geq y^0$, and $\delta_{y^0} = 0$ for all $y < y^0$. When the sample is all piled up at y^0 its mean is y^0, i.e., $\bar{y}\{\delta_{y^0}\} = y^0$. The estimator $\bar{y}$ applied to the contaminated distribution function: $(1-\epsilon)F + \epsilon\delta_{y^0}$, for $0 < \epsilon < 1$ will yield: $\bar{y}\{(1-\epsilon)F + \epsilon\delta_{y^0}\} = (1-\epsilon)\mu + \epsilon y^0$ as the new estimate of the mean of the contaminated distribution.

Hampel's influence function for measuring the effect of contamination at any point y on the estimator y^* is defined by

$$\psi(y^*, F, y) = \lim_{\epsilon \to 0} \frac{[y^*\{(1-\epsilon)F + \epsilon\delta_y\} - y^*\{F\}]}{\epsilon} \, .$$

For the particular estimator $\bar{y}$, the theoretical influence function becomes:

$$\begin{aligned} \psi(\bar{y}, F, y) &= \lim_{\epsilon \to 0} \epsilon^{-1}[(1-\epsilon)\mu + \epsilon y - \mu] \\ &= \lim_{\epsilon \to 0} \epsilon^{-1}[\epsilon(y - \mu)] \\ &= y - \mu \, . \end{aligned}$$

If we plot y on the horizontal axis and ψ on the vertical axis $\psi(\bar{y}, F, y)$ is the 45° line through the origin. Since this function increases indefinitely as $y \to \infty$, $\bar{y}$ is said to be non-robust. The median is robust if and only if $F^{-1}(½)$ contains no more than one point.

We will not discuss this topic in further detail in this book. However ψ has many useful properties. Expected value of its square gives the asymptotic variance. The largest absolute value of ψ measures "gross error sensitivity". Effect of perturbation can be measured by a difference between two values of ψ.

2.6.6 Outlier Detection: t test

We have noted that "influencial" points may be regarded as "outliers". After omitting t^{th} observation and estimating the $b_{(-t)}$ regression coefficients based on the remaining observations we can "plug in" the tth observation in the model to predict y_t:

$$\hat{y}_{(-t)} = x_t' \, b_{(-t)} \, .$$

If the tth point is not influential, the expectation $E\,\hat{y}_{(-t)}$ should equal to the observed value y_t of the dependent variable. Now the variance is

$$V[\hat{y}_{(-t)}] = \sigma^2(x_t'[X_{(-t)}'X_{(-t)}]^{-1}\, x_t) \ ,$$

where $X_{(-t)}$ denotes the X matrix without the tth row. Hence a Students' t test for outlier $out\,(t)$ is given by dividing $[y_t - \hat{y}_{(-t)}]$ by its standard error. It can be shown that this t statistic can be written in terms of the h_{tt} and hence in terms of studentized residuals as

$$out\,(t) = Student\,(\hat{u}_t)\left[\frac{T-p-1}{T-p-[Student\,(\hat{u}_t)]^2}\right]^{½} . \qquad (22c)$$

(See Exercise 2.5 at the end of this chapter).

2.7 ADMISSIBILITY PROPERTIES (UNBEATEN SOMEWHERE)

Admissibility is an important concept borrowed from statistical decision theory. A decision maker in our context is a person choosing between various estimators based on a certain loss function; e.g. MSE. Roughly speaking, an "admissible" estimator is one which cannot be beaten by another estimator at all points of the parameter space. The phrase "unbeaten somewhere" is sufficient to ensure admissibility (but not necessary). It is intended to aid the memory of the student. However, similar phrase is not found in textbooks because it can be misleading for certain pathological cases, where one can construct admissible estimators which are not unbeaten.

2.7.1 Definition: Better Estimator (Under MSE Criterion Similar to Section 2.2.4)

An estimator b_1 is better than estimator b_2 if $MSE(b_1) \leq MSE(b_2)$ for all β, and $MSE(b_1) < MSE(b_2)$ (strict inequality) for at least one point in the parameter space.

Clearly, a similar definition can be stated in terms of a "weighted" MSE criterion.

2.7.2 Definition: Admissible Estimator

Let a $p \times 1$ vector b be an estimate of β, and let WMSE(b) $= E(b-\beta)'D(b-\beta)$ be the "weighted" MSE for a specified positive definite matrix D. The estimator b is said to be admissible under WMSE(b) if there exists no other estimator b^* such that $WMSE(b^*) \leq WMSE(b)$ with strict inequality for at least one value of β. (That is, there is no estimator b^* "better" than b everywhere.)

An Example Consider regressing y on x_1

$$y = \beta_1 x_1 + u \ . \tag{23}$$

Now, let $x_1 \equiv 1$ for all observations. In matrix notation,

$$y = \iota\beta_1 + u \ , \tag{23a}$$

where ι is a vector of all ones. The OLS estimator is

$$b_1 = (\iota'\iota)^{-1}\iota' y = \frac{1}{T}\, T\bar{y} = \bar{y} \ . \tag{23b}$$

Clearly, this is the arithmetic mean of T values of y in the sample. For this model it is well known that the OLS estimator is "admissible". Stein [1956] gives earlier references to the result that $b_1 = \bar{y}$ is an "admissible" estimator of β_1 in the above model for an MSE-type criterion function.

2.7.3 Inadmissible Estimator ("Better" Estimator Exists)

If there is another estimator in the set of all possible estimators such that it is "better" than our estimator when evaluated at all possible values of the parameter, then our estimator is said to be "inadmissible". When an estimator is not "inadmissible" it is admissible.

In Figure 2.3 we have three MSE functions associated with three estimators. We assume that these three lines represent the set of all possible estimators, and that the plot includes all possible values of the unknown parameter.

Note that the estimator #1 is inadmissible because #3 is as good or better than #1 everywhere. Also, note that #2 is "admissible" because it is better than the other two in some regions of the above Figure 2.3 assuming we have plotted all possible estimators and all possible values of the parameter.

2.7.4 Limitation of the Admissibility Criterion

The admissibility criterion suggests that all inadmissible estimators be left out of consideration. This leaves the class of all admissible estimators, which however is generally very large, and admissibility says nothing as to how a choice should be made from this large class. The choice is therefore made, in practice, by applying other statistical principles such as unbiasedness, minimaxity, etc.; or by taking into consideration the researcher's prior beliefs regarding the weight to be attached to the different possible values of the parameter. In the last mentioned case, if the prior beliefs about β are expressed as a probability distribution, the MSE integrated or averaged over β gives the Bayesian MSE or "risk". The appropriate estimator, called the minimum average risk estimator, is then the one which minimizes Bayes risk (See Section 2.10).

Another point to be noted is that if a decision rule is inadmissible, then there exists another rule which should be used in preference to it. But this does not mean that every admissible rule is to be preferred to any inadmissible rule. It is easy to construct examples of rules which are admissible but which it would be absurd to use. An example in point estimation is an estimator which is equal to some constant c whatever be the observations. The MSE then vanishes for $\beta = c$ and the estimator is an admissible one. Makani [1977] provides a more sophisticated example.

2.8 MINIMAX PROPERTIES

Consider a pessimistic decision maker who believes that the "worst that can happen will happen". In the context of parameter estimation "worst that can

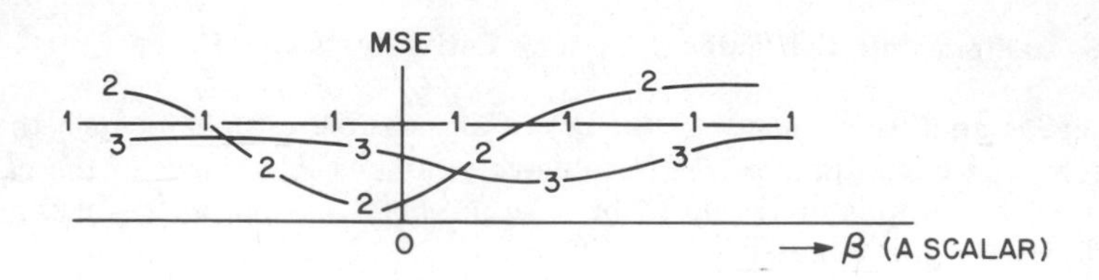

Figure 2.3 (#1 Inadmissible, #2 Admissible, #3 Admissible)

happen" means that the parameter estimates are completely unacceptable. For example, the estimator might have infinite WMSE.

2.8.1 Definition: Minimax Estimator Under MSE Criterion

For a criterion function defined by MSE a minimax estimator is one that minimizes the maximum MSE.

In Figure 2.4 we plot a new set of three estimators which are assumed to be all possible estimators.

In Figure 2.4 estimators #1 and #2 have the same maximum MSE, and the maximum MSE of #3 is larger than that of the other two. Therefore both are minimax. The minimaxity criterion will not help us in choosing between them. It is interesting to note that #3 is better than #1 and #2 in most regions of the parameter space. The minimaxity criterion being pessimistic, it will not recommend #3 because in a small region near x it is worse than the other two. In fact, by the minimaxity criterion #3 will be regarded as the worst of the three. On the other hand, #2 is admissible because near x it is unbeaten.

In Figure 2.5, the MSE for #3 is better than #2 everywhere, hence #2 is inadmissible. Since #3 is better than others at least somewhere, #3 is admissible, but not minimax because its maximum MSE is somewhat larger than the maximum MSE of #1 at x.

2.8.2 Digression: Minimaxity and Ridge Regression

In the following chapters we will note that some of the ridge regression estimators are generally not minimax, except for certain choices of the biasing parameter k in a range close to zero. The mathematical statisticians warn the users

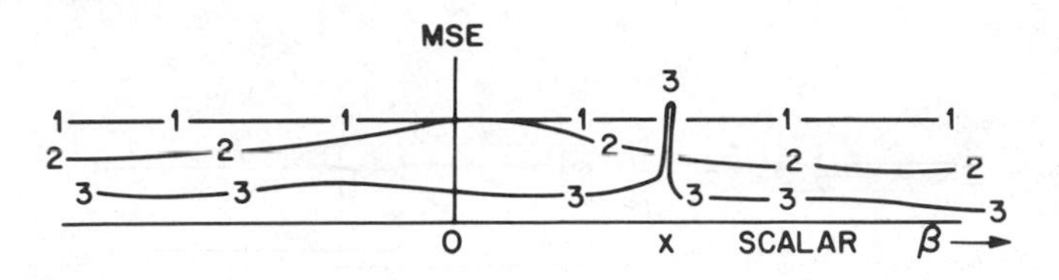

Figure 2.4 (#1 Minimax Inadmissible, #2 Minimax Admissible, #3 is Admissible)

of ridge regression against using larger k values because of this lack of minimaxity. We feel that blind reliance on minimaxity may not be always good. A "skilled" use of estimators that may not be minimax can sometimes help lead the researcher closer to the truth about the model, especially when the data are multicollinear.

2.8.3 Proposition: Admissible Estimator with Constant MSE Must Be Minimax

It can be verified that any admissible estimator which has constant MSE must also be minimax. In Figure 2.5 the constant risk estimator #1 is minimax because it is admissible (unbeaten) near x. In Figure 2.4 the constant MSE estimator #1 is not minimax because it is inadmissible (beaten everywhere by #2).

The OLS estimator has constant MSE, hence whenever OLS is admissible it must be minimax also.

2.9 ACCEPTABLE RANGE OF SHRINKAGE FACTOR FOR REDUCING THE MSE OF OLS

In the univariate model $(y, \iota\beta_1, \sigma^2)$ of Eq. (23) the OLS estimator $b_1 = \bar{y}$ is unbiased and

$$MSE(b_1) = V(b_1) = \sigma^2(\iota'\iota)^{-1} = \frac{\sigma^2}{T}. \tag{24}$$

Now, we consider MSE of a certain biased estimator $b_1\delta$ of β_1 where the shrinkage factor δ is a constant.

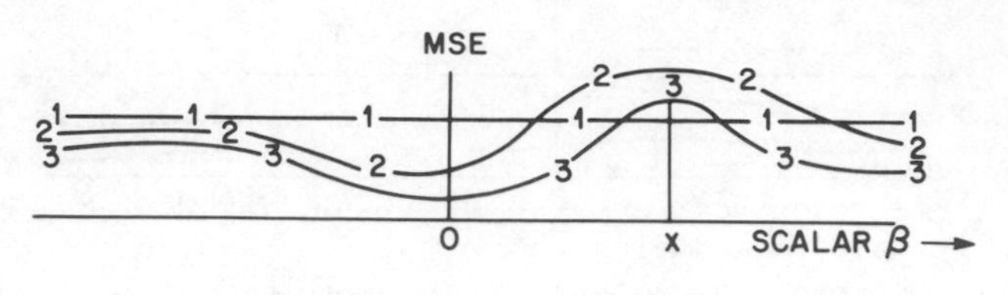

Figure 2.5 (#1 Minimax Admissible, #2 Inadmissible, #3 Admissible but Not Minimax)

$$
\begin{aligned}
MSE(b_1\delta) &= E(b_1\delta - \beta_1)^2 \\
&= E(b_1\delta - \beta_1\delta + \beta_1\delta - \beta_1)^2 \\
&= E[\delta(b_1 - \beta_1)]^2 + (\beta_1\delta - \beta_1)^2 \\
&= \delta^2 V(b_1) + [Bias(b_1\delta)]^2 , \qquad (25)
\end{aligned}
$$

where we have used the fact that $Eb_1 = \beta_1$.

Next, we study the conditions under which the biased estimator reduces the MSE of OLS

$$
\begin{aligned}
MSE(b_1) - MSE(\delta b_1) &= \frac{\sigma^2}{T} - \frac{\delta^2\sigma^2}{T} - (\delta - 1)^2\beta_1^2 \\
&= (1 - \delta)(1 + \delta)\frac{\sigma^2}{T} - (1 - \delta)^2\beta_1^2 . \qquad (26)
\end{aligned}
$$

This expression is strictly positive provided both terms above are strictly positive. The first term being positive implies the following inequality: $1 > \delta > -1$. For the entire expression to be positive, the second term implies $(1 + \delta) > (1 - \delta)Q_1 = Q_1 - \delta Q_1$, where $Q_1 = T\beta_1^2/\sigma^2$ is the unknown noncentrality parameter or signal to noise ratio (SNR) in engineering terms. Rearranging terms, we require $\delta + \delta Q_1 > Q_1 - 1$, that is,

$$
\delta > \frac{Q_1 - 1}{Q_1 + 1} . \qquad (27)
$$

Collecting the two inequalities, we write the following proposition.

Proposition The univariate $MSE(b_1) = \sigma^2/T$ is reduced by a biased estimator, δb_1 with a constant shrinkage factor δ in the following "acceptable" range

$\dfrac{Q_1 - 1}{Q_1 + 1} < \delta < 1$. If we have prior information about Q_1, we can uniformly reduce the MSE of OLS. Q_1 involves knowing the length of the regression parameter and its variance. A conservative approach would be to use δ closer to unity, that is staying close to the OLS estimator. Let an upper bound on Q_1 be denoted by Q_1^{up}. Now we can write

$$\frac{Q_1 - 1}{Q_1 + 1} < \frac{Q_1^{up} - 1}{Q_1^{up} + 1} < \delta < 1 \, . \tag{28}$$

A general result in terms of coefficient of variation is found in Blight [1971]. Blight also suggests replacing the variance in the coefficient of variation by its Cramér-Rao lower bounds. However, these methods do not seem to be applicable to regression analysis.

2.9.1 Novel Estimation of Acceptable Range of δ Using Noncentral F Distribution

The result in (28) shows that prior knowledge of Q_1^{up} will suggest an "acceptable range" of the shrinkage factor δ. The admissibility of OLS in the univariate case discussed above implies that it is impossible to dominate OLS everywhere in the parameter space. However, we may be willing to specify that extremely large values of β_i^2 or extremely small values of σ^2 are unrealistic for our problem. Furthermore, we may be willing to accept a small (say, 5%) chance that the unrealistic values are in fact true, rather than an ironclad guarantee.

The OLS estimate of β_1 is $b_1 = \bar{y}$, and an unbiased estimate of σ^2 is s^2 defined by

$$(T - 1)s^2 = (y - \iota b)'(y - \iota b) = \sum_{t=1}^{T} (y_t - \bar{y})^2 \, . \tag{29}$$

An OLS estimate of Q_1 is $Q_1^{OLS} = Tb_1^2/s^2$, which is a random variable. Recall that $b_1 \sim N(\beta_1, \sigma^2/T)$ implies that $\sqrt{T}\,(b_1 - \beta_1)/\sigma \sim N(0,1)$. Furthermore, since $y_t \sim N(0, \sigma^2)$, $(T - 1)s^2/\sigma^2 = \sum_{t=1}^{T} (y_t - \bar{y})^2/\sigma^2$ is a sum of squares of unit normal variables. It is known that $(T - 1)s^2/\sigma^2$ is distributed as a central chi-square variable with $T - 1$ degrees of freedom. We write $(T - 1)s^2/\sigma^2 \sim \chi_c^2(T - 1)$ with subscript c to suggest "central". Now, Tb_1^2/σ^2 is distributed as the square of a normal variable with non-zero mean and one degree of freedom (d.f.). We write $Tb_1^2/\sigma^2 \sim \chi_{nc}^2(1, T\beta_1^2/\sigma^2) = \chi_{nc}^2(1, Q_1)$, a "non-central" chi-square variable with Q_1 as non-centrality parameter. The ratio $(T - 1)\chi_{nc}^2(1, Q_1)/\chi_c^2(T - 1)$ is a non-central F variable with 1 d.f. in the numerator, $T-1$ in the denominator and Q_1 as the non-centrality parameter. We write after cancelling the σ^2

$$\frac{(T-1)\chi^2_{nc}(1,Q_1)}{\chi^2_c(T-1)} \sim \frac{(T-1)Tb_1^2}{(T-1)s^2} = Q_1^{OLS} \sim F_{nc}(1,T-1,Q_1) \ . \qquad (30)$$

Under the null hypothesis $\beta_1 = 0, Q_1^{OLS}$ is a central F variable and the usual F table can be used to find the level of significance. For example, if $f_1 \sim F_c(1,T-1)$ we can find tabulated value $F^{Tab}(1,T-1,0)$ so that $Pr[f_1 > F^{Tab}(1,T-1,0)] = 0.05$. In our context, we can define $\hat{Q}_1^{up} = F^{Tab}(1,T-1,0)$ to satisfy $Pr[Q_1^{OLS} > \hat{Q}_1^{up}] = \alpha$ (where $\alpha = .05$, say).

Note that a non-zero non-centrality parameter shifts the curve for the central F distribution to the right hand side as it were. Hence the probability content to the left side of $F^{Tab}(1,T-1,0)$ for a non-central F will be smaller. In Figure 2.6 the area (probability content) marked by + is larger than the area marked by 0. Hence we have the following inequality proposition:

Proposition Let $Q_1^{OLS} \sim F_{nc}(1,T-1,Q_1)$. Now, define the tabulated value $F_1^{Tab}(1,T-1,Q_1)$ by the relation $Pr[Q_1^{OLS} > F_1^{Tab}(1,T-1,Q_1)] = \alpha$.

We have the following inequality among tabulated values

$$\hat{Q}_1^{up} = F_1^{Tab}(1,T-1,0) < F_1^{Tab}(1,T-1,Q_1) \ . \qquad (31)$$

Instead of giving a formal proof we refer the reader to the accompanying Figure 2.6.

Thus, $\hat{Q}_1^{up}$ from the central F tables will lead to a liberal estimate of the acceptable range (28) of shrinkage factors δ. That is, it may permit excessive shrinkage because $(\hat{Q}_1^{up} - 1)/(\hat{Q}_1^{up} + 1)$ is small (close to zero) when $\hat{Q}_1^{up}$ is small.

This suggests that we have to use the non-central F distribution based on estimated Q_1^{OLS}:

$$Q_1^{OLS} \sim F_{nc}(1,T-1,Q_1^{OLS}) \ . \qquad (32)$$

Since tabulated values of F_{nc} are not readily available, we may use the following approximation by the "central" F distribution:

$$F_{nc}(1,T-1,Q_1^{OLS}) = (1 + Q_1^{OLS})\, F_c\left[\frac{1+Q_1^{OLS}}{1+2Q_1^{OLS}}\, ,\ T-1\right], \qquad (33)$$

where $Q_1^{OLS} \geq 0$. We can write the probability

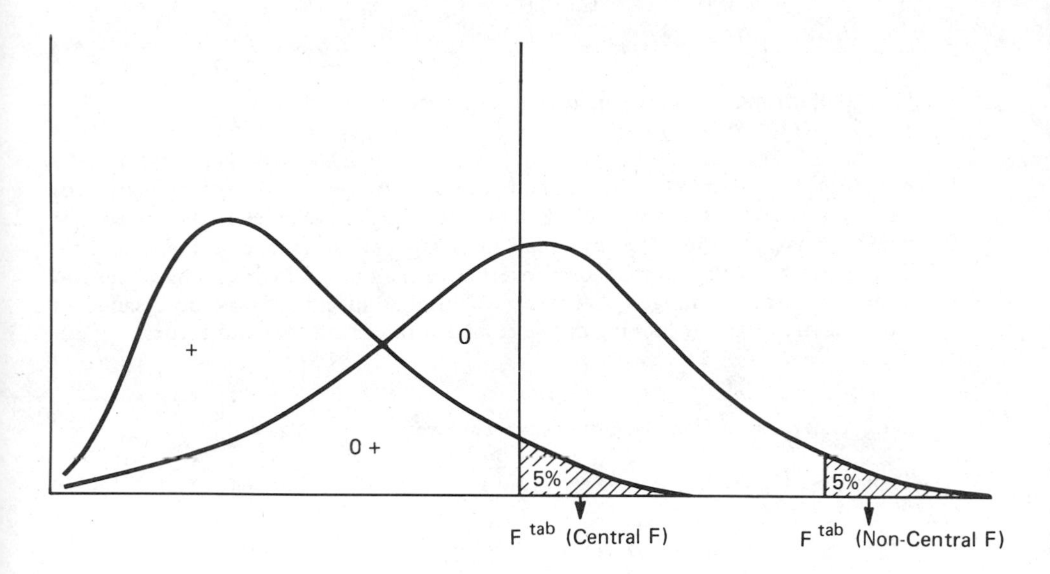

Figure 2.6

$$Pr\,[Q_1^{OLS} < F_{nc}^{Tab}(1,T-1,Q_1^{OLS})]$$

$$= Pr\left[Q_1^{OLS} < (1 + Q_1^{OLS})\; F_c^{Tab}\left(\frac{1 + Q_1^{OLS}}{1 + 2Q_1^{OLS}}\,,\; T-1\right)\right] = \alpha\,. \tag{34}$$

For specified α (say, .05) we can find the value of

$$F_c^{Tab}\left(\frac{1 + Q_1^{OLS}}{1 + 2Q_1^{OLS}}\,,\; T-1\right)$$

from the usual F tables for the central F distribution. This leads to a new estimate of the upper bound denoted by

$$\tilde{Q}_1^{up} = (1 + Q_1^{OLS}) F_c^{Tab} \left[\frac{1 + Q_1^{OLS}}{1 + 2Q_1^{OLS}}, T-1 \right].$$

Now the estimate (approximate) of the acceptable range for shrinkage δ is $1 > \delta > (\tilde{Q}_1^{up} - 1)/(\tilde{Q}_1^{up} + 1)$.

As long as δ lies in this acceptable range, we have $MSE(\delta\, b_1) < MSE(b_1)$, except for the $100 \times \alpha\%$ (e.g., 5%) approximate probability (cases) that the true Q_1 is larger than $\tilde{Q}_1^{up}$. Whenever the true Q_1 is larger than $\tilde{Q}_1^{up}$, the $MSE(b_1)$ may be small and shrinkage is inappropriate. The researcher who wants to be conservative may do so by choosing δ close to unity in the above range. A practical choice of specific δ may be decided in data analytic terms by looking at the results with certain trial and error.

2.9.2 Selection of Specific Shrinkage Factors

Thompson [1968] has compared, among others, the following specific choices of δ:

$$i)\ 1; \quad ii)\ Q_1/(1 + Q_1); \quad iii)\ Q_1^{OLS}/(1 + Q_1^{OLS}),$$

where $Q_1 = T\beta_1^2/\sigma^2$, and $Q_1^{OLS} = Tb_1^2/s^2$.

The motivation for considering the choice ii) is as follows. If we minimize

$$MSE(\delta\, b_1) = \delta^2 \frac{\sigma^2}{T} + (1 - \delta)^2 \beta_1^2 \tag{35}$$

by differentiating it with respect to δ, and set the derivative equal to zero (first-order condition for a minimum), we have

$$2\delta \frac{\sigma^2}{T} - 2\beta_1^2(1 - \delta) = 0\,.$$

Upon dividing by $2\sigma^2/T$ we have $\delta = Q_1 - Q_1\delta$ leading to $\delta = Q_1/(1 + Q_1)$. Thus $Q_1/(1 + Q_1)$ is in some sense an optimal choice of δ. However, it is not operational because it involves the unknown non-centrality parameter Q_1. From our discussion of the non-central F distribution of Q_1^{OLS} we suggest the new shrinkage factor

$$iv)\ \tilde{Q}_1^{up}/(1 + \tilde{Q}_1^{up}), \tag{36}$$

where

$$\tilde{Q}_1^{up} = (1 + Q_1^{OLS})\, F_c^{Tab}\left[\frac{1 + Q_1^{OLS}}{1 + 2Q_1^{OLS}}\,,\, T-1\right], \tag{37}$$

where F_c^{Tab} denotes the tabulated values (upper tail) of the central F distribution with the indicated degrees of freedom in the numerator and the denominator respectively.

The properties of the new shrinkage factor (37) can be studied graphically as in Thompson [1968]. (See Exercise 2-6).

2.10 BAYES THEOREM

Bayes Theorem dates back to the year 1763 (reprinted in Bayes [1958]) before the American revolution. It is also revolutionary in some sense, and continues to be a useful tool. Suppose β is any scalar unknown parameter. Since it is unknown, its values are uncertain, hence β can be treated as a random variable. To have some information about β we can begin with the some subjective information or belief about β. We can then go back and look for sample evidence y about β and combine this with our prior knowledge to obtain a revised information or opinion about it. This is essentially the process involved in a Bayesian analysis. Usually, the prior possibilities about the values of β are formulated in terms of a probability density, called a prior probability density (pd). The random sample vector y is expressed in terms of probability of sample data and called a likelihood function. And the revised information, called the posterior probability density, is obtained by combining the prior probability distribution with the likelihood function. Specifically, using the definition of conditional probabilities we have

$$p(y\,|\,\beta) = \frac{p(y,\beta)}{p(h)} \quad \text{and} \quad p(\beta|y) = \frac{p(y,\beta)}{p(y)}\,.$$

Now the first equation gives $p(y,\beta) = p(y\,|\,\beta)p(b)$, which may be substituted in the numerator of the second to yield

$$p(\beta|y) = \frac{p(\beta)\,p(y\,|\,\beta)}{p(y)} \tag{38}$$

with $p(y) \neq 0$.

The unconditional probability of observing the data, $p(y)$, is equal to $\Sigma_\beta\, p(\beta)\, p(y\,|\,\beta)$ in the discrete case and equal to $\int_\beta p(\beta)\, p(y\,|\,\beta)d\beta$ in the continuous case, and therefore it is considered as the reciprocal of the normalizing constant for the $p(\beta|y)$. We can thus write

$$p(\beta|y) = e^* \, p(\beta) \, p(y|\beta) \, , \quad e^* = \frac{1}{p(y)}$$

$$\propto p(\beta) \, p(y|\beta)$$

$$\propto p(\beta) \, L(\beta) \, ,$$

where $p(\beta|y)$ is the posterior *pd* for the parameter β given the data vector y,$p(\beta)$ is the prior *pd* for β and $p(y|\beta)$, viewed as a function of β, is the familiar likelihood function $L(\beta)$. Equation (38) is a statement of Bayes theorem. All the above equations remain the same if β is a vector of unknown parameters in the regression model. According to a folklore Mr. Bayes did not publish this theorem while he was alive, because he was unsure about its subtle implications for inductive science.

To illustrate Bayes theorem consider an experiment with two Bags- the first has 1 black ball and 5 white balls and the second has 2 black and 3 white balls. A bag is chosen at random and a ball is picked. It turns out to be white. What is the probability that the first bag was chosen?

In the above experiment the unknown parameter β takes values 1 or 2, depending on which is the chosen bag and this is what we want to know. *A priori,* we know that there are two possible values for β, that is, Bag 1 and Bag 2 each can be assigned a 50% chance of being chosen. Thus our prior for β can be formulated in terms of a discrete *pd* given as

β	$p(\beta)$
1	1/2
2	1/2 .

Now the sample evidence of size one represents a white ball. The probability of this data, given that $\beta = 1$ is true is $p(y|\beta = 1) = L(\beta = 1) = 5/6$. Therefore the posterior probability of $\beta = 1$ given data is

$$p(\beta = 1|y) \propto \frac{1}{2} \times \frac{5}{6} = \frac{5}{12} \, .$$

Similarly, $p(\beta = 2|y) \propto 1/2 \times 3/6 = 3/12$. To obtain the exact posterior probability we note that the probability of data $p(y) = \Sigma_\beta \, p(\beta) \, p(y|\beta) = 8/12$. Thus the exact posterior probabilities are $p(\beta = 1|y) = 5/12 \times 12/8 = 5/8$ and $p(\beta = 2|y) = 3/8$. The results can be summarized in a table as

β $\quad p(\beta) \quad L(\beta) = p(y|\beta) \quad p(\beta) \, L(\beta) \quad p(\beta|y) =$ *Posterior*

1	1/2	5/6	5/12	5/8
2	1/2	3/6	3/12	3/8
	1		$p(y) = 8/12 = \sum_{\beta} p(\beta) L(\beta)$	1

These calculations indicate that the posterior odds for Bag 1 to Bag 2 are 5/3, which is greater than the prior odds ratio.

Consider another example in which a researcher wants to analyze the unknown average income β of a population. In this case β can have infinite number of possibilities, and thus can take any value in $-\infty < \beta < \infty$. Any specification of prior *pd* in this case will be in terms of a continuous *pd*, say normal. The probability of a random sample of incomes for given β can be written. We can then obtain posterior *pd* of β. If we are also interested in the variation in income σ^2, then the marginal posterior density of each parameter β and σ^2 is obtained by integrating out the joint posterior density $p(\beta, \sigma^2|y)$ over the remaining parameters.

Note that the sampling method of inference uses only the likelihood function, which according to the "likelihood principle" constitutes the entire information about the parameter β. In the Bayesian approach, however, the posterior pd is used to make inferences about β. If β is a vector then the marginal distribution of each parameter is obtained by integrating the joint posterior probability over the remaining parameters. The Bayesian inference entails (though not always) an analysis of the mean and/or mode (especially if the posterior distribution is asymmetric) and variance of the posterior marginal distributions.

The functional form of the joint posterior distribution of the parameters depends on the form of the prior probability density function which can be taken as normal, gamma, beta, etc. It is often convenient to select the priors such that the functional form of the posterior is the same as that of the prior so that the task of integration to obtain marginal distributions is easily performed. A class of such priors for the Bayesian approach is called the "conjugate priors". The advantage of these priors is that every time new sample information is observed the revision of the opinion about the parameter of interest and its posterior distribution can easily be obtained from the previous stage, since the posterior distribution has the same functional form as the prior. A disadvantage of these priors is that we may not have such prior knowledge about the parameters, in which case a "noninformative" or "diffuse" prior can be used. Such priors recommended by Jeffreys [1961] are also known as "improper" priors because their probabilities do not add to one. The improper prior about a parameter, which can take any value from $-\infty$ to ∞, is represented by the uniform probability density function. An advantage of noninformative priors is that the posterior density is simply proportional to the likelihood function.

Hence the mode (mean) of the posterior marginal distribution of a regression coefficient is also the Maximum Likelihood (ML) estimator if the likelihood function is asymmetric (symmetric). Of course, the interpretation of the Bayesian and classical estimates are different even though the noninformative priors yield the Bayesian estimates which are similar to the classical estimates.

2.10.1 Minimum Expected Loss (MELO) Bayesian Estimator

As pointed out earlier, in Bayesian analysis we determine and study the complete posterior density of the parameter β. One could therefore obtain summary measures such as central tendency, dispersion and skewness. If desired, a measure of central tendency can be considered as a point estimate for β. To help decide upon a particular measure of central tendency, let us consider a quadratic loss function $L^\circ(\beta,\hat{\beta}) = (\beta - \hat{\beta})'D(\beta - \hat{\beta})$, where D is a known nonstochastic positive definite symmetric matrix and $\hat{\beta}$ is a point estimate based on the given data vector y. Note that L° is random because β is random. Now we choose that value of $\hat{\beta}$ for which the posterior expected loss, that is

$$E\ L^\circ(\beta,\hat{\beta}) = E(\beta - \hat{\beta})'D(\beta - \hat{\beta}) = \int_\beta L^\circ(\beta,\hat{\beta})\ p(\beta|y)d\beta \tag{39}$$

$$= E[(\beta - E\beta) - (\hat{\beta} - E\beta)]'D[(\beta - E\beta) - (\hat{\beta} - E\beta)]$$

$$= E(\beta - E\beta)'D(\beta - E\beta) + (\hat{\beta} - E\beta)'D(\hat{\beta} - E\beta)$$

is minimum. It is easy to verify that if the $E\ L^\circ(\beta,\hat{\beta})$ exists, then it is minimized for $\hat{\beta} = E\beta$. Thus for (positive definite) quadratic loss functions the mean of the posterior density (if it exists) is an optimal point estimate. Such an estimator, $\hat{\beta}$ in the sampling theory framework, is the minimum average risk estimator.

This chapter has discussed the idea of a good regression estimator from various criteria. There is no general agreement on the subject, and one has to be aware of the properties of these criteria to make a judicious choice for a particular scientific estimation problem.

EXERCISES

2.1

(a) Let u be a $T \times 1$ vector with zero mean and covariance matrix Ω. If A is any nonstochastic matrix of order $T \times T$, show that $Eu'Au = TrA\ \Omega$.

(b) In the linear regression model $y = X\beta + u$ with p regressors, and where $Eu = 0$ and $Euu' = \sigma^2 I$, show that

$$s^2 = \frac{1}{T-p}\,\hat{u}'\hat{u}\,, \quad \hat{\sigma}^2 = \frac{1}{T}\,\hat{u}'\hat{u}\,, \quad \ddot{\sigma}^2 = \frac{1}{T-p+1}\,\hat{u}'\hat{u}$$

are all consistent estimators of σ^2. Show that the first is unbiased, and the second is biased but asymptotically unbiased. Show that the third has the lowest MSE here. (See Rao [1973], p. 316).

2.2

(a) Define consistency, unbiasedness and efficiency of an estimator.

(b) Consider a scalar parameter β and let $\hat{\beta}$ be its estimator with the following probability distribution.

$\hat{\beta}$	$P(\hat{\beta})$
β	$1 - \frac{1}{T}$
T	$\frac{1}{T}$

Show that the estimator is consistent and asymptotically unbiased but biased in small samples.

(c) Consider the following case

$\hat{\beta}$	$P(\hat{\beta})$
$\beta + \frac{\alpha_1}{T}$	$1 - \frac{1}{T}$
$\beta - \alpha_2 T$	$\frac{1}{T}$

Show that $\hat{\beta}$ is consistent, but as long as $\alpha_2 \neq 0$ this $\hat{\beta}$ is neither unbiased nor asymptotically unbiased.

(d) If $\hat{\beta} = \beta + \alpha_1 T$ with probability 1/3 and $\hat{\beta} = \beta - \alpha_2 T$ with probability 2/3, then show that $\hat{\beta}$ is not consistent, but if $\alpha_1 = 2\alpha_2$ then $\hat{\beta}$ is unbiased.

(Reference Srinivasan [1970]).

2.3

(a) Suppose an estimator $\hat{\beta}$ of β in $y = X\beta + u$ is such that $MSE(\hat{\beta}) < MSE(b)$ for all β; where b is the least squares estimator. Then, is it true that $\hat{\beta}$ is minimax?

(b) If an estimator is admissible it can still be beaten by another estimator in some parameter space. Comment.

2.4 Let $\hat{\beta}_1$ and $\hat{\beta}_2$ be the two estimators of the parameter vector β. Show that the criterion $MtxMSE(\hat{\beta}_2)$ "$\geq$" $MtxMSE(\hat{\beta}_1)$ is the same as (i) the criterion $WMSE(\hat{\beta}_1) - WMSE(\hat{\beta}_2) \leq 0$ for all W (the weight matrix) in the WMSE. (ii) the criterion $MSE(\eta'\hat{\beta}_1) \leq MSE(\eta'\hat{\beta}_2)$ for all nonzero column vectors η. Also show that $WMSE(\hat{\beta}) = E(\hat{\beta} - \beta)'W(\hat{\beta} - \beta) = MSE(\eta'\hat{\beta})$ for $W = \eta\eta'$, where $\hat{\beta}$ is an estimator of β.

2.5 Explain the distinction between the variance covariance matrix for u and $\hat{u}$. Show that $(X'X)^{-1}$ and $(X'_{(-t)}\, X_{(-t)})^{-1}$ (where the subscript $(-t)$ denotes that t^{th} row is omitted) are related to each other by the following identity, Cook [1977, e.g. (6)]:

$$[X'_{(-t)}\, X_{(-t)}]^{-1} = (X'X)^{-1} + \frac{(X'X)^{-1}\, x_t x_t'(X'X)^{-1}}{(1 - \hat{h}_{tt})} ,$$

where

$$\hat{h}_{tt} = x_t'(X'X)^{-1}\, x_t .$$

Use this identity to prove (22b) and (22c).

2.6 Write the theoretical influence function for statistic $\int (y-\mu)^2 dF$ for F which has a finite variance σ^2 and known mean μ. Show that it equals $(y-\mu)^2 - \sigma^2$.

2.7 Following Thompson [1968] use the shrinkage factors of Section 2-9-2 to evaluate numerically the ratio $MSE(b_1\delta)/MSE(b_1)$. Denote $w = (T-1)/2$ and show that this ratio can be written as the following double integral

$$2^{-w}(2\pi)^{-1/2}w^{-1/2}\int_0^{\infty}\int_{-\infty}^{\infty}(\delta b_1-\beta_1)^2\exp[-(1/2)(b_1-\beta_1)^2 - w]v^w\, db_1 dv .$$

Plot the numerical values of the integrals as in Thompson.

3

Restricted Least Squares and Bayesian Regression

In this chapter we study the regression model with both stochastic and nonstochastic restrictions on the regression coefficients. The Bayesian analysis is also expounded. The relationship between Bayesian estimators and various restricted estimators is analyzed.

3.1 RESTRICTED LEAST SQUARES ESTIMATOR (RLS)

Consider the normal linear regression model having T observations and p regressors as

$$y = X\beta + u, \tag{1}$$

where u is distributed as multivariate normal with the mean vector zero and variance covariance matrix $\sigma^2 I$, i.e., $u \sim N(0, \sigma^2 I)$. Further, suppose there are m linear equality restrictions on the $p \times 1$ parameter vector β as

$$R\beta = r, \tag{2}$$

where R is an $m \times p$ matrix of rank m and r is an $m \times 1$ vector, both consisting of known nonstochastic numbers. The problem is to estimate the parameter vector β in (1) subject to the restrictions given in (2). A rank condition on R to make sure that the constraints are "consistent" is that Rank $(R) = m \leq p$.

We can use the maximum likelihood (ML) method of estimation, which will yield the restricted maximum likelihood or restricted least squares (RLS) estimator below. We maximize the likelihood function:

$$L(\beta, \sigma^2 | y, X) = L = (2\pi\sigma^2)^{-T/2} \exp\left[-\frac{1}{2\sigma^2}(y - X\beta)'(y - X\beta)\right], \tag{3}$$

by minimizing the error sum of squares (SSE)

$$(y - X\beta)'(y - X\beta)$$

subject to $R\beta = r$ of (2) by the Lagrangian expression:

$$S^* = (y - X\beta)'(y - X\beta) + \mu'(r - R\beta)$$

$$= y'y - 2\beta'X'y + \beta'X'X\beta + \mu'(r - R\beta), \tag{4}$$

where μ is an $m \times 1$ vector of m Lagrangian multipliers associated with the m constraints along the m rows of $R\beta = r$. We differentiate the Lagrangian with respect to β and μ and equate these to zero. This gives for $\beta = b_R$ the normal equations as:

$$\frac{\partial S^*}{\partial \beta} = 0 = -2X'y + 2X'Xb_R - R'\mu, \tag{5}$$

$$\frac{\partial S^*}{\partial \mu} = 0 = r - Rb_R, \tag{6}$$

where the subscript R in b_R represents the "restricted" solution for β and should not be confused with the matrix R in (2).

To solve (5) and (6) for b_R and μ, we first premultiply (5) with $(X'X)^{-1}$ and write

$$-2b + 2b_R - (X'X)^{-1}R'\mu = 0, \tag{7}$$

where

$$b = (X'X)^{-1}X'y \tag{8}$$

is the OLS estimator of β. Next, premultiplying (7) by R and using (6), we can get

$$\mu = 2[R(X'X)^{-1}R']^{-1}(r - Rb),$$

provided the inverse of $R(X'X)^{-1}R'$ exists. A necessary (but not sufficient) condition for the existence of this inverse is that Rank $(R) = m \leq p$, which was assumed earlier. Substituting this value of μ in (7) and solving for b_R we obtain

$$b_R = b - (X'X)^{-1}R'[R(X'X)^{-1}R']^{-1}(Rb - r). \tag{9}$$

This solution for b_R is called the restricted least squares estimator of β.

The restricted maximum likelihood (ML) estimator of σ^2 is

$$\hat{\sigma}_R^2 = \frac{1}{T}(y - Xb_R)'(y - Xb_R). \tag{10}$$

3.2 GENERAL LINEAR HYPOTHESIS TESTS

The equality restriction $R\beta - r = \phi = 0$ may be thought to be a null hypothesis H_0 which may or may not be supported by the data. This is a well-known null hypothesis called the general linear hypothesis. To test this hypothesis against the alternative hypothesis H_1: $R\beta - r = \phi \neq 0$ by the likelihood ratio test, we will maximize the likelihood function (3) subject to no constraint. The well-known solution obtained by Lagrangian methods leads to $b = (X'X)^{-1}X'y$, the OLS estimator of β, and

$$\hat{\sigma}^2 = \frac{1}{T}(y - Xb)'(y - Xb). \tag{11}$$

An intuitive description of the likelihood ratio test is as follows. If the ratio of the likelihood function evaluated at b and $\hat{\sigma}^2$ to the same function evaluated at b_R and $\hat{\sigma}_R^2$ is large, then b is in some sense "more likely" than b_R, and we reject the null hypothesis which leads to the b_R estimator.

Intuitively, it is clear that 'likelihood' is inversely related to the sum of squared errors (SSE) in model (1). Hence the likelihood ratio rejection region defined by

$$\frac{L(b, \hat{\sigma}^2)}{L(b_R, \hat{\sigma}_R^2)} > c^*, \tag{12}$$

is large when the following ratio is large:

$$SSE(b_R)/SSE(b) > c, \tag{13}$$

where c^* and c are the constants of rejection or simply critical points. We can see that b is now moved to the denominator. The term $SSE(b_R)$ represents SSE due to the estimation of β by b_R, i.e., $(y - Xb_R)'(y - Xb_R)$.

Now we will write SSE(b_R) in terms of the error in the null hypothesis:

$$e = Rb - r = \hat{\phi} \tag{14}$$

evaluated by replacing β by the OLS estimator b. Recall from (9) that

$$b_R = b - Pe, \tag{15}$$

where

$$P = (X'X)^{-1}R'Q^{-1} \tag{16}$$

and

$$Q = R(X'X)^{-1}R'. \tag{17}$$

Thus

$$
\begin{aligned}
SSE(b_R) &= (y - Xb_R)'(y - Xb_R) \\
&= [y - X(b - Pe)]'[y - X(b - Pe)] \\
&= (y - Xb + XPe)'(y - Xb + XPe) \\
&= (y - Xb)'(y - Xb) + e'P'X'y - e'P'X'Xb + y'XPe \\
&\quad -b'(X'X)Pe + e'P'X'XPe \\
&= SSE(b) + e'P'R'Q^{-1}e \\
&= SSE(b) + e'Q^{-1}e,
\end{aligned} \tag{18}
$$

where we have used $X'Xb = X'y$, $X'XP = R'Q^{-1}$ and $P'R' = Q^{-1}R(X'X)^{-1}R' = Q^{-1}Q = I$.

Since the likelihood ratio test is equivalent to checking whether the ratio (13) of SSE's is large, we have the statistic:

$$\frac{SSE(b) + e'Q^{-1}e}{SSE(b)} > c$$

or equivalently:

$$\frac{e'Q^{-1}e}{SSE(b)} > c - 1. \tag{19}$$

Since the rejection constant may be redefined by the -1, there is no loss of generality in considering the statistic

$$w = \frac{e'Q^{-1}e}{SSE(b)} = \frac{(Rb - r)'[R(X'X)^{-1}R']^{-1}(Rb - r)}{(y - Xb)'(y - Xb)}. \tag{19a}$$

Verify that under the null hypothesis the numerator is a sum of squares of m normal variables, which is a central χ_m^2 a Chi-square variable with m degrees of freedom (d.f.). Also, the denominator is a χ_{T-p}^2 with $T - p$ degrees of freedom. It is well-known that a ratio of independent χ^2 variables follows an F distribution. Hence to test the general linear hypothesis:

$$
\begin{aligned}
H_0 &: R\beta - r = 0 \\
H_1 &: R\beta - r \neq 0
\end{aligned}
$$

we use the test statistic

$$w = \frac{(Rb - r)'[R(X'X)^{-1}R']^{-1}(Rb - r)/m}{(y - Xb)'(y - Xb)/T - p} \sim F(m, T - p). \tag{20}$$

The null hypothesis H_0 is rejected if

$$w > c, \tag{21a}$$

where the critical value c is determined, for a given level of the test α, by

$$\int_{c}^{\infty} dF(w) = \alpha. \tag{21b}$$

This can be found from tables of the F distribution with numerator $d.f. = m$ and denominator $d.f. = T - p$.

3.3 PROPERTIES OF RESTRICTED ESTIMATORS (RLS)

If the hypothesis $R\beta = r$ is true, that is, $R\beta - r = \phi = 0$, then the restricted estimator b_R is the best linear unbiased estimator. However, if $R\beta = r$ is not true, that is, $R\beta - r \neq 0$, then the estimator b_R is biased. This can be shown as below. From (9) we write

$$E(b_R) = \beta - (X'X)^{-1}R'[R(X'X)^{-1}R']^{-1}\phi, \tag{22}$$

where we have used $E(b) = \beta$. Since the second term on the right-hand side of (22) is (in general) not zero, the estimator b_R is biased. In the particular case when the restrictions are exact, ϕ=0 and the second term is zero.

One motivation for imposing restrictions despite the bias is to reduce the variance. Since the estimator b_R is a vector, we need to consider its variance covariance matrix:

$$\begin{aligned} V(b_R) &= E(b_R - Eb_R)(b_R - Eb_R)' \\ &= E[b - \beta - PR(b - \beta)][b - \beta - PR(b - \beta)]' \\ &= E(b - \beta)(b - \beta)' - E(b - \beta)(b - \beta)'R'P' \\ &\quad - PR\, E(b - \beta)(b - \beta)' + PR\, E(b - \beta)(b - \beta)'R'P' \\ &= V(b) - \sigma^2 PQP', \end{aligned} \tag{23}$$

where we use $(b_R - Eb_R) = b - \beta - PR(b - \beta)$ from (9) and (22), and P and Q are as defined in (16) and (17), respectively. Further, we have

$$\begin{aligned} V(b) &= E(b - \beta)(b - \beta)' \\ &= (X'X)^{-1}X'(Euu')X(X'X)^{-1} \\ &= \sigma^2(X'X)^{-1} \end{aligned} \tag{24}$$

is the variance covariance matrix of the OLS estimator b.

Since the matrix PQP' is non-negative definite, we have $V(b) - V(b_R) =$ a positive semi-definite matrix, and hence the following matrix inequality:

$$V(b_R) \text{ “}\leq\text{” } V(b), \tag{25}$$

where the quotations around the inequality are reminders that it is not the usual scalar inequality. Thus, we have shown that the variance covariance matrix has been "reduced" by imposing equality restrictions.

3.3.1 Matrix MSE of RLS

Now we look at the matrix MSE criterion which has been defined in Chapter 2. For the estimator b_R

$$\begin{aligned} MtxMSE(b_R) &= E(b_R - \beta)(b_R - \beta)' \\ &= V(b_R) + [Bias(b_R)][Bias(b_R)]' \qquad (26) \\ &= \sigma^2(X'X)^{-1} - \sigma^2 PQP' + P\phi\phi'P', \end{aligned}$$

where $Bias(b_R) = Eb_R - \beta = -P\phi$ from (22). Thus

$$\begin{aligned} \Delta = MtxMSE(b) - MtxMSE(b_R) &= \sigma^2 PQP' - P\phi\phi'P' \qquad (27) \\ &= P[\sigma^2 Q - \phi\phi']P'. \end{aligned}$$

As noted in Chapter 2, for Δ to be non-negative definite we need to show that $\eta'\Delta\eta \geq 0$ for all $\eta \neq 0$, where η is any $p \times 1$ nonzero constant vector. From (27) we have

$$\eta'\Delta\eta = \eta'P[\sigma^2 Q - \phi\phi']P'\eta \geq 0$$

iff (if and only if)

$$\sigma^2 \geq \frac{\eta'P\phi\phi'P'\eta}{\eta'PQP'\eta}$$

or

$$\theta_0 = \frac{\xi'\phi\phi'\xi}{\sigma^2\xi'Q\xi} \leq 1, \qquad (28)$$

where $\xi = P'\eta$. The result (28) holds for all nonzero ξ iff

$$\theta = \sup_{\xi}(\theta_0) = \frac{\phi'Q^{-1}\phi}{\sigma^2} \leq 1,$$

where *sup* respresents *supremum* of θ_0 or alternatively, iff

$$\theta = \frac{\phi'Q^{-1}\phi}{2\sigma^2} \leq \frac{1}{2}, \qquad (29)$$

where $\phi = R\beta - r$. In writing (29) we have used the algebraic result that the supremum of $(\phi'\xi)^2/\xi'Q\xi$ is $\phi'Q^{-1}\phi$, Rao [1973,p.60].

Thus we have proved the following result due to Toro-Vizcarrondo and Wallace [1968].

Proposition A necessary and sufficient condition for $MtxMSE(b)$ "$\geq$" $MtxMSE(b_R)$, that is, b_R to be preferred to b, is

$$\theta = \frac{(R\beta - r)'[R(X'X)^{-1}R']^{-1}(R\beta - r)}{2\sigma^2} \leq \frac{1}{2}. \tag{30}$$

Although this proposition states a general "necessary and sufficient" condition, the argument for "necessity" needs a further comment. The proposition is asymmetric in that when $\theta > (1/2)$ there are η vectors for which $\eta' b_R$ is always preferred to $\eta' b$. Giles and Rayner [1979] consider an example in which $\eta = \overline{\eta}$ is orthogonal to the bias vector $bias(b_R)$, in the sense that $bias(\overline{\eta}' b_R) = 0$. It is obvious that in this case the MSE equals the covariance matrix, and since that variance of b_R and its linear combinations is smaller, b_R is preferred to b for this special choice of $\overline{\eta}$. Of course, the special choice of η is possible only for $m \geq 2$.

To clarify the situation consider the difference defined in (27) in a reverse order. We write

$$\Delta_1 = MtxMSE(b_R) - MtxMSE(b) = P\phi\phi'P' - \sigma^2 PQP'. \tag{27a}$$

Note that the conditions based on (27) and (27a) are equivalent for $m=1$. The condition for Δ_1 to be positive semidefinite is $\eta'\Delta_1\eta \geq 0$ for all $\eta \neq 0$. Pre-multiply (27a) by η' and post-multiply by η to yield:

$$\eta'\Delta_1\eta = \eta' P(\phi\phi' - \sigma^2 Q)P'\eta \geq 0,$$

if and only if

$$\frac{\eta' P\phi\phi' P'\eta}{\sigma^2 \eta' PQP'\eta} = \frac{\xi'\phi\phi'\xi}{\sigma^2 \xi' Q\xi} \geq 1, \tag{30a}$$

where $\xi = P'\eta$. When there is only one restriction (i.e., $m=1$) the vectors ξ and ϕ on the left hand side of (30a) become scalars, and the iff condition becomes $\phi^2/(\sigma^2 Q) \geq 1$. It is obviously possible to choose the values of the parameters β and σ^2 for which this condition is satisfied when $m=1$. (Hint: Q is a scalar for $m=1$).

For $m \geq 2$, the situation is asymmetric in the following sense. The condition (30a) is satisfied iff the infimum of the left hand side is ≥ 1. From Rao [1973, p.60] the infimum of $(\xi'\phi\phi'\xi/\sigma^2\xi'Q\xi)$ over ξ is simply zero. In other words, there are no parameter values for which OLS is preferred to the restricted estimator when $m \geq 2$.

By considering (27a) in addition to (27) we have noted that the advantages associated with using the restricted estimator may not have been fully realized until very recently.

We note that the condition (30) depends on unknown β and σ^2, and thus it is not possible to check this condition in practice. Toro-Vizcarrondo and Wallace [1968] suggested a statistical F test for $\theta \leq \frac{1}{2}$ by substituting OLS estimators of σ^2 and β in (30) and dividing by the appropriate degrees of freedom. Compared to the $F(m, T-p)$ defined for the general linear hypothesis test in (20), this test for $\theta \leq \frac{1}{2}$ is seen to be somewhat weaker. We are no longer testing the truth of the $R\beta = r$ constraint per se. We are testing whether the imposition of this constraint is helping the estimation by "reducing" the MtxMSE(b). Thus $R\beta \neq r$ is permissible for the null hypothesis involving $\theta \leq \frac{1}{2}$. Therefore, the quadratic form on the right side of (30) will not be zero even under the null hypothesis, implying that the mean of the underlying normal variable $Rb - r$ need not be zero. The sum of squares of such a normal variable having a nonzero mean is distributed as a "non-central" Chi-square variable. Therefore, the Toro-Vizcarrondo and Wallace [1968] test statistic w in (20) is distributed as a non-central F, that is,

$$w \sim F'(m, T-p, \theta), \tag{31}$$

with d.f. m and $T-p$, respectively, and non-centrality parameter

$$\theta = \frac{\phi' Q^{-1} \phi}{2\sigma^2}. \tag{32}$$

The null hypothesis $\theta \leq \frac{1}{2}$ (b_R, the restricted estimator, is better than OLS) is rejected when the estimated statistic satisfies the relevant inequality with respect to the tabulated non-central F distribution, Toro-Vizcarrondo and Wallace [1968].

3.3.2 Weighted MSE of RLS

The b_R and b can be compared with respect to a "wieghted MSE" criterion (see Chapter 2),

$$\begin{aligned} WMSE(b_R) &= E(b_R - \beta)' W (b_R - \beta) \\ &= Tr[WE(b_R - \beta)(b_R - \beta)'] \\ &= Tr[W \; MtxMSE(b_R)], \end{aligned} \tag{33}$$

where W is a positive definite weight matrix. We can write

$$\begin{aligned} \Delta^* = WMSE(b_R) - WMSE(b) &= Tr[W \; MtxMSE(b_R) - W \; MtxMSE(b)] \\ &= TrW\Delta, \end{aligned} \tag{34}$$

where $\Delta = P[\sigma^2 Q - \varphi\phi']P'$ from (27).

In a special case when $W = X'X$, we can simplify (34) as

$$\Delta^* = \sigma^2 m - \phi'[R(X'X)^{-1}R']^{-1}\phi, \tag{35}$$

where we have used $TrR'P' = TrP'R' = TrI = m$, and $Tr\phi' Q^{-1}\phi = TrQ^{-1}\phi\phi'$.

Using (35), Wallace [1972] gives the following condition which is "weaker" than (30), for b_R to be better than b in the sense of having a low WMSE with the weight matrix $W = (X'X)$:

$$\theta = \frac{(R\beta - r)'[R(X'X)^{-1}R']^{-1}(R\beta - r)}{2\sigma^2} \le \frac{m}{2}. \tag{36}$$

This condition is "weaker" because it will allow a larger right hand side than (30) before rejecting the null hypothesis. The relevant statistic follows the same non-central F as in (31) for the MtxMSE criterion.

In another special case, when $W = I$, the WMSE becomes MSE. We write (27) after taking the trace as follows:

$$TrW\Delta = Tr\Delta = Tr[\sigma^2 PQP' - P\phi\phi'P'] \tag{37}$$

$$= Tr[\sigma^2(X'X)^{-1}R'Q^{-1}R(X'X)^{-1}] - \phi'P'P\phi,$$

where $Tr(P\phi\phi'P') = \phi'P'P\phi$. Unfortunately (37) does not simplify without additional algebra.

If the second term in (37) had $(X'X)$ in the middle, there would be a cancellation shown below: First we use $P = (X'X)^{-1}R'Q^{-1}$ to write

$$\phi'P'P\phi = \phi'Q^{-1}R(X'X)^{-1}(X'X)^{-1}R'Q^{-1}\phi$$

$$\phi'P'(X'X)P\phi = \phi'Q^{-1}\phi = 2\sigma^2\theta. \tag{38}$$

Recall that the eigenvalues of $X'X$ are denoted by $\lambda_1 \ge \lambda_2 \cdots \ge \lambda_p$. From matrix algebra we know

$$\lambda_p = \lambda_{\min} \le \frac{\phi'P'(X'X)P\phi}{\phi'P'P\phi} \le \lambda_{\max} = \lambda_1. \tag{39}$$

Reciprocals of all terms and use of (38) leads to

$$\lambda_{\max}^{-1} \le \frac{\phi'P'P\phi}{2\sigma^2\theta} \le \lambda_{\min}^{-1}. \tag{40}$$

From (34) $Tr\Delta \ge 0$ is the condition for $WMSE(b_R)$ to be lower than $WMSE(b)$ with $W = I$. Hence the condition from (37) is

$$\phi'P'P\phi \le Tr[\sigma^2(X'X)^{-1}R'Q^{-1}R(X'X)^{-1}].$$

If we replace the left hand side by its upper bound $(2\lambda_{\min}^{-1}\sigma^2\theta)$ from (40), the inequality is harder to satisfy. Since the upper bound can be reached, we must use it if we want an iff condition. The condition becomes

$$\theta \le \frac{1}{2}\lambda_{\min} Tr[(X'X)^{-1}R'[R(X'X)^{-1}R']^{-1}R(X'X)^{-1}] \tag{41}$$

given by Wallace [1972]. The right hand side containing known non-stochastic matrices can be shown to be less than $p/2$. An appropriate significance test for this needs tables of a non-central F distribution, but do not involve any conceptual difficulties. Generally speaking we expect (41) to be a "somewhat weaker" condition than (30) or (36). Unfortunately, this cannot be guaranteed.

3.4 MIXED REGRESSION STOCHASTIC RESTRICTIONS MODEL

The "mixed regression" model developed by Theil and Goldberger [1961] uses the linear restrictions $R\beta = r$ of the previous section stochastically. We write the linear restrictions as

$$r = R\beta + v, \tag{42}$$

where r is an $m \times 1$ vector of $m \leq p$ random variables. The $m \times p$ matrix R specifies the restrictions and v is an $m \times 1$ vector of normally distributed errors with mean zero and covariance matrix $\sigma_v^2 \Omega$, i.e., $v \sim N(0, \sigma_v^2 \Omega)$. Also, v and u are independent.

Now mixing $y = X\beta + u$, $u \sim N(0, \sigma^2 I_T)$ with (42) means adding m observations represented by (42). This can be written as

$$\begin{bmatrix} y \\ r \end{bmatrix} = \begin{bmatrix} X \\ R \end{bmatrix} \beta + \begin{bmatrix} u \\ v \end{bmatrix} \tag{43}$$

or

$$y_M = X_M \beta + u_M, \tag{44}$$

where M represents "mixed". The model (43) also represents the combination of prior (42) and sample ($y = X\beta + u$) information. We note that $u_M \sim N(0, \Omega_M)$ where

$$\Omega_M = \begin{bmatrix} \sigma^2 I_T & 0 \\ 0 & \sigma_v^2 \Omega \end{bmatrix}, \text{ and } \Omega_M^{-1} = \begin{bmatrix} \frac{1}{\sigma^2} I_T & 0 \\ 0 & \frac{1}{\sigma_v^2} \Omega^{-1} \end{bmatrix} \tag{45}$$

is a positive definite matrix such that $P' \Omega_M P = I$, $\Omega_M^{-1} = PP'$; P is a $(m + T) \times (m + T)$ matrix of transformations and should not be confused with the P matrix given in (16). Thus $P' u_M \sim N(0, I)$ and we can apply OLS estimation on the transformed model

$$P' y_M = P' X_M \beta + P' u_M, \tag{46}$$

that is, we minimize

$$(P' y_M - P' X_M \beta)'(P' y_M - P' X_M \beta) = (y_M - X_M \beta)' \Omega_M^{-1} (y_M - X_M \beta)$$

$$= \frac{(y - X\beta)'(y - X\beta)}{\sigma^2} + \frac{1}{\sigma_v^2}(r - R\beta)' \Omega^{-1} (r - R\beta)$$

with respect to β. This gives

$$b_M = [(P' X_M)'(P' X_M)]^{-1} (P' X_M)' P' y_M \tag{47}$$

$$= (X_M' \Omega_M^{-1} X_M)^{-1} X_M' \Omega_M^{-1} y_M,$$

the Aitken generalized least squares (GLS) estimator,which is the BLUE estimator for known Ω_M. In our context we call it a "mixed" estimator. Since

$$X'_M \Omega_M^{-1} X_M = [X'R'] \begin{bmatrix} \frac{1}{\sigma^2} I_T & 0 \\ 0 & \frac{1}{\sigma_v^2}\Omega^{-1} \end{bmatrix} \begin{bmatrix} X \\ R \end{bmatrix}$$

$$= \frac{1}{\sigma^2} X'X + \frac{1}{\sigma_v^2} R'\Omega^{-1}R,$$

an alternative form of b_M, the mixed estimator, is

$$\begin{aligned} b_M &= \left[\frac{1}{\sigma^2}X'X + \frac{1}{\sigma_v^2}R'\Omega^{-1}R\right]^{-1}\left[\frac{X'y}{\sigma^2} + \frac{R'\Omega^{-1}r}{\sigma_v^2}\right], \qquad (48) \\ &= (X'X + \lambda R'\Omega^{-1}R)^{-1}(X'y + \lambda R'\Omega^{-1}r) \\ &= (X'X + \lambda R'\Omega^{-1}R)^{-1}(X'Xb + \lambda R'\Omega^{-1}Rb_p), \end{aligned}$$

where $\lambda = \sigma^2/\sigma_v^2$, is the ratio of variances (not to be confused with eigenvalues) $b = (X'X)^{-1}X'y$ is the OLS estimator of β in $y = X\beta + u$ and $b_p = (R'\Omega^{-1}R)^{-1}R'\Omega^{-1}r$ is the prior estimator of β from the restrictions $r = R\beta + v$. The form in (48) expresses the mixed estimator as the weighted matrix combination of the sample estimator b and the prior estimator b_p, i.e.,

$$b_M = W_1 b + W_2 b_p, \qquad (49)$$

where $W_1 = (X'X + \lambda R'\Omega^{-1}R)^{-1}X'X$ and $W_2 = \lambda(X'X + R'\Omega^{-1}R)^{-1}$ $R'\Omega^{-1}R$ such that $W_1 + W_2 = I$.

Another useful form of b_M follows from the third equality of (48) as

$$\begin{aligned} b_M &= (X'X + \lambda R'\Omega^{-1}R)^{-1}[X'X(b - b_p) + (\lambda R'\Omega^{-1}R + X'X)b_p] \qquad (50) \\ &= b_p + (X'X + \lambda R'\Omega^{-1}R)^{-1}X'X(b - b_p). \end{aligned}$$

From (48) or (50) it is clear that if prior information is useless, i.e., $\sigma_v^2 \to \infty$ or $\lambda \to 0$, then $b_M = b$. On the other hand, if $\lambda \to \infty$, i.e., $\sigma_v^2 \to 0$, then $b_M = b_p$.

A special case of interest is when the prior restrictions $r = R\beta + v$ are such that $r = R\beta_0$ where β_0 is a vector of known values. Then $b_p = (R'\Omega^{-1}R)^{-1}R'\Omega^{-1}R\beta_0 = \beta_0$ and (50) becomes

$$b_M = \beta_0 + (X'X + \lambda R'\Omega^{-1}R)^{-1}X'X(b - \beta_0). \qquad (51)$$

For example, $\beta_0 = 0$ may be substituted in (51). If $r = 0$ and $R = I$, then

$$b_M = (X'X + \lambda I_p)^{-1} X'y. \tag{52}$$

We shall see in Chapters 7 and 8 that the estimators (51) and (52) are similar to the "generalized" and "ordinary" ridge estimators, respectively, proposed by Hoerl and Kennard [1970].

In practice we use the usual unbiased estimator $S^2 = (y - Xb)'(y - Xb)(T - p)^{-1}$ for σ^2 assuming that the number p of regressors includes one for the intercept. Now we assume that $\sigma_v^2 \Omega$ is known. Then we have the estimator

$$\hat{b}_M = \left[\frac{1}{S^2} X'X + \frac{1}{\sigma_v^2} R'\Omega^{-1} R\right]^{-1} \left[\frac{1}{S^2} X'y + \frac{1}{\sigma_v^2} R'\Omega^{-1} r\right]. \tag{53}$$

Though this estimator is unbiased its variance covariance matrix will be different (compared to b_M) because S^2 appearing in (53) is stochastic.

3.4.1 Properties of Mixed Estimators and Tests for Stochastic Restrictions

First, substituting $y = X\beta + u$ and $r = R\beta + v$ in (48) we write the sampling error of b_M as

$$b_M - \beta = (X'X + \lambda R'\Omega^{-1} R)^{-1} (X'u + \lambda R'\Omega^{-1} v), \tag{54}$$

where $\lambda = \sigma^2/\sigma_v^2$ is the ratio of variances. Thus, since both $EX'u$ and $ER'\Omega^{-1}v$ are zero it is easily verified that b_M is unbiased, that is,

$$E(b_M - \beta) = 0,$$

and

$$\begin{aligned} V(b_M) &= E(b_M - Eb_M)(b_M - Eb_M)' \qquad (55) \\ &= (X'X + \lambda R'\Omega^{-1}R)^{-1}(\sigma^2 X'X + \sigma_v^2 \lambda^2 R'\Omega^{-1} R)(X'X + \lambda R'\Omega^{-1}R)^{-1} \\ &= \sigma^2 (X'X + \lambda R'\Omega^{-1} R)^{-1}. \end{aligned}$$

Now the sampling error of $\hat{b}_M$ in (53) is

$$\hat{b}_M - \beta = (X'X + \hat{\lambda} R'\Omega^{-1} R)^{-1} (X'u + \hat{\lambda} R'\Omega^{-1} v) \tag{56}$$

where $\hat{\lambda} = S^2/\sigma_v^2$ is now stochastic. However, S^2 is a consistent estimator of σ^2, $\hat{\lambda} \to \lambda$ as $T \to \infty$. Thus, asymptotically, $\hat{b}_M$ is unbiased and $Asy \cdot V(\hat{b}_M)$ is the same as $V(b_M)$. When the sample is small, we evaluate

$$E(\hat{b}_M - \beta) = E_{(S^2)}[E(\hat{b}_M - \beta)|S^2], \tag{57}$$

where $E_{(S^2)}$ represents the expectation with respect to S^2 of the term $E(\hat{b}_M - \beta)|S^2$, that is, $E(\hat{b}_M - \beta)$ given S^2. Since $E(\hat{b}_M - \beta)|S^2 = 0$, $E(\hat{b}_M - \beta) = 0$, that is, $\hat{\beta}_M$ is unbiased.

Alternatively, note that

$$S^2 = \frac{(y - Xb)'(y - Xb)}{T - p} = \frac{y'My}{T - p} = \frac{u'Mu}{T - p}; \; M = I - X(X'X)^{-1}X' \quad (58)$$

is distributed independently of $X'u$ because $MX = 0$. Also S^2 and v are independent because $Euv' = 0$. Thus

$$E(\hat{b}_M - \beta) = E(X'X + \hat{\lambda}R'\Omega^{-1}R)^{-1}E(X'u) \quad (59)$$

$$+ E\hat{\lambda}(X'X + \hat{\lambda}R'\Omega^{-1}R)^{-1}E(R'\Omega^{-1}v)$$

$$= 0$$

provided $E(X'X + \hat{\lambda}R'\Omega^{-1}R)^{-1}$ exists. One can also use the procedure in Kakwani [1967] to show the unbiasedness of $\hat{b}_M$. Because $S^2 = u'Mu(T - p)^{-1}$, observe that $\hat{b}_M - \beta = g(u,v)$, where $g(u,v)$ is an odd function of normal error vectors u and v, that is, $g(-u,-v) = -g(u,v)$. Now, using the property that the moments of any odd function of symmetrically distributed normal errors are zero, it is obvious that $E(\hat{b}_M - \beta) = Eg(u,v) = 0$.

To evaluate the variance covariance matrix of $\hat{b}_M$ we write

$$V(\hat{b}_M) = E(\hat{b}_M - E\hat{b}_M)(\hat{b}_M - E\hat{b}_M)' = E_{S^2}[E(\hat{b}_M - E\hat{b}_M)(\hat{b}_M - E\hat{b}_M)'|S^2] \quad (60)$$

$$= E_{S^2}[\hat{A}^{-1}(\sigma^2X'X + \sigma_v^2\hat{\lambda}^2R'\Omega^{-1}R)\hat{A}^{-1}],$$

where $\hat{A} = (X'X + \hat{\lambda}R'\Omega^{-1}R)$ and $\hat{A}^{-1}(\sigma^2X'X + \sigma_v^2\hat{\lambda}^2R'\Omega^{-1}R)\hat{A}^{-1}$ is the conditional covariance matrix. To obtain unconditional $V(\hat{b}_M)$ we need to evaluate the second expectation in (60) with respect to S^2. This has been evaluated by Swamy and Mehta [1976] and Charette [1978]. However, the mathematical expression is too complicated to draw useful conclusions.

Recall that the mixed estimator b_M is unbiased only when the stochastic restriction $r = R\beta + v$ is unbiased, that is, when $Er - R\beta = Ev = \phi = 0$. If the stochastic restriction is biased with $Ev = Er - R\beta = \phi \neq 0$, then the mixed estimator b_M is biased with

$$Eb_M = \beta + \lambda A^{-1}R'\Omega^{-1}\phi \quad (61)$$

and

$$MtxMSE(b_M) = E(b_M - \beta)(b_M - \beta)' \quad (62)$$

$$= V(b_M) + [Bias(b_M)][Bias(b_M)]'$$

$$= \sigma^2A^{-1} + \lambda^2A^{-1}R'\Omega^{-1}\phi\phi'\Omega^{-1}RA^{-1},$$

where $A = (X'X + \lambda R'\Omega^{-1}R)$. Note that ϕ here is used for $Er - R\beta$ instead of $r - R\beta$. As in the nonstochastic restriction case of Sections 3.2 and 3.3, we now look at two questions: (i) test for the unbiasedness of stochastic restrictions and (ii) test for $MtxMSE(b_M) - MtxMSE(b)$ "$\leq$" 0 or $MSE(\eta'b_M) \leq MSE(\eta'b)$ for any $\eta \neq 0$, when the restriction is not true.

Note that Toro-Vizcarrondo and Wallace tests used in the nonstochastic restrictions case are not directly applicable to the stochastic restrictions case. To use their tests, in order to answer questions in (i) and (ii), we reformulate below the model $y = X\beta + u$, with $r = R\beta + v$ in terms of a non-stochastic restrictions model. Within this context, we observe that the mixed estimator b_M for the stochastic restricted model

$$\begin{bmatrix} y \\ r \end{bmatrix} = \begin{bmatrix} X \\ R \end{bmatrix}\beta + \begin{bmatrix} u \\ v \end{bmatrix} \tag{63}$$

is equivalent to the restricted estimator for β in

$$\begin{bmatrix} y \\ r \end{bmatrix} = \begin{bmatrix} X & O \\ O & I \end{bmatrix}\begin{bmatrix} \beta \\ \mu + \phi \end{bmatrix} + \begin{bmatrix} u \\ v \end{bmatrix}, \tag{64}$$

under the nonstochastic restriction $R\beta = \mu$ written as

$$[R \ -I]\begin{bmatrix} \beta \\ \mu + \phi \end{bmatrix} = -\phi. \tag{65}$$

The main idea (due to Judge and Bock [1978]) here is to write (63) as Zellner's [1962] seemingly unrelated regressions (SUR) model (to be discussed in Chapter 10). Note that the parameters β in the two equation of the restricted model (63) are the same, whereas the parameters of the SUR model in (64) are respectively β and $\mu + \phi \equiv R\beta + \phi$. The definitional relation $R\beta = \mu$ is now the non-stochastic constraint implied by (65). The formulation (63) to (65) does not allow for the general variance covariance matrix of v. Recall from (42) that $v \sim N(0, \sigma_v^2 \Omega)$. In terms of the variance ratio $\lambda = \sigma^2/\sigma_v^2$, and for our context,we have: $v \sim N(\phi, \sigma^2\lambda^{-1}\Omega)$.

In order to reduce the matrix Ω to the conventional case, we rewrite the model (64) and (65) as

$$\begin{bmatrix} y \\ Gr \end{bmatrix} = \begin{bmatrix} X & O \\ O & G \end{bmatrix}\begin{bmatrix} \beta \\ \mu + \phi \end{bmatrix} + \begin{bmatrix} u \\ Gv \end{bmatrix} \tag{66}$$

subject to

$$G[R \ -I]\begin{bmatrix} \beta \\ \mu + \phi \end{bmatrix} = -G\phi \quad or \quad [R \ -I]\begin{bmatrix} \beta \\ \mu + \phi \end{bmatrix} = -\phi, \tag{66a}$$

where G is such that $G(\lambda^{-1}\Omega)G' = I$. The model given by (66) and (66a) may be written in a compact form as

$$y^* = X^*\beta^* + u^*, \tag{67}$$

subject to

$$R^*\beta = r^*, \tag{68}$$

where

$$y^* = \begin{bmatrix} y \\ Gr \end{bmatrix}, \ X^* = \begin{bmatrix} X & O \\ O & G \end{bmatrix}, \ \beta^* = \begin{bmatrix} \beta \\ \mu + \phi \end{bmatrix} = \begin{bmatrix} \beta_1 \\ \beta_2 \end{bmatrix} \tag{69}$$

$$u^* = \begin{bmatrix} u \\ Gv \end{bmatrix}, \ R^* = [R \ -I], \ r^* = -\phi. \tag{70}$$

If $\phi = 0$, then $r^* = 0$ is obvious from (70).

Using this notation, the restricted least squares estimator of β^* is

$$b_R^* = \begin{bmatrix} b_{1R} \\ b_{2R} \end{bmatrix} = b^* - (X^{*\prime}X^*)^{-1}R^{*\prime}[R^*(X^{*\prime}X^*)^{-1}R^{*\prime}]^{-1}R^*b^* \tag{71}$$

where we have used $r^* = 0$, and

$$b^* = (X^{*\prime}X^*)^{-1}X^{*\prime}y^* = \begin{bmatrix} (X'X)^{-1}X'y \\ r \end{bmatrix} = \begin{bmatrix} b \\ r \end{bmatrix}$$

is the unrestricted least squares estimator. Simplifying (71), we can write the following (Notation λ comes from the variance of $v = \sigma^2\lambda^{-1}\Omega$, and should not be confused with the eigenvalues of $X'X$).

$$b_{1R} = b - (X'X)^{-1}R'[R(X'X)^{-1}R' + \lambda^{-1}\Omega]^{-1}(Rb - r)$$

$$b_{2R} = r + \lambda^{-1}\Omega[R(X'X)^{-1}R' + \lambda^{-1}\Omega]^{-1}(Rb - r).$$

Finally, using certain matrix inversions (see Exercise 3.1) we note that

$$b_M = (X'X + \lambda R'\Omega^{-1}R)^{-1}(X'y + \lambda R'\Omega^{-1}r)$$

$$= [(X'X)^{-1} - (X'X)^{-1}R'(R(X'X)^{-1}R' + \lambda^{-1}\Omega)^{-1}R(X'X)^{-1}](X'y + \lambda R'\Omega^{-1}r)$$

$$= b_{1R}$$

In view of the above relations all the results of Sections 3.2 and 3.3 relating to test statistics and conditions for the restricted estimator to be superior to the OLS estimator directly apply for corresponding comparisons between the "stochastic restricted" and the OLS estimators. For example, the test statistic for testing the null hypothesis that the stochastic restriction is unbiased, that is,

$$H_0: r^* = -\phi = 0$$

against the alternative hypothesis:

$$H_1: r^* = -\phi \neq 0$$

is

$$w^* = \frac{(Rb - r)'(R(X'X)^{-1}R' + \lambda^{-1}\Omega)^{-1}(Rb - r)/m}{(y - Xb)'(y - Xb)/(T - p)} \sim F(m, T-p), \tag{72}$$

which has a central F distribution under H_O.

If H_O is not true, then b^* and hence b_M is biased. In this case the condition under which $MtxMSE(b) - MtxMSE(b_M) = \Delta$ is positive definite, or $\text{MSE}(\eta'b) \geq \text{MSE}(\eta'b_M)$ for all $\eta \neq O$, is

$$\theta^* = \frac{\phi'(R(X'X)^{-1}R' + \lambda^{-1}\Omega)^{-1}\phi}{2\sigma^2} \leq \frac{1}{2}. \tag{73}$$

To test for $\theta^* \leq 1/2$ one uses a new w^* which will now be a noncentral $F(m, T-p, \theta^*)$ with θ^* as the noncentrality parameter.

3.5 PRELIMINARY TEST ESTIMATOR (PT)

Most practitioners of regression start their analysis with a test of the hypothesis H_O: $R\beta = r$, or $\phi = O$, against H_1: $R\beta \neq r$, or $\phi \neq O$, by using the statistic w for an F test given in eq. (20) above. We reject H_O if $w > c$, where c is the critical value determined by $\int_c^\infty dF(w) = \alpha$. If we accept H_O, we use the restricted least squares estimator b_R as our estimator for β in $y = X\beta + u$; otherwise, we use the usual OLS estimator b. In this "testing followed by estimation" procedure, the estimation part is dependent on a "preliminary" test of significance, and therefore the following 'preliminary test estimator' results. Recall that b_R denotes a "restricted" estimator.

$$b_{PT} = I(w)b_R + (1 - I(w))b$$

$$= b - I(w)\,(X'X)^{-1}R'[R(X'X)^{-1}R']^{-1}(Rb - r)\,, \tag{74}$$

where PT represents preliminary test, and $I(w)$ is the "indicator function" such that:

$$I(w) = 1, \text{ if } w \leq c,$$

$$= 0, \text{ if } w > c.$$

If $w > c$, that is, when the hypothesis $R\beta = r$ is rejected, then $I(w) = 0$ and $b_{PT} = b$.

If the restrictions are stochastic, that is, $r = R\beta + v$, then a stochastic version of the preliminary test estimator can be written as

$$\tilde{b}_{PT} = I(w)b_M + (1 - I(w))b, \tag{75}$$

where b_M denotes our "mixed" estimator. The bias and various MSE criteria of both b_{PT} and its stochastic version $\tilde{b}_{PT}$ have been analyzed in the work of Bock et al. [1973], and Judge et al. [1973]. It follows from these references that the estimator b_{PT} is to be preferred to OLS in the sense that

$$\Delta = MtxMSE(b) - MtxMSE(b_{PT})$$

is positive semi-definite, or MSE$(\eta' b) \geq$ MSE$(\eta' b_{PT})$ for all $\eta \neq O$, when the noncentrality parameter θ in (32) is $\leq 1/4$. For $\tilde{b}_{PT}$ to be preferred to b we require $\theta^* \leq \frac{1}{4}$ in (73). The noncentral F statistics w and w^* can be used to test for the null hypotheses $\theta \leq 1/4$ and $\theta^* \leq 1/4$, respectively. It is obvious from (74) that as the significance level α, (the size of the test), approaches zero and thus c approaches infinity, one would expect $MtxMSE(b_{PT})$ to be close to $MtxMSE(b_R)$. Conversely, when α approaches 1, $MtxMSE(b_{PT})$ will be close to $MtxMSE(b)$. Thus the significance level α plays an important role in studying the MSE matrix of the preliminary test estimators. Sclove, Morris and Radhakrishnan [1972] prove that a Stein-Rule (see Chapter 6) modification of these preliminary test estimators is uniformly superior to traditional pre-test estimator. Therefore *PT* estimator is inadmissible in terms of an MSE criterion. If the criterion is modified to incorporate a penalty for incorporating too many regressors, the *PT* estimator can be shown to be admissible.

3.6 BAYESIAN ANALYSIS OF REGRESSION

An alternative way of combining the prior and sample informatior is by using Bayes' theorem, given in Chapter 2. We will discuss the Bayesian analysis of the model $y = X\beta + u$ with (i) conjugate priors and (ii) diffuse priors. A conjugate prior is a prior that, when combined with the likelihood function, provides a posterior distribution that has the same functional form as the prior. When we know nothing about the parameters we use a noninformative or diffuse prior. This is normally represented by a uniform probability distribution.

3.6.1 Conjugate Priors (Normal and Multivariate t)

Consider $u \sim N(0, \sigma^2 I)$ such that y is distributed as a multivariate normal with the mean vector $X\beta$ and variance covariance matrix $\sigma^2 I$, i.e., $y \sim N(X\beta, \sigma^2 I)$. The density function (the probability of data) of y can therefore be written as

$$f(y|\beta, \sigma, X) = \frac{1}{(2\pi)^{T/2}\sigma^T} \exp[-\frac{1}{2\sigma^2}(y - X\beta)'(y - X\beta)].$$

For a given data set on y and X this is regarded as the likelihood function for β, σ, i.e.,

$$L(\beta, \sigma) = L(\beta, \sigma|y) = f(y|\beta, \sigma, X).$$

Let us write the quadratic form in the likelihood function as

$$(y - X\beta)'(y - X\beta) = (y - Xb)'(y - Xb) + (\beta - b)'X'X(\beta - b)$$

$$= SSE(b) + (\beta - b)'X'X(\beta - b).$$

Then the likelihood function can be rewritten as

$$L(\beta,\sigma) = \frac{1}{(2\pi)^{T/2}\ \sigma^T} \exp\left[\frac{1}{2\sigma^2}\left\{SSE(b) + (\beta - b)'X'X(\beta - b)\right\}\right].$$

$$\propto \sigma^{-T} \exp\left[\frac{1}{2\sigma^2}\left\{SSE(b) + (\beta - b)'X'X(\beta - b)\right\}\right]. \quad (76)$$

where $\propto$ denotes propotionality and permits a simplification by omitting the constants.

Consider the following cases:

Case 1. σ known. In this case we can write (76) as

$$L(\beta) \propto exp[-\frac{1}{2\sigma^2}(\beta - b)'X'X(\beta - b)],$$

because $SSE(b)/2\sigma^2$ does not involve β. This form of likelihood suggests that one may use the normal distribution as a prior for β. Let it be

$$p(\beta) = \frac{1}{(2\pi)^{p/2}|\sigma^2\Omega|^{1/2}} \exp\left[\frac{1}{2\sigma^2}(\beta - \beta_0)'\Omega^{-1}(\beta - \beta_0)\right]$$

$$\propto \exp[-\frac{1}{2\sigma^2}(\beta - \beta_0)'\Omega^{-1}(\beta - \beta_0)], \quad (77)$$

which is multivariate normal with the prior mean vector β_0 and prior variance covariance $\sigma^2\Omega$. Then using Bayes theorem (see Chapter 2) the posterior density is propotional to the prior times the likelihood, and it is written as:

$$p(\beta|y) \propto p(\beta)L(\beta)$$

$$\propto \exp\left[-\frac{1}{2\sigma^2}\left\{(\beta - b)'X'X(\beta - b) + (\beta - \beta_0)'\Omega^{-1}(\beta - \beta_0)\right\}\right]. \quad (78)$$

Now write

$$(\beta - b)'X'X(\beta - b) + (\beta - \beta_0)'\Omega^{-1}(\beta - \beta_0)$$

$$= \beta'(X'X + \Omega^{-1})\beta - 2\beta'(X'Xb + \Omega^{-1}\beta_0) + c \quad (79)$$

$$= (\beta - b^*)'\Omega^{*-1}(\beta - b^*) + c^*,$$

where $c = b'X'Xb + \beta_0'\Omega^{-1}\beta_0$ does not involve β. Further,

$$b^* = (X'X + \Omega^{-1})^{-1}(X'Xb + \Omega^{-1}\beta_0)$$

$$\Omega^* = (X'X + \Omega^{-1})^{-1}, \quad (79a)$$

and $c^* = c - b^{*\prime}\Omega^{*-1}b^*$ does not contain β. Using (79) we can write the posterior density:

$$p(\beta|y) \propto exp[-\frac{1}{2\sigma^2}(\beta - b^*)'\Omega^{*-1}(\beta - b^*)], \tag{79b}$$

where the normalizing constant is $(2\pi)^{-T/2}|\sigma^2\Omega^*|^{-1/2}$. This density is again a multivariate normal with the mean vector b^* and variance covariance matrix Ω^*, that is, in the usual notation: the posterior density of $\beta \sim N(b^*, \sigma^2\Omega^*)$.

Since both the prior and posterior are normal, the prior for β given above is called a "conjugate" prior. More generally, we can easily verify that, if the prior of $\beta \sim N(\beta_0, \sigma_\beta^2\Omega)$ then the posterior density is $\beta \sim N(b_1^*, \sigma^2\Omega_1^*)$, where

$$b_1^* = \left(\frac{X'X}{\sigma^2} + \frac{\Omega^{-1}}{\sigma_\beta^2}\right)^{-1}\left(\frac{X'Xb}{\sigma^2} + \frac{\Omega^{-1}\beta_0}{\sigma_\beta^2}\right)$$

$$\Omega_1^* = \left(\frac{X'X}{\sigma^2} + \frac{\Omega^{-1}}{\sigma_\beta^2}\right)^{-1} \tag{79c}$$

For $\sigma_\beta^2 = \sigma^2$, we have $b_1^* = b^*$ and $\Omega_1^* = \Omega^*$.

Case 2. σ unknown. In this case consider a prior similar in form to the likelihood function in (76). The conjugate prior is a joint distribution

$$p(\beta,\sigma) = p(\beta|\sigma)p(\sigma) \propto \sigma^{-(l+p)} exp[-\frac{1}{2\sigma^2}[(l-1)a^2 + (\beta - \beta_0)'\Omega^{-1}(\beta - \beta_0)]], \tag{80}$$

where $a > 0$, $l > 0$. Now consider the conditional distribution

$$p(\beta|\sigma) \propto |\Omega|^{-1/2}\sigma^{-p} \exp\left[\frac{1}{2\sigma^2}(\beta - \beta_0)'\Omega^{-1}(\beta - \beta_0)\right],$$

which is the multivariate normal distribution of β given σ, and where

$$p(\sigma) \propto \sigma^{-l} \exp\left[\frac{(l-1)a^2}{2\sigma^2}\right],$$

is an inverted gamma distribution with $l > 1$ and $a > 0$. Verify that $p(\beta,\sigma) = p(\beta|\sigma)p(\sigma)$ which is similar to the familar relation for joint probability. The prior in (80) is called a gamma-normal prior. It can be verified that if we integrate $p(\beta,\sigma)$ over $\sigma > 0$ then the resulting marginal distribution of β will be multivariate student t.

We now multiply the likelihood function (76) with the prior (80) and using (79) write the joint posterior distribution of β and σ as

$$p(\beta,\sigma|y) \propto \sigma^{-n} \exp\left[\frac{1}{2\sigma^2}\left\{c_0 + (\beta - b^*)'\Omega^{*-1}(\beta - b^*)\right\}\right], \tag{81}$$

where $n = (T + l + p)$ and $c_0 = SSE(b) + (l - 1)a^2 + c^*$. Integrating (81) with respect to σ we get the marginal posterior distribution of β as

$$p(\beta|y) \propto [c_0 + (\beta - b^*)'\Omega^{*^{-1}}(\beta - b^*)]^{-(n-1)/2}. \tag{82}$$

This is a multivariate t density with the mean vector b^* and the variance covariance matrix $\dfrac{c_0}{n-p-1}\Omega^*$, that is,

$$\text{posterior density of } \beta \sim \text{multivariate } t\left(b^*, \frac{c_0}{n-p-1}\Omega^*\right). \tag{83}$$

Since both the prior (80) and posterior (81) after integrating out σ have the same (multivariate t) distribution, the prior is said to be "conjugate."

3.6.2 Diffuse (Improper) Priors

When we know nothing about the parameters β and σ we consider the joint prior of β and σ as

$$\begin{aligned} p(\beta,\sigma) &= p(\beta|\sigma)p(\sigma) \\ &\propto \frac{1}{\sigma}, \end{aligned} \tag{84}$$

where $p(\beta|\sigma)$ is a constant, and $p(\sigma)$ is proportional to $1/\sigma$. This can be obtained by letting the variances of prior distributions of β and σ in (80) tend to infinity, that is, $\Omega^{-1} \to 0$ and $a \to 0$ in (80). The prior (84) is referred to as a "diffuse" prior, or "improper" prior. The word improper is used because it does not integrate to 1. It is also called Jeffrey's prior in honor of its originator Jeffrey.

Using the diffuse prior we can get the posterior as

$$p(\beta,\sigma|y) \propto \sigma^{-(T+1)} \exp\left[-\frac{1}{2\sigma^2}\left\{SSE(b) + (\beta - b)'X'X(\beta - b)\right\}\right]. \tag{85}$$

Thus given σ the posterior of $\beta \sim N(b, \sigma^2(X'X)^{-1})$. However, integrating with respect to σ, the marginal posterior distribution of β is

$$p(\beta|y) \propto [SSE(b) + (\beta - b)'X'X(\beta - b)]^{-T/2}. \tag{86}$$

This is multivariate t with the mean vector b and variance covariance matrix $\dfrac{SSE(b)}{T-p}(X'X)^{-1}$.

To make inferences about σ we integrate the joint posterior distribution with respect to β.

3.6.3 Bayesian Regression Estimator

We have shown before that in the case of conjugate and diffuse joint priors on β and σ, the marginal posterior distribution of β is a multivariate t distribution. Thus the parameter β can be analyzed by using the well known properties of the multivariate t density function. The mean of this symmetric posterior density is also its mode and provides the Bayes estimator for the parameter. The Bayes estimator is a minimum expected loss (MELO) estimator with respect to a quadratic loss function (see Chapter 2). Thus from (86), assuming the diffuse prior on β and σ, Bayes estimator of β is identical with the maximum likelihood (or OLS) estimator, $b = (X'X)^{-1}X'y$. This should be expected, because under the diffuse prior, the posterior density is dominated by the likelihood function.

In case of conjugate priors on β and σ, Bayes estimator of β from (79a) is

$$b^* = (X'X + \Omega^{-1})^{-1}(X'Xb + \Omega^{-1}\beta_0) \tag{87}$$

$$= (X'X + \Omega^{-1})^{-1}X'Xb + (X'X + \Omega^{-1})^{-1}\Omega^{-1}\beta_0. \tag{87a}$$

The estimator b^* is a weighted matrix combination of the ML estimator b, based on data, and the prior mean β_0. Verify that the weight matrices in (87a) sum to the identity matrix. Also, the weight matrices of b and β_0 are the normalized inverse variance covariance matrices (or precision matrices) of b and β, respectively. Since $V(b) = \sigma^2(X'X)^{-1}$ and $V(\beta) = \sigma^2\Omega$, we have $V(b)^{-1} + V(\beta_0)^{-1} = \frac{1}{\sigma^2}(X'X + \Omega^{-1})$. Thus the normalized weight for b is given by its precision matrix $[V(b)^{-1} + V(\beta)^{-1}]^{-1}[V(b)]^{-1} = (X'X + \Omega^{-1})^{-1}X'X$. Similarly, the weight for β_0 is its precision matrix. Note that $V(b)^{-1} + V(\beta)^{-1}$ is $\sigma^2\ \Omega^{*-1}$, which is the precision matrix for the posterior of β.

We can also write b^* the Bayesian estimator from (87) by adding and subtracting $X'X\beta_0$ inside the second parentheses as:

$$b^* = (X'X + \Omega^{-1})^{-1}[X'X(b - \beta_0) + (X'X + \Omega^{-1})\beta_0] \tag{88}$$

$$= \beta_0 + (X'X + \Omega^{-1})^{-1}X'X(b - \beta_0). \tag{88a}$$

Alternatively, using $X'X = X'X + \Omega^{-1} - \Omega^{-1}$, we can write

$$b^* = \beta_0 + (I - (X'X + \Omega^{-1})^{-1}\Omega^{-1})\ (b - \beta_0). \tag{89}$$

Both the forms (87) and (88a) of the Bayesian estimator b^* correspond to the forms of the mixed estimator (b_M) in (48) and (50), respectively, for $R = I$, $\lambda = 1$ and $r = \beta_0$. The interpretation of this is that the prior $p(\beta|\sigma)$ is multivariate normal with mean β_0 and variance covariance matrix $\sigma^2\Omega$. Therefore one can write $\beta = \beta_0 + v$ where $Ev = 0$ and $Evv' = \sigma^2\Omega$. This implies that we are dealing with the model $y = X\beta + u$ with the stochastic restrictions $r = R\beta + v$, where $r = \beta_0$ and $R=I$. In fact, conditional on β_0, the minimization of

$$\frac{(y - X\beta)'(y - X\beta)}{\sigma^2} + \frac{1}{\sigma^2}(\beta - \beta_0)'\Omega^{-1}(\beta - \beta_0)$$

with respect to β gives the GLS estimator, called the mixed estimator, which will be identical to b^*. This minimization corresponds to the minimization in (46).

Although the Bayesian results are identical to the mixed regression results from Theil-Goldberger, the interpretations are different. In the Bayesian model β is a random variable, whereas in a mixed regression model it is not. For further discussion of this point see Swamy [1980].

3.6.4 Properties of the Bayesian Estimators

Normally, the classical sampling properties of Bayesian estimators are not required by a strict Bayesian. Nonetheless, since many sampling estimators discussed in following chapters will be compared with the Bayesian estimator, it is interesting to study their sampling properties. Specifically we shall analyze the properties of estimators b and b^*.

First we can easily show that as $T \to \infty$, $(X'X + \Omega^{-1})^{-1}X'X \equiv (\frac{X'X}{T} + \frac{\Omega^{-1}}{T})^{-1}\frac{X'X}{T}$ tends to the identity matrix, and $(X'X + \Omega^{-1})^{-1} \equiv (\frac{X'X}{T} + \frac{\Omega^{-1}}{T})^{-1}\frac{1}{T}$ tends to zero. This is because $\frac{X'X}{T}$ is a positive definite matrix and $\frac{\Omega^{-1}}{T}$ is a zero matrix when $T \to \infty$. Thus, from (87), for large samples b^* and b are identical, and so b^* is consistent, asymptotically unbiased and asymptotically efficient. Note that b^* and b will not be identical if the prior information is totally "dogmatic," that is, $\Omega = 0$. In this case b^* from (87) will be identical with β_0.

Since b^* is unbeaten (optimal) in the region of the parameter space where the prior is true, it is "admissible" in the sense defined in Section 2.7.2 of Chapter 2. Also, b^* can be shown to be a minimum expected loss (MELO) estimator (See Eq. (39) of Chapter 2) with respect to the conjugate prior and squared (error) loss function. On the other hand, under diffuse prior and squared loss, although b is MELO, it is inadmissible for models with three or more regressors. This will be shown in Chapter 6. With respect to the property of unbiasedness, as is well known the estimator b is BLUE. On the other hand, b^* is a linear non-homogeneous estimator and it is biased. Taking expectations on both sides of (87), the bias is given by:

$$Eb^* - \beta = -(X'X + \Omega^{-1})^{-1}\Omega^{-1}(\beta - \beta_0)$$

$$= -\Omega^*\Omega^{-1}\beta^*, \qquad (90)$$

where $\Omega^* = (X'X + \Omega^{-1})^{-1}$ as in (79a) and $\beta^* = \beta - \beta_0$.

Now we consider the variance covariance matrix and MtxMSE comparisons of b and b^*. Subtracting (87) from (90) we can write

$$b^* - Eb^* = -\beta^* + (I - \Omega^*\Omega^{-1})(b - \beta_0) + \Omega^*\Omega^{-1}\beta^*$$

$$= (I - \Omega^*\Omega^{-1})(b - \beta) \tag{91}$$

and

$$(b^* - Eb^*)(b^* - Eb^*)' = (I - \Omega^*\Omega^{-1})(b - \beta)(b - \beta)'(I - \Omega^{-1}\Omega^*). \tag{92}$$

Thus taking expectations on both sides

$$V(b^*) = (I - \Omega^*\Omega^{-1})V(b)(I - \Omega^{-1}\Omega^{*\prime})$$

$$= \sigma^2(I - \Omega^*\Omega^{-1})(X'X)^{-1}(I - \Omega^{-1}\Omega^{*\prime})$$

$$= \sigma^2\Omega^* X'X\,\Omega^*, \tag{93}$$

because $\Omega^{*-1} = (X'X + \Omega^{-1})$, so that $\Omega^{-1} = \Omega^{*-1} - X'X$, and $\Omega^*\Omega^{-1} = (I - \Omega^* X'X)$.

Let $\Delta = V(b) - V(b^*) = \sigma^2[(X'X)^{-1} - \Omega^* X'X\,\Omega^*]$. Then for b^* to have smaller variance we require to show that Δ is a positive definite matrix. This can be shown by proving that $V(\eta' b) \geq V(\eta' b^*)$ or that $\eta'\Delta\eta \geq 0$ for all $\eta \neq 0$. Therefore write

$$\eta'\Delta\eta = \sigma^2[\eta'(X'X)^{-1}\eta - \eta'\Omega^* X'X\,\Omega^*\eta] \tag{94}$$

or

$$\xi'\Omega^{*-1}\Delta\,\Omega^{*-1}\xi = \sigma^2\xi'[\Omega^{*-1}(X'X)^{-1}\Omega^{*-1} - X'X]\xi\ ,$$

where $\xi = \Omega^*\eta$. But, since

$$\Omega^{*-1}(X'X)^{-1}\Omega^{*-1} = \Omega^{-1}(X'X)^{-1}\Omega^{-1} + X'X + 2\Omega^{-1}$$

we can write the right hand side of (94) as $\sigma^2\xi'(\Omega^{-1}(X'X)^{-1}\Omega^{-1} + 2\Omega^{-1})\xi$. Now Ω^{-1} and $\Omega^{-1}(X'X)^{-1}\Omega^{-1}$ are both positive semi-definite so $\sigma^2\xi'(\Omega^{-1}(X'X)^{-1}\Omega^{-1} + 2\Omega^{-1})\xi > 0$, and hence $\eta'\Delta\eta \geq 0$ for all $\eta \neq 0$. Thus $V(\eta' b) > V(\eta' b^*)$ and in particular $V(b_i) > V(b_i^*)$, $i = 1, \ldots, p$. This result implies that though b^* is biased it has smaller variance compared to b.

Now consider the MtxMSE criterion according to which b^* will be preferred to b if $\Delta^* = MtxMSE(b) - MtxMSE(b^*)$ is a positive definite matrix, or alternatively $\text{MSE}(\eta' b) \geq \text{MSE}(\eta' b^*)$ for all $\eta \neq 0$. Let us write

$$MtxMSE(b) = V(b) + [Bias(b)][Bias(b)]' = \sigma^2(X'X)^{-1}$$

$$MtxMSE(b^*) = V(b^*) + [Bias(b^*)][Bias(b^*)]'$$

$$= \sigma^2\Omega^* X'X\,\Omega^* + \Omega^*\Omega^{-1}\beta^*\beta^{*\prime}\Omega^{-1}\Omega^{*\prime}. \tag{95}$$

Then Δ^* is positive semi-definite if $\eta'\Delta^*\eta \geq 0$ for all $\eta \neq 0$, or using $\xi = \Omega^*\eta$

the iff condition for the superiority of the Bayesian estimator b^* over OLS is

$$\xi'[\Omega^{*-1}(X'X)^{-1}\Omega^{*-1} - X'X - \frac{\Omega^{-1}\beta^*\beta^{*\prime}\Omega^{-1}}{\sigma^2}]\xi \geq 0.$$

The above is true if

$$\theta_0 = \frac{(\xi'\Omega^{-1}\beta^*\beta^{*\prime}\Omega^{-1}\xi)}{2\sigma^2\xi'[\Omega^{-1}(X'X)^{-1}\Omega^{-1} + 2\Omega^{-1}]\xi} \leq \frac{1}{2}. \tag{96}$$

Finally, (96) holds for all nonzero ξ if

$$\theta = \operatorname*{Sup}_{\xi} \theta_0$$

$$= \frac{\beta^{*\prime}[(X'X)^{-1} + 2\Omega]^{-1}\beta^*}{2\sigma^2} \leq \frac{1}{2}, \tag{97}$$

where $\beta^* = \beta - \beta_0$. Thus b^* is preferred to b if $\theta \leq \frac{1}{2}$. Note that this condition is a special case of the condition in (30) for $R = I, r = \beta_0$ and $\lambda = 1$. In principle one could use the Toro-Vizcarrondo and Wallace test to test the null hypothesis H_0: $\theta \leq \frac{1}{2}$. However, in this case the test statistic w^* with $R = I$, $\lambda = 1$ and $r = \beta_0$ is biased (toward rejection) as indicated by Giles and Rayner [1979].

3.6.5 Lindley's Bayesian Regression Model

In this section we shall consider the Bayesian analysis of the model $y = X\beta + u$ when the parameter vector β itself has a structure which would influence the choice of a prior. For example, consider a simple case where this structure on β is given by

$$E(\beta) = A\beta_0, \tag{98}$$

where A is a $p \times p_0$ known matrix (not to be confused with the A matrix of (61) above) and β_0 is a $p_0 \times 1$ vector of hyperparameters. Let β have a multivariate normal distribution with known variance covariance matrix Ω_0. Then the above structure implies the following normal prior distribution of $\beta \sim N(A\beta_0, \sigma^2\Omega_0)$, instead of the one with the mean β_0 discussed in the previous section. For example, the structure of this kind can appear in "polynomial distributed lag" models and in the multi-regression models discussed later in Chapter 9.

At this point, the Bayesian analysis conditional on β_0 can be carried out exactly as was done in the earlier section. However, the process could be repeated by incorporating additional structure on β_0. In most applications the process can be concluded at a stage which expresses the structure of the hyperparameters by supposing

$$E(\beta_0) = A_1\mu \ , \tag{99}$$

where A_1 and μ, of orders $p_0 \times p_1$ and $p_1 \times 1$ respectively, are both known. Normally, p_1 is much smaller than p and in many applications it is just one. The hyperparameters β_0 are again assumed to be distributed as normal variables with a known variance covariance matrix Ω_1, that is, $\beta_0 \sim N(A_1\mu, \sigma^2\Omega_1)$. Our problem is then to analyze the general Bayesian regression model $y = X\beta + u$, where we assume that:

$$\begin{aligned} y &\sim N(X\beta, \sigma^2 I), \\ \beta &\sim N(A\beta_0, \sigma^2\Omega_0), \\ \beta_0 &\sim N(A_1\mu, \sigma^2\Omega_1). \end{aligned} \tag{99a}$$

The general Bayesian regression model in (99a) essentially consists of linear regression model with a (three stage) hierarchical form of prior structures due to Lindley (1971) and Lindley and Smith (1972). A two stage hierarchy of priors is one in which the first stage prior describes the relationship between parameters in the linear regression and the second stage prior describes the knowledge about the form of this relationship. When only the first stage prior is used, with $A=I$, one gets the Bayesian regression model described earlier in Sections 3.6.1 to 3.6.4

Since the structures in (98) and (99) are linear and the distributions are normal, the above model reduces to

$$y \sim N(X\beta, \ \sigma^2 I)$$

$$\beta \sim N(A_2\mu, \ \sigma^2\Omega_2); \ A_2 = AA_1; \ \Omega_2 = A\,\Omega_1 A' + \Omega_0.$$

This is because we can write $\beta = A\beta_0 + v$ and $\beta_0 = A_1\mu + w$, where $v \sim N(0, \Omega_0)$, $w \sim N(0, \Omega_1)$ and $Evw' = 0$. Thus $\beta = AA_1\mu + v + Aw$ is normal with the mean $AA_1\mu$ and variance covariance matrix $\sigma^2(A\,\Omega_1 A' + \Omega_0)$. Now, from (79b), the posterior density of β, conditional on σ^2, Ω_0, Ω_1 and the y data, can be written as

$$p(\beta | y, \sigma^2, \Omega_0, \Omega_1) \propto exp[-\frac{1}{2\sigma^2}(\beta - b_L^*)'\Omega_L^{*-1}(\beta - b_L^*)], \tag{100}$$

where

$$\begin{aligned} b_L^* &= (X'X + \Omega_2^{-1})^{-1}(X'Xb + \Omega_2^{-1}A_2\mu), \\ \Omega_L^* &= (X'X + \Omega_2^{-1})^{-1}, \end{aligned} \tag{101}$$

where L stands for Lindley, and b is the OLS estimator. An alternative form of b_L^* is

$$b_L^* = A_2\mu + [I - (\Omega_2^{-1} + X'X)^{-1}\Omega_2^{-1}]\,(b - A_2\mu). \tag{101a}$$

This posterior density is multivariate normal with the mean vector b_L^* and variance covariance matrix Ω_L^*. Thus Lindley's Bayesian estimator of β from (101) is

$$b_L^* = [X'X + (A\Omega_1 A' + \Omega_0)^{-1}]^{-1}[X'Xb + (A\Omega_1 A' + \Omega_0)^{-1}AA_1\mu], \quad (102)$$

where we have rewritten (101a) with

$$\Omega_2^{-1} = (A\Omega_1 A' + \Omega_0)^{-1} \quad (102a)$$

$$= \Omega_0^{-1} - \Omega_0^{-1}A(A'\Omega_0^{-1}A + \Omega_1^{-1})^{-1}A'\Omega_0^{-1}$$

and used a formula for matrix inversion. (See Exercise 3.1 at the end of this chapter). The sampling properties of b_L^* follow directly from the results of an earlier section.

For the case when $\Omega_1 \to \infty$, that is, when we have a noninformative or vague prior on β_0, we write (102a) without the term involving Ω_1 as:

$$\Omega_2^{-1} = \Omega_0^{-1} - \Omega_0^{-1}A(A'\Omega_0^{-1}A)^{-1}A'\Omega_0^{-1}, \quad (102b)$$

and hence we have the following cancellation:

$$\Omega_2^{-1}A = \Omega_0^{-1}A - \Omega_0^{-1}A(A'\Omega_0^{-1}A)^{-1}(A'\Omega_0^{-1}A) = 0.$$

Thus $\Omega_2^{-1}A_2 = 0$, and b_L^* of (101) simplifies to

$$b_L^* = (X'X + \Omega_2^{-1})^{-1}X'y$$

$$= [I + (\Omega_2 X'X)^{-1}]^{-1}b, \quad (103)$$

where $(\Omega_2 X'X)^{-1}$ essentially provides a correction term to the OLS estimator b. A similar correction of this form occurs for Stein-rule and ridge estimators discussed in Chapters 6, 7, and 8. These estimators were obtained using non-Bayesian methods by James and Stein [1961] and Hoerl and Kennard [1970]. Thus Lindley's hyperparameter model provides a Bayesian view of these non-Bayesian estimators.

If the parameter of interest is β_0 in (99a) instead of β and $\Omega_1 \to \infty$, then one can obtain the posterior density of β_0 given σ^2, Ω_0, Ω_1 and y. This will be multivariate normal with the following mean (Bayes estimator)

$$b_0^* = \left\{A'((X'X)^{-1} + \Omega_0)^{-1}A\right\}^{-1} A'((X'X)^{-1} + \Omega_0)^{-1}b. \quad (104)$$

One can also obtain b_0^* by noting from (99a) that $y \sim N(XA\beta_0, \sigma^2 + \sigma^2 X\Omega_0 X')$. Thus if β_0 has a diffuse prior distribution, it follows from the analysis in section 3.6.1 that the posterior mean of β_0 will simply be its generalized least squares estimator. Thus

$$b_0^* = [A'X'(I + X\Omega_0 X')^{-1}XA]^{-1} A'X'(I + X\Omega_0 X')^{-1}y,$$

which is identical with (104) by using the matrix inversion of $(I + X\Omega_0 X')^{-1}$, (See Exercises 3.1(b) and 3.1(c) at the end of this chapter).

Now we can write an alternative form of the Lindley Bayes estimator b_L^* by obtaining the matrix inversion of $X'X + \Omega_2^{-1} = (X'X + \Omega_0^{-1}) - (\Omega_0^{-1}A)(A'\Omega_0^{-1}A)^{-1}(A'\Omega_0^{-1})$ from (102b). Notice that the right hand side is of the same form as Exercise 3.1(b). Using this we can get

$$b_L^* = (X'X + \Omega_0^{-1})^{-1}(X'Xb + \Omega_0^{-1}Ab_0^*)$$

$$= Ab_0^* + [I + (\Omega_0 X'X)^{-1}]^{-1}(b - Ab_0^*). \tag{105}$$

In this form b_L^* appears as the weighted matrix combination of the estimators b and Ab_0^*. Also, the second line shows that the estimator b_L^* shrinks the least squares estimator b towards Ab_0^*. Thus, even though there is an apparent similarity in b_L^* and ridge estimators ($\Omega_2 = I$) from (103), it is clear from (105) that they differ with respect to the direction of shrinkage. We will note in Chapter 7 that in ridge regression the direction of shrinkage is usually towards zero, whereas in b_L^* it is towards Ab_0^*. Finally, note that the form of b_L^* in (105) appears as if we have obtained the posterior mean of β in the linear regression $y = X\beta + u$ with the prior $\beta \sim N(Ab_0^*, \Omega_0)$. It is therefore true that b_L^* in (105) comes very close to the form of b^* in (87). However, the interpretations of these two are different as they belong to different prior structures.

3.7 INEQUALITY RESTRICTED (IR) LEAST SQUARES ESTIMATOR

In the earlier sections we considered the linear regression model with exact stochastic and nonstochastic restrictions on its coefficients. However, in applied work one often deals with the models that are subject to inequality restrictions on the coefficients. These restrictions can be represented as

$$R\beta \geq r$$

or

$$R\beta + \phi = r,$$

where R is an $m \times p$ matrix and r is an $m \times 1$ vector, both consisting of known constants, and ϕ is a $m \times 1$ unknown vector. The estimation problem which results from combining $R\beta \geq r$ with the sample information $y = X\beta + u$ can be specified as a quadratic programming problem for which a number of solution algorithms exists in Operations Research literature, (Liew [1976]).

Two Regressor Case Consider a simple case of linear regression with two regressors as

$$y = x_1\beta_1 + x_2\beta_2 + u = X\beta + u,$$

where

$$X = [x_1 \; x_2], \text{ and } \beta = \begin{bmatrix} \beta_1 \\ \beta_2 \end{bmatrix} \geq 0.$$

The above inequality restrictions are in the form of $R\beta \geq r$ where R is a 2X2 identity matrix, and r is a 2X1 vector of zero elements.

The least squares estimator of β is decomposed as:

$$b = \begin{bmatrix} b_1 \\ b_2 \end{bmatrix} = (X'X)^{-1}X'y.$$

Since the estimators b_1 and b_2 may or may not violate the constraint, the "inequality restricted estimator" can be expressed as

$$b_{IR} = \begin{cases} b & \text{if } b_1 \text{ and } b_2 \textit{ are both } \geq 0 \\ b^{(1)} & \text{if } b_1 \geq 0 \text{ and } b_2 \leq 0 \\ b^{(2)} & \text{if } b_1 \leq 0 \text{ and } b_2 \geq 0 \\ b^{(0)} & \text{if } b_1 \text{ and } b_2 \textit{ are both } \leq 0, \end{cases}$$

where IR represents inequality restricted, and

$$b^{(1)} = \begin{bmatrix} b_1 \\ 0 \end{bmatrix}, \; b^{(2)} = \begin{bmatrix} 0 \\ b_2 \end{bmatrix}, \; b^{(0)} = \begin{bmatrix} 0 \\ 0 \end{bmatrix}.$$

These four are all the possible outcomes for b_{IR}. It can be shown that the estimator b_{IR} satisfies the Kuhn-Tucker conditions for the minimization of $(y - X\beta)'(y - X\beta)$, subject to $\beta_1 \geq 0$, and $\beta_2 \geq 0$ (Schmidt [1978]).

The sampling properties of the inequality restricted least squares estimator have not been fully explored yet. In special cases, some work regarding this can be found in Judge and Yancey [1978a and 1978b], Thomson [1979] and Thomsòn and Schmidt [1979]. The results of these authors indicate, that the inequality restricted estimator will, in general, be biased. Also, if the direction of the inequality is correctly known, IR estimator will be better than the estimators b and b_R (in Section 3.1) in the MSE sense.

3.8 NEED FOR JUDGMENT IN EMPIRICAL RESEARCH

The discussion in this chapter lays certain foundations for the results in the following chapters. It is not appropriate to impose restrictions or use Bayesian methods without thinking about the specific problem at hand. We will see that the job of a researcher is becoming more and more skilled, requiring some judgement and experience. In the (good) old days when ordinary least squares was the only available estimator, any high school student could be taught to do scientific empirical research. The recent advances have made researchers' task complicated, but exciting.

EXERCISES

3.1 Show that the following matrix identities are true:

(a) $(D + AB)^{-1} = D^{-1} - D^{-1}A(I + BD^{-1}A)^{-1}BD^{-1}$

(b) $(D + EFE')^{-1} = D^{-1} - D^{-1}E(E'D^{-1}E + F^{-1})^{-1}E'D^{-1}$

(c) $(D + B)^{-1} = D^{-1} - D^{-1}(D^{-1} + B^{-1})^{-1}D^{-1}$

(d) $(D + B)^{-1}B = I - (D + B)^{-1}D$

where it is assumed that all matrices are of appropriate (conformable) dimensions, and that all the stated inverses exist.

3.2 Derive the posterior distribution of the parameters β and β_0 in the model: $y \sim N(X\beta, \sigma^2\Omega)$; $\beta \sim N(A\beta_0, \sigma_\beta^2\Omega_0)$ and $\beta_0 \sim N(A_1\beta_1, \Omega_1)$, with $\Omega_1 \to \infty$. Then show that the posterior mean of β is

$$\beta^* = \left[X'X + k\,\Omega_0^{-1}\right]^{-1}\left[X'Xb + k\,\Omega_0^{-1}A\beta_0^*\right]$$

where $k = \sigma^2/\sigma_\beta^2$ and $\beta_0^* = (A'\Omega_0^{-1}A)^{-1}A'\Omega_0^{-1}\beta^*$ is the posterior mean of β_0.

3.3 Suppose you consider a true model: $y = X\beta + u$, with no intercept and apply the least squares method to obtain an estimate of β. Another researcher does use your model, but before obtaining the least square estimates he takes deviations from the sample means for the variable y and those in the X matrix. What would be the effect on the efficiency of the estimates of β due to these two approaches. Explain your model and estimators in the framework of restricted and unrestricted least squares theory. What would be your suggestion to an applied researcher on the basis of your results?

3.4 The model: $y_t = \beta_0 + \beta_1 x_{1t} + \beta_2 x_{2t} + \beta_3 x_{3t} + u_t$, was estimated by ordinary least squares from 26 observations yielding:

$$y_t = 2 + \underset{(1.9)}{3.5}x_{1t} - \underset{(2.2)}{0.7}x_{2t} + \underset{(1.5)}{2}\,x_{3t} + \hat{u}_t$$

with t-ratios in parentheses and $R^2 = 0.9832$.

The same model was estimated with the restriction $\beta_2 = \beta_1$. Estimates were

$$y_t = 1.5 + 3(x_{1t} + x_{3t}) - 0.6x_{2t} + \hat{u}_t \quad ; \quad R^2 = 0.876$$

(a) Test the significance of the vector $\beta = [\beta_1\ \beta_2\ \beta_3]'$ from the unrestricted estimates.

(b) Test the significance of the restriction: $\beta_2 = \beta_1$. Clearly state any assumptions utilized (Univ. of London MSc (Econ.) examinations, 1977).

4

Autoregressive Moving Average (ARMA) Regression Errors and Heteroscedasticity

The consequences of autoregressive (AR) and/or moving average (MA) errors, non-stationarity and heteroscedasticity are discussed in many textbooks: Box and Jenkins [1976], Johnston [1972], Dhrymes [1978], etc. In this chapter, our discussion of such generally available material is very brief.

If $\epsilon_t \sim N(0,\sigma^2)$ denotes independent and identically distributed (iid) normal variable with zero mean and common variance σ^2, the process: $u_t = \phi_1 u_{t-1} + \epsilon_t$ is called autoregressive (u_t regressed on lagged values of itself) processes of the first order, AR(1). First order moving average process MA(1) is defined by: $u_t = \epsilon_t + \theta_1 \epsilon_{t-1}$, where the right hand side contains a weighted (moving) sum of the iid variables. In general mixed ARMA(p,q) process is defined by $u_t = \phi_1 u_{t-1} + ... + \phi_p u_{t-p} + \theta_1 \epsilon_t + ... + \theta_q \epsilon_{t-q}$. Regression errors can satisfy general ARMA processes which may not be stationary i.e., may not belong to a process which remains in equilibrium about a constant mean level.

4.1 DURBIN-WATSON-VINOD-WALLIS TESTS

Testing for the presence of autocorrelated errors of order j is made by the Durbin-Watson-type (DW) statistic $\hat{d}_j$ of order j defined by

$$\hat{d}_j = \sum_{t=j+1}^{T} (u_t - \hat{u}_{t-j})^2 \Big/ \sum_{t=1}^{T} \hat{u}_t^{\,2} . \tag{1}$$

This is tabulated for j=1 in Durbin and Watson [1950, 1951]. For j = 2, 3 and 4 Tables are given by Vinod [1973], and reproduced in Dhrymes [1978, p. 404-415]. For j=4 Wallis (see [1972], "Added in proof footnote" p. 617) using quarterly data for regressions has independently developed a test similar to

Vinod's. Roughly speaking $\hat{d}_j$ close to 2 suggests that the errors are not jth order serially correlated. By contrast, when d_j is close to zero or four, it suggests that ρ_j the serial correlation of order j is "significant" in some sense. The Durbin-Watson type test is often used by practitioners to check whether some important variable has been omitted in the basic model specification. (See Savin and White [1978] for interesting results.)

The DW test statistic is

$$d = \hat{d}_1 = \frac{\hat{u}'A\hat{u}}{\hat{u}'\hat{u}} = \frac{u'MAMu}{u'Mu}\ , \tag{1a}$$

where $M = I - X(X'X)^{-1}X'$, $\hat{u} = y - Xb$ and

$$A = \begin{bmatrix} 1 & -1 & 0 & \cdots & 0 \\ -1 & 2 & -1 & \cdots & 0 \\ 0 & -1 & 2 & \cdots & 0 \\ \vdots & \vdots & \vdots & \cdots & \vdots \\ 0 & 0 & 0 & \cdots & 1 \end{bmatrix} \cdot \tag{1b}$$

The probability distribution of $\hat{d}_1$, under $\rho = 0$, depends on X. Hence DW derived upper ($\hat{d}_{1UP}$) and lower ($\hat{d}_{1LO}$) limits for the significance level of $\hat{d}_1$. Their tables are to test the hypothesis $H_0 : \rho = 0$ against the one sided alternative $H_1 : \rho > 0$. They suggested rejecting the null hypothesis of no positive autocorrelation (AR(1) error) if $\hat{d}_1 < \hat{d}_{1LO}$, accepting it if $\hat{d}_1 > \hat{d}_{1UP}$ and treating the test as inconclusive if $\hat{d}_{1LO} < \hat{d}_1 < \hat{d}_{1UP}$. For testing the hypothesis $\rho = 0$ against $\rho < 0$, (negative autocorrelation) one considers $4 - d_1$ and uses the DW tables as though one were testing for positive autocorrelation.

Note that the bounds for the DW test statistic $\hat{d}_1$ are given only for the regression model with an intercept. Kramer [1971], Farebrother [1980,81] and King [1980a] have published tables of bounds appropriate for testing AR(1) disturbances in regressions without an intercept (forced through the origin). King [1981] further extends the DW results to regressions (i) with a full set of quarterly seasonal dummy variables (ii) with an intercept and a linear trend variable, and (iii) regressions with a full set of quarterly seasonal dummy variables and a linear trend regressor. King notes that in each of these cases the size of the inconclusive region for $\hat{d}_1$ becomes smaller. Intuitively, this follows, because by including more regressors explicitly, we reduce the uncertainty regarding the form of the X matrix in $y = X\beta + u$.

There are some other methods in the literature which deal with the problem of the inconclusive DW test region. Some of them are: using $\hat{d}_{1UP}$ instead of $\hat{d}_1$ (Hannan and Terrell [1968]), the fitting of beta distribution on approximate moments (Theil and Nagar [1961]), fitting of a beta distribution (Durbin and Watson [1951], Henshaw [1966]), and Durbin and Watson's [1971] more recent method.

Durbin and Watson [1950] have shown that their statistic $\hat{d}_1$ is approximately uniformly most powerful, if X is spanned by p of the characteristic vectors of the matrix A. Kariya [1977] generalized this result to the case where u follows any non-normal symmetric distribution. He also found that with the same restriction on the X matrix, the two-sided DW test is an approximately "uniformly most powerful" unbiased (UMPU) test of $H_0{:}\rho = 0$ against $H_1{:}\rho \neq 0$ for a restrictive class of symmetric distributions of the error vector u. King [1980b] has studied the invariance property of the DW test. He found that, under the restriction on X mentioned above, the one sided DW test is an approximately UMP invariant test of H_0 against H_1. He also showed that for general X, the one sided DW test is approximately locally best invariant in the neighborhood of $\rho = 0$. The invariance of $\hat{d}_1$ here implies the invariance with respect to the transformation $y^* = \alpha_0 y + X\alpha$ or $My^* = \alpha_0 My$ or $g(u) = \alpha_0 u$ ($My = MX\beta + Mu = Mu$ because $MX = 0$) where α_0 is a positive scalar and α is a $p \times 1$ vector. For details on invariance see Lehmann [1959].

The above results will similarly hold for the higher order DW-type statistics studied by Wallis [1972] and Vinod [1973].

It was noted earlier that the distribution of the OLS residuals depends on X. Also note that these residuals are not independent. Thus an alternative to the DW test is to use a test based on a new set of fewer iid residuals. Theil [1971, Chapter 5] derived a _b_est _l_inear _u_nbiased residual vector with _s_calar covariance matrix (BLUS), and suggested its use in the von Neumann [1941] ratio (ratio of mean-square successive differences to the variance) to test the independence of residuals. Another alternative has been proposed by Harvey and Phillips [1974]. They suggested the use of recursive residuals which also are LUS. Further alternatives have been suggested by Abrahamse and Koerts [1971], Abrahamse and Louter [1971], Golub and Styan [1973], Neudecker [1977], Sims [1975] Dent and Styan [1978], and King [1980c]. It should be noted that all these alternatives provide a set of T-p independent residuals. This is because the T OLS residuals are subject to p linear restrictions.

The power properties of many of these alternatives to DW have been investigated by Koerts and Abrahamse [1969], Abrahamse and Koerts [1969, 1971], Dubbelman, Louter and Abrahamse [1978], Dent and Styan [1978], Dent and Cassing [1978], L'Esperance and Taylor [1975] among others. The basic conclusion from all these papers is that the DW test is a "most powerful" test, provided we deal with the inconclusive region accurately. Thus it appears that a search in the recent literature for an alternative to DW, which is more powerful for a general X, and whose critical regions are free from the matrix X, has not yet produced a better alternative to DW (King [1980c]).

It is often not emphasized in the textbooks (e.g., Daniel and Wood [1971]) that the behavior of observed residual $\hat{u}_t$ may not indicate the behavior of true unknown residuals u_t. Note that $\hat{u} = (I - X(X'X)^{-1}X')y = Mu$ $(MX=0)$ implies that $\hat{u}_t$ are a linear transformation of u_t, and it is possible to construct examples where $\hat{u}_t$ may "look" random on a plot of residuals and yet may have arisen from serially correlated u_t. Statistical literature contains a great body of graphical tools for looking at observed residuals. Their use should be made with an awareness that the true errors may look quite different. The Durbin-

Watson type tests have the important advantage that serial correlation among the true unknown u_t values is correctly tested. The main disadvantage of these tests is the presence of an "indeterminate range" within the tabulated bounds, where no definite conclusion is reached.

We have noted that the distribution of $\hat{d}_j$ can be approximated by a beta distribution by equating the first two (or four) moments of $\hat{d}_j$ with the corresponding (theoretically known) moments of a beta variable. Hence, it should be possible to use the power of modern computers to directly obtain the significance level itself for any given matrix of regressors, without the need to use bounding significance values from published tables. It is somewhat surprising that such computer programs are not widely available, because this procedure detailed in Section 4.6 would be more accurate, and would not have any "indeterminate range".

4.2 TESTS FOR REGRESSION COEFFICIENTS

Apart from testing the serial correlations of various orders, one is usually more interested in tests for regression coefficients β_i, or their linear combinations. The usual F tests for the null hypothesis $R\beta = r_0$, are sometimes severely affected by the presence of ARMA errors.

The equality restricted least squares model can be used to study the effect of low-order autoregressive moving average (ARMA) errors on significance tests for the OLS regression coefficients. Watson [1955], Watson and Hannan [1956], Vinod [1976i] and recently Kiviet [1980] have developed the theory which is discussed in this and the following two sections.

Consider the model of Chapter 3 equations (1) and (2) with a more general error covariance matrix:

$$y = X\beta + u \quad , \quad R\beta = r \quad , \quad u \sim N(0, \sigma^2\Omega) \; ,$$

where R is $m \times p$, y and u are $T \times 1$, and Ω is $T \times T$.

If Ω is known, Aitken's generalized least squares (GLS) estimator of β is

$$b^{GLS} = (X'\Omega^{-1}X)^{-1} X'\Omega^{-1}y. \tag{2}$$

If Ω is unknown, the GLS estimator must be based on "consistent" estimates of the T^2 elements of Ω. The number of unknowns in Ω is reduced to only one if we are willing to assume that the errors follow a first order AR or MA process. The corresponding matrices are given below in (29) and (42) respectively. Some practitioners do not prefer GLS over OLS, because the GLS fit looks distorted in a graphical sense and/or the forecasting performance of GLS is poor.

If the investigator is mainly interested in significance tests for regression vector b, it may be possible to use the simpler OLS estimator instead of a GLS estimator based on "estimated" Ω. A continued reliance on significance tests based on Students' t values or F values computed in usual OLS computer programs despite ARMA errors may be dangerous. It requires some assurance that the final conclusions from these tests will not be reversed, if the GLS method was used. A purpose of the following discussion is to devise some bounds on t or F values, such that the conclusions (eg. reject a null hypothesis) from OLS computations will not be reversed, even if we had undertaken the additional computation involved in specifying various values of ARMA parameters defining an estimated Ω, and then used significance tests based on GLS computations.

4.3 DERIVATION OF BOUNDS ON F TESTS BASED ON EIGENVALUES OF Ω

In this section we derive bounds on Snedecor's F values, such that whenever the observed F is larger (smaller) than the tabulated upper (lower) bound, GLS will not reverse the conclusions based on OLS. The main result is given in equation (20). The tabulation of bounds is discussed in the following section and illustrated in a subsection.

Recall the test statistic of equation (20) in Chapter 3 for an F test $F(m,T-p)$ which is proportional to

$$F^* = \frac{(Rb-r)'[R(X'X)^{-1}R']^{-1}(Rb-r)}{(y-Xb)'(y-Xb)},$$

where $b = (X'X)^{-1}X'y$ is the OLS estimator. We may write this in terms of the errors u

$$F^* = \frac{u'X(X'X)^{-1}R'Q^{-1}R(X'X)^{-1}X'u}{u'Mu}, \tag{3}$$

where

$$Q = R(X'X)^{-1}R' \quad , \quad \text{and} \quad M = I - X(X'X)^{-1}X' . \tag{4}$$

It is convenient to introduce a slight change of notation here. The $m \times p$ matrix R of rank m (new notation, R_m) will now be augmented by another $(p-m)\times p$ matrix R_{p-m} of rank $p-m$ such that the augmented $p\times p$ matrix R_p is nonsingular. Thus the subscripts denote number of rows, as well as, the rank of the corresponding matrix. Now we will use certain algebraic results to write (3) in a form which is suitable for application of a Lemma given in Section 4.3.1 below.

Let a matrix having rank p be defined by

$$S_p = X(X'X)^{-1}R_p' . \tag{5}$$

We decompose S_p into two parts, one of rank m denoted by S_m and another of rank $p-m$ denoted by S_{p-m} as:

$$S_p = [S_m \colon S_{p-m}] \ , \ \textit{where} \ \ S_m = X(X'X)^{-1}R_m' \ ,$$

and $S_{p-m} = X(X'X)^{-1}R_{p-m}'$. Now, from (5) write $X = S_p[(X'X)^{-1}R_p']^{-1}$ and $X(X'X)^{-1} = S_p[R_p']^{-1}$. Hence

$$\begin{aligned} X(X'X)^{-1}X' &= S_p[R_p']^{-1}[S_p(R_p')^{-1}(X'X)]' \\ &= S_p[R_p']^{-1}X'X[R_p]^{-1}S_p' \\ &= S_p[R_p(X'X)^{-1}R_p']^{-1}S_p' \\ &= S_p[R_p(X'X)^{-1}X'X(X'X)^{-1}R_p']^{-1}S_p' \\ &= S_p(S_p'S_p)^{-1}S_p' \ . \end{aligned} \tag{6}$$

Also, we have

$$\begin{aligned} Q &= R_m(X'X)^{-1}R_m' \\ &= R_m(X'X)^{-1}X'X(X'X)^{-1}R_m' \\ &= S_m'S_m \ . \end{aligned} \tag{7}$$

Upon substitution in F^* of (3) we have

$$F^* = \frac{u'S_m(S_m'S_m)^{-1}S_m'u}{u'[I - S_p(S_p'S_p)^{-1}S_p']u} \ . \tag{8}$$

Now we will simplify the expression $S_m(S_m'S_m)^{-1}S_m'$ by using the "singular value decomposition" $S_m = H_m\Lambda_m^{1/2}G_m'$, where H_m is a $T \times m$ matrix Λ_m is a diagonal $m \times m$ matrix, and G_m is an orthogonal $m \times m$ matrix, satisfying $G_m' = G_m^{-1}$, $S_m'S_m = G_m\Lambda_mG_m'$ and $H_m'H_m = I_m$.

$$\begin{aligned} S_m(S_m'S_m)^{-1}S_m' &= H_m\Lambda_m^{1/2}G_m'G_m\Lambda_m^{-1}G_m'G_m\Lambda_m^{1/2}H_m' \\ &= H_mH_m' \ . \end{aligned} \tag{9}$$

Similarly, $S_p(S_p'S_p)^{-1}S_p' = H_pH_p'$ in an analogous notation. Substituting in (8) we have

$$F^* = \frac{u'H_m H_m' u}{u'[I - H_p H_p']u} . \tag{10}$$

Now, let H_{T-p} be such that $H_T = [H_m : H_{p-m} : H_{T-p}]$ is a $T \times T$ orthogonal matrix: $H_T' H_T = H_T H_T' = (H_p H_p' + H_{T-p} H_{T-p}')$. Substituting in (10) we have

$$F^* = \frac{u'H_m H_m' u}{u'H_{T-p} H_{T-p}' u} . \tag{11}$$

Now consider a $T \times T$ symmetric square-root matrix of the $T \times T$ matrix Ω defined by $\sigma^2 \Omega = PP'$, and note that $\mathrm{v} = Pu^{-1} \sim N(0,I)$. Thus $u = P\mathrm{v}$, and we have

$$F^* = \frac{\mathrm{v}'P'H_m H_m' P \mathrm{v}}{\mathrm{v}'P'H_{T-p} H_{T-p}' P \mathrm{v}} . \tag{12}$$

Now we need to find some bounds on F^*, which in turn involve eigenvalues of

$$P'H_m H_m' P \quad \text{and} \quad P'H_{T-p} H_{T-p}' P$$

appearing in the numerator and the denominator of F^*. Let us concentrate on the former since the latter is analogous. Note that $H_m H_m'$ is an idempotent matrix of rank m: $H_m H_m' H_m H_m' = H_m I_m H_m' = H_m H_m'$, whose eigenvalues are ones or zeroes.

The matrix $\Omega = PP'$, is proportional to the covariance matrix of regression errors. If the errors are assumed to follow a low order autoregressive moving average ARMA type stochastic process, the eigenvalues of Ω are known, except for a scale factor. Since $H_m H_m'$ is idempotent, its eigenvalues are either zero or unity. The nonzero eigenvalues of $P'H_m H_m' P$ are the same as the nonzero eigenvalues of $H_m H_m' PP' H_m H_m' = H_m H_m' \Omega H_m H_m'$.

Thus the eigenvalues ν_i of the expression $P'H_m H_m' P$ in the numerator of (12) are the same as the eigenvalues of $(P'H_m H_m')(H_m H_m' P)$ a product of two $T \times T$ matrices and equal the eigenvalues of $H_m H_m' \Omega H_m H_m'$ (we have used the well-known matrix theory result that: the eigenvalues of AB are the same as the eigenvalues of BA provided A and B are square matrices of the same order).

It is desirable to find bounds on the eigenvalues ν_i from the numerator of (12) in terms of the eigenvalues μ_i of Ω which are more readily known up to a constant of proportionality from low order ARMA specifications. For this, we use the following Lemma.

4.3.1 Durbin-Watson-Anderson (DWA) Lemma

Using algebraic results similar to those in Durbin and Watson [1950, p. 416] and T. W. Anderson [1971, p. 611, Lemma 104.3] the reader can verify the following result.

LEMMA Let $\mu_1 \geq \mu_2 ... \geq \mu_T$ be the eigenvalues of a symmetric matrix Ω. Define an idempotent matrix M^* of rank m, and let the m nonzero eigenvalues of $M^* \Omega M^*$ be denoted by $\nu_1 \geq ... \geq \nu_m$ where the $(T-m)$ of the zero eigenvalues are omitted. Then we have the inequality

$$\mu_i > \nu_i \geq \mu_{i+(T-m)} \; , \quad (i=1,...,m) \; . \tag{13}$$

The proof of this result will be a major digression, and hence omitted. This Lemma is useful for bounding the eigenvalues of $M^* \Omega M^*$ in terms of known eigenvalues of Ω.

4.3.2 Application of the DWA Lemma

First, write the numerator of (12) in terms of an eigenvalue eigenvector decomposition as a weighted sum of squares

$$\mathrm{v}'P'H_m H_m' P \mathrm{v} = \mathrm{v}'G_\nu \Lambda_\nu G_\nu' \mathrm{v} = \sum_{i=1}^{m} \nu_i \eta_i^2 \; , \tag{14}$$

where G_ν is the matrix of eigenvectors of $P'H_m H_m' P$, Λ_ν is the diagonal matrix of its eigenvalues ν_i (in a descending order of magnitude), and $\eta = G_\nu' \mathrm{v}$, is a vector of unit normals, $N(0,I)$.

The inequalities (13) of the Lemma above continue to hold true even after a summation sign and our weights η_i^2 are introduced. Thus the bounds for the numerator are

$$\sum_{i=1}^{m} \mu_i \eta_i^2 \geq \sum_{i=1}^{m} \nu_i \eta_i^2 \geq \sum_{i=1}^{m} \mu_{i+T-m} \eta_i^2 \; . \tag{15}$$

The matrix $H_{T-p} H_{T-p}'$ in the denominator of (12) is also idempotent, and completely analogous argument can be made. Hence we can write

$$F_{UP}^* \geq F^* \geq F_{LO}^* \; , \tag{16}$$

where

$$F_{UP}^{*} = \left[\sum_{i=1}^{m} \mu_i \eta_i^2\right]\left[\sum_{i=p+1}^{T} \mu_i \eta_i^2\right]^{-1} , \tag{17}$$

$$F_{LO}^{*} = \left[\sum_{i=1}^{m} \mu_{i+T-m} \eta_i^2\right]\left[\sum_{i=p+1}^{T} \mu_{i-p} \eta_i^2\right]^{-1} , \tag{18}$$

which do not involve the design matrix X or restriction matrix R, and which are respectively the upper and lower bounds on F^*. We can now explain a Theorem due to Kiviet [1980, p. 354].

Recall that the ratio of the quadratic forms in normal variables involved in a so-called "general linear hypothesis test," from Chapter 3 equation (20) leads to Snedecor's test statistic $F(m,T-p) = \left[\frac{T-p}{m}\right]F^*$. Assume that the eigenvalues μ_i of the matrix Ω of error covariances of the regression model based on a low order autoregressive or moving average (ARMA) structure are known (at least numerically), upto a constant of proportionality. (Analytical expressions for μ_i are available for the asymptotic, $T \to \infty$, case).

Assume also that η_i are independent and identically distributed unit normal variables with zero mean and unit variance, $\eta_i \sim N(0,1)$. We use the definitions of F_{LO}^* and F_{UP}^* from (18) and (17) respectively which do not depend on a specific knowledge of the design matrix X, or of the restrictions $R\beta = r_0$. Then we write the following bounds on Snedecor's test statistic

$$\frac{T-p}{m} F_{LO}^* \le F(m,T-p) \le \frac{T-p}{m} F_{UP}^* , \tag{19}$$

where the bounding statistics depend only on the number of observations T, number of regressors p, the number of linearly independent restrictions m, and the eigenvalues of Ω which is the matrix of error covariances for OLS regression, known upto a constant of proportionality (i.e., σ^2 is not assumed known). The bounds are attained when the restrictions R coincide with appropriate linear combinations of the eigenvectors of Ω.

Kiviet's theorem implies that the "rejection constant" (also known as the "critical value") as in Chapter 3 equation (19) for a 100α percent, $(0<\alpha<1)$, test can be bounded. Denoting the rejection constant by $F^{\alpha}(m,T-p)$ with a superscript α, we can write the following bounds on the rejection constants:

$$F_{LO}^{*\alpha}\left[\frac{T-p}{m}\right] \le F^{\alpha}(m,T-p) \le F_{UP}^{*\alpha}\left[\frac{T-p}{m}\right] , \tag{20}$$

where the bounding values do not depend on X or R, and where we define the bounding rejection constants $F_{UP}^{*\alpha}$ and $F_{LO}^{*\alpha}$ by following probabilities:

$$Pr[F_{UP}^* \ge F_{UP}^{*\alpha}] = \alpha , \tag{21}$$

$$Pr[F_{LO}^* \ge F_{LO}^{*\alpha}] = \alpha . \tag{22}$$

Evaluations of the bounding rejection constants for appropriately specified numbers α, $1 \leq m \leq p \leq T$, and for an arbitrary Ω with known eigenvalues can be found by solving (21) and (22) for $F_{UP}^{*\alpha}$ and $F_{LO}^{*\alpha}$ respectively. This amounts to "inverting" the cumulative distribution functions of the underlying quadratic forms in normal variables from (17) and (18). We wish to specify α, the cumulative probability and find the value of the random variable at which it is attained.

Although analytic expressions for inverted distributions are complicated, numerical inversions can be obtained by following the approximate techniques developed by Imhof [1961]. On a modern computer Imhof's inversion procedure is generally quite accurate. Computer programs are available in Koerts and Abrahams [1969].

For $m = 1$ one usually considers a Students' t test, which involves a square root of the corresponding expressions for Snedecor's F test. Some bounding significance points for a t test were first given by Vinod [1976i].

4.4 EIGENVALUES OF Ω FOR ARMA PROCESSES AND TABULATION OF BOUNDS

Consider a first order autoregressive process, AR(1) defined by

$$u_t = \rho u_{t-1} + \epsilon_t \quad , \quad |\rho| < 1 , \tag{23}$$

where $t = 1,2,...,T$, denotes the tth value such that $\{\epsilon_t : 0, \pm 1, \pm 2, ...,\}$ is a sequence of independent and identically distributed (iid) random variables with zero mean and finite variance, σ^2.

When regression errors follow this simple stochastic process they are seen to be serially correlated. Consider a starting point t^* of the stochastic process (23). Now we have

$$u_{t^*+1} = \rho u_{t^*} + \epsilon_{t^*+1} ,$$

$$u_{t^*+2} = \rho u_{t^*+1} + \epsilon_{t^*+2} .$$

Now substituting the first equation into the second rewrite the second as

$$u_{t^*+2} = \rho u_{t^*} \rho u_{t^*+1} + \epsilon_{t^*+2} .$$

Verify that, in general

$$u_t = \rho^{t-t^*} u_{t^*} + \sum_{i=0}^{t-t^*-1} \rho^i \epsilon_{t-i} . \tag{24}$$

Since the process AR(1) is assumed to have started a long time ago, $t^* \to -\infty$, we have

$$u_t = \sum_{i=0}^{\infty} \rho^i \epsilon_{t-i} \tag{25}$$

because the first term will vanish under the assumption $|\rho| < 1$ from (23).

From $\epsilon_t \sim N(0,\sigma^2)$ and (25) we have $E(u_t) = 0$, and covariance of u_t and u_{t+l} is the expectation of an infinite double sum of products, $E\left[\sum_i \sum_j \rho^i \rho^j \epsilon_{t+l-i}\epsilon_{t-j}\right]$. Whenever the subscripts are unequal the expectation vanishes because ϵ_t are independent and identically distributed random variables.

The expectations of the nonzero terms when $l = 0$ are σ^2, $\sigma^2\rho^2, \sigma^2\rho^4, \ldots$, respectively. The infinite sum of these nonzero terms can be compactly written as $\sigma^2(1-\rho^2)^{-1}$. When $l > 0$

$$cov(u_{t+l}, u_t) = \sigma^2\rho^l(1-\rho^2)^{-1} . \tag{26}$$

Now the covariance matrix for u is

$$cov(u) = \frac{\sigma^2}{1-\rho^2}\begin{vmatrix} 1 & \rho & \rho^2 & .. & \rho^{T-1} \\ \rho & 1 & \rho & .. & \rho^{T-2} \\ . & . & . & & . \\ . & . & . & .. & . \\ \rho^{T-1} & \rho^{T-2} & \rho^{T-3} & .. & 1 \end{vmatrix} \tag{27}$$

$$= \sigma^2\Omega .$$

In the infinite dimensional case ($T \rightarrow \infty$) spectral theory is used to state the eigenvalues μ_t of Ω to be given by

$$\mu_t = 1 + \rho^2 - 2\rho \cos \omega, \tag{28}$$

where $t = 1, \ldots, T$ and where ω is one of the T equidistant values in the range $-\pi$ to $+\pi$ (roots of unity).

The numerical evaluation of the eigenvalues of Ω from a ARMA process may be made as in Vinod [1976i] from an analytically known Ω^{-1} which is a tri-diagonal matrix:

$$\Omega^{-1} = \begin{vmatrix} 1 & -\rho & 0 & 0 & .. & 0 & 0 \\ -\rho & 1+\rho^2 & -\rho & 0 & .. & 0 & 0 \\ 0 & -\rho & 1+\rho^2 & -\rho & .. & 0 & 0 \\ . & . & . & . & .. & . & . \\ . & . & . & . & .. & . & . \\ 0 & 0 & 0 & 0 & .. & 1+\rho^2 & -\rho \\ 0 & 0 & 0 & 0 & .. & -\rho & 1 \end{vmatrix} . \tag{29}$$

It can be verified by direct multiplication that $\Omega^{-1}\Omega = I_T$. For tri-diagonal matrices (nonzero terms only along three diagonals centered at the main diagonal) the so-called "QR algorithms" with desirable numerical stability properties are available for computing the eigenvalues.

From numerically known eigenvalues one can readily compute the eigenvalues $\mu_t\,(t=1,...,T)$, F_{UP}^*, F_{LO}^* from (17) and (18). The bounding rejection constants $F_{UP}^{*\alpha}$ and $F_{LO}^{*\alpha}$ from (19) and (20) can be directly determined on a modern computer for any specified α values.

4.4.1 An Illustrative Example Using Bounds for t Tests

This example is not particularly good for illustrating the autocorrelation problem. It is discussed simply because it is already discussed in Chapter 1. Consider the illustrative example having 9 observations plotted in the three dimensional pillar chart of Figure 1.1. The fitted OLS solution when the variables are measured from their means is:

$$y = \underset{(4.8568)}{12.397} \; \underset{(0.4542)}{-0.1737}x_1 + \underset{(5.6035)}{1.1169}x_2 \,, \tag{30}$$

where the numbers in parentheses represent the absolute values of Students' t statistics. The value of the Durbin-Watson (DW) statistic (1a) d based on the residuals of (30) is 1.3547 and the corresponding first-order serial correlation coefficient ($\hat{\rho}$) is 0.1523. Now the estimated regression coefficient $b_2 = 1.1169$ of (30) is statistically significant at the 95% level, and the null hypothesis, $\beta_2 = 0$, is rejected.

Since there is some evidence of serial correlation among errors, we may wonder whether the conclusion $\beta_2 \neq 0$ above may be reversed by the presence of autocorrelated errors. In the attached Table 4.1 for the true unknown $\rho = 0.5$ and $T = 10$, $p = 3$, the value of the upper bound t^U is 5.30. For our illustrative example, $T = 9$ we extrapolate linearly to obtain $t^U = 5.43$ as follows. From $T = 15$, $t^U = 4.64$ the value of $t^U = 4.64$ is increased by 0.66 when the number of observations decrease by 5 from 15 to 10. From 10 to 9, there is a decrease of 1 implying an increase of $(1/5)0.66 = 0.132$. Hence t^U for $T = 9$ is approximately $5.30 + 0.13 = 5.43$. Since the observed t statistic for β_2 is 5.6035, which is larger than this $t^U = 5.43$, we are assured that the conclusion $\beta_2 \neq 0$ is not reversed, even if the true unknown first order serial correlation coefficient ρ was as high as 0.50. Since the estimated $\hat{\rho}$ is only 0.1523, we may conclude that $\beta_2 \neq 0$ is a fairly safe conclusion.

The test for $\beta_1 = 0$ is accepted for (30) because the $|t| = 0.4542$ is very small. From the lower bounds of Table 4.1 it is clear that this conclusion is not reversed by the presence of $\rho \leq 0.9$.

If we suspected that the errors follow a first order moving average (MA) process with the MA parameter θ it is clear from Table 4.2 that the conclusions $\beta_1 = 0$, $\beta_2 \neq 0$ are not reversed by MA processes with parameters as high as $\theta = 0.9$.

There are many practical problems where the main interest is focussed on the significance of certain regression coefficients. The presence of correlated errors which follow an autoregressive (AR) or moving average (MA) process are simply nuisance parameters. The researcher would much rather forget about these nuisance parameters rather than estimate them, change the whole specification, and re-estimate the regression coefficients by generalized least squares (GLS). Tables 4.1 and 4.2 are helpful in these situations. If the observed $|t|$ values are large (small) enough, the researcher may not have to worry about first order AR or MA errors. If the $|t|$ values lie between the bounds t^L and t^U of Tables 4.1 and 4.2 the results are indeterminate, and the researcher may well have to worry about AR or MA errors. If a mixed ARMA process of first order is suspected, or if two or more regression parameters have to be jointly tested by an F test, Kiviet's [1980] F Tables, which have not been reproduced here, should be used.

4.1 Bounds on t Values at 5% Level for AR(1) Disturbances

			$\rho = .3$		$\rho = .5$		$\rho = .7$		$\rho = .9$	
p	n	$\rho = 0$	t^L	t^U	t^L	t^U	t^L	t^U	t^L	t^U
1	10	2.2622	1.66	3.21	1.34	4.29	1.05	6.40	.73	13.33
	15	2.1448	1.57	3.02	1.27	4.01	.98	5.93	.66	12.76
	25	2.0639	1.51	2.87	1.21	3.78	.92	5.56	.59	12.03
	50	2.0096	1.47	2.77	1.17	3.59	.87	5.16	.53	10.96
2	10	2.3060	1.64	3.42	1.31	4.78	1.02	7.58	.70	17.12
	15	2.1604	1.56	3.13	1.24	4.32	.95	6.82	.64	16.00
	25	2.0687	1.50	2.94	1.19	3.95	.90	6.09	.58	14.64
	50	2.0106	1.47	2.80	1.16	3.67	.87	5.42	.53	12.65
3	10	2.3646	1.63	3.66	1.28	5.30	.98	8.71	.66	20.26
	15	2.1788	1.54	3.26	1.22	4.64	.93	7.62	.62	18.67
	25	2.0739	1.49	3.00	1.18	4.13	.89	6.62	.57	16.91
	50	2.0117	1.46	2.83	1.15	3.76	.86	5.70	.52	14.28
4	10	2.4469	1.63	3.94	1.25	5.85	.94	9.80	.63	23.12
	15	2.2010	1.52	3.40	1.19	4.95	.90	8.36	.60	20.92
	25	2.0796	1.48	3.07	1.16	4.32	.87	7.14	.56	18.87
	50	2.0129	1.46	2.86	1.15	3.84	.85	5.98	.52	15.79
5	10	2.5706	1.65	4.28	1.24	6.47	.91	10.97	.59	26.07
	15	2.2281	1.51	3.54	1.17	5.27	.88	9.04	.58	22.91
	25	2.0860	1.47	3.14	1.15	4.51	.86	7.63	.55	20.58
	50	2.0141	1.45	2.89	1.14	3.93	.84	6.27	.51	17.16

4.2 Bounds on t Values at 5% Level
for MA(1) Disturbances

			$\theta = .2$		$\theta = .3$		$\theta = .5$		$\theta = .9$	
p	n	$\theta = 0$	t^L	t^U	t^L	t^U	t^L	t^U	t^L	t^U
1	10	2.2622	1.77	2.72	1.53	2.92	1.07	3.25	.48	3.55
	15	2.1448	1.68	2.56	1.44	2.74	.99	3.02	.34	3.25
	25	2.0639	1.61	2.45	1.38	2.61	.93	2.85	.23	3.04
	50	2.0096	1.57	2.38	1.35	2.52	.90	2.74	.17	2.90
2	10	2.3060	1.77	2.84	1.51	3.09	1.04	3.51	.46	3.92
	15	2.1604	1.67	2.62	1.42	2.83	.96	3.15	.32	3.44
	25	2.0687	1.61	2.48	1.37	2.65	.92	2.92	.23	3.13
	50	2.0106	1.57	2.39	1.34	2.54	.89	2.76	.17	2.94
3	10	2.3646	1.77	2.99	1.50	3.30	1.02	3.85	.44	4.44
	15	2.1788	1.65	2.69	1.40	2.92	.94	3.31	.31	3.66
	25	2.0739	1.60	2.51	1.36	2.70	.90	2.99	.22	3.23
	50	2.0117	1.56	2.40	1.33	2.56	.89	2.79	.17	2.97
4	10	2.4469	1.79	3.18	1.50	3.57	1.01	4.30	.44	5.18
	15	2.2010	1.65	2.77	1.39	3.03	.92	3.49	.31	3.94
	25	2.0796	1.59	2.54	1.34	2.74	.89	3.06	.22	3.34
	50	2.0129	1.56	2.41	1.32	2.58	.88	2.82	.17	3.01
5	10	2.5706	1.84	3.44	1.53	3.92	1.02	4.91	.44	6.30
	15	2.2281	1.64	2.85	1.37	3.16	.91	3.72	.30	4.30
	25	2.0860	1.58	2.58	1.33	2.80	.88	3.15	.22	3.46
	50	2.0141	1.55	2.42	1.32	2.60	.87	2.85	.17	3.05

4.5 ESTIMATION ALLOWING FOR AR(1) ERRORS

If the parameter ρ of a first order autoregressive AR(1) process (23) is somehow known, we can write

$$\Omega^{-1} = B'B, \tag{31}$$

where

$$B = \begin{vmatrix} (1-\rho^2)^{1/2} & 0 & 0 & \cdots & 0 & 0 \\ -\rho & 1 & 0 & \cdots & 0 & 0 \\ 0 & -\rho & 1 & \cdots & 0 & 0 \\ . & . & . & \cdots & . & . \\ . & . & . & \cdots & . & . \\ . & . & . & \cdots & . & . \\ 0 & 0 & 0 & \cdots & -\rho & 1 \end{vmatrix}. \tag{32}$$

The result in (31) may be verified by direct multiplication. Now the GLS estimator is obtained by applying OLS to the transformed model:

$$By = BX\beta + Bu \tag{33}$$

with the estimator of β given by

$$b^{GLS} = (X'B'BX)^{-1}X'B'By$$

$$= (X'\Omega^{-1}X)^{-1}X'\Omega^{-1}y. \tag{2a}$$

The maximum likelihood (ML) estimator can be obtained as follows. Let $u \sim N(0, \sigma^2\Omega)$. Then, given X, the likelihood function is

$$L = L(\beta, \sigma^2, \Omega \mid y) = (2\pi)^{-T/2} \mid \sigma^2\Omega \mid^{1/2} \exp\left[-\frac{(y-X\beta)'\Omega^{-1}(y-X\beta)}{2\sigma^2} \right] \tag{34}$$

is the likelihood function and therefore the log likelihood, excluding constant, is

$$\log L = -\frac{T}{2}\log \sigma^2 - \frac{1}{2}\log \mid \Omega \mid - \frac{1}{2\sigma^2}(y-X\beta)'\Omega^{-1}(y-X\beta). \tag{34a}$$

Conditional on β and Ω, the ML estimator for σ^2 is

$$\hat{\sigma}^2 = (y-X\beta)'\Omega^{-1}(y-X\beta)/T.$$

Substituting this into $\log L = \log L\ (\beta, \sigma^2, \Omega \mid y)$ we can write the concentrated log likelihood function as

$$\log L_c = constant - \frac{T}{2}\log(y-X\beta)'\Omega^{-1}(y-X\beta) - \frac{1}{2}\log \mid \Omega \mid, \tag{34b}$$

where the subscript c in L_c represents concentrated.

Note that $\mid \Omega \mid = (1-\rho^2)$. Thus the ML estimators for β and ρ are those values which maximize $\log L_c$. Also notice that the maximization of $\log L_c$ is the same as the minimization of

$$\mid \Omega \mid^{1/T}(y-X\beta)'\Omega^{-1}(y-X\beta). \tag{34c}$$

Since this expression contains the determinant of Ω which depends on ρ, minimization of (34c) will yield different estimator than GLS of (2a) based on the minimization of $(y-X\beta)'\Omega^{-1}(y-X\beta)$ when ρ is unknown (non-constant). Of course, if ρ is a known constant the ML and GLS procedures yield the same estimator for β. If ρ needs to be estimated the iterative GLS and ML estimators will not, in general, coincide.

4.5.1 Improvements to Cochrane-Orcutt Estimator

Cochrane and Orcutt (CO) [1949] did not include the first row of matrix of B, and considered the transformed model (33) with $T-1$ observations and the following estimate

$$\hat{\rho}_{CO} = \sum_{t=2}^{T} \hat{u}_t \, \hat{u}_{t-1} \,/ \sum_{t=1}^{T-1} \hat{u}_t^2 \tag{35}$$

to estimate their Ω. The CO estimator of β is based on substituting their Ω in the GLS estimator (2a).

Later, Prais and Winston (PW) [1954] suggested the correct expression for B in (32) and the following estimate

$$\hat{\rho}_{PW} = \sum_{t=2}^{T} \hat{u}_t \, \hat{u}_{t-1} \,/ \sum_{t=2}^{T-1} \hat{u}_t^2, \tag{36}$$

where the denominator has one fewer term than the number of terms in the numerator.

The simple two-stage CO or PW estimator is to pretend that the first-stage estimates of ρ from (35) or (36) are true, insert them in B, pre-multiply by B similar to (33) and use OLS on the transformed model. Intuitively, this seems to be better than OLS, because it does "correct for" serially correlated errors, even if the first stage estimate of ρ may not be perfect.

As a further refinement, we recognize that ρ is an unknown parameter (nuisance), and its estimate from the available T observations in the sample should be treated as a random variable. Therefore we cannot afford to be dogmatic about sample estimates of ρ, and should consider a range of values of ρ, $|\rho| \leq 1$. The basic statistical problem then is one of estimating all the regression coefficients, variance σ^2 and ρ jointly by using the least squares criterion of minimizing residual sum of squares, (SSRes) or maximizing the likelihood (MaxL) if normality is assumed.

Certain iterative versions of *CO* or *PW* estimators have been used in the literature where a range of ρ values is considered, and SSRes is recorded for each choice of ρ. The final choice of ρ is the one that minimizes these recorded values of SSRes.

Beach and McKinnon [1978] propose a full fledged maximum likelihood estimator for joint estimation of all parameters (see (34b) above). Computationally, their estimator is similar to iterative PW estimator except that they use a different estimate of ρ.

Certain simulations in Park and Mitchell [1980] suggest that iterative PW estimator is the best choice. The full fledged maximum likelihood estimator is close to it in performance. The simulation discourages the use of any form of CO estimator where the first observation is omitted (without good reason). They report a somewhat surprising result that OLS without any correction of ρ yields estimates that are closer to true values than any form of the CO estimator based on $T-1$ observations. When iterative PW estimator (using all T observations) is used for a model having large ρ it is considerably more efficient than the OLS.

4.5.2 Analytical Study of Cochrane-Orcutt Type Estimators

Several analytical studies assume that ρ is known, and analyze the behavior of various estimators discussed above. In general it has been noted that, when the sample is large, the effect of dropping the first row of B in the CO estimator is negligible on its efficiency. However, Maeshiro [1976] indicates in his calculations that if the model has an intercept and a trended variable, then for all positive values of ρ, the GLS estimator computed by the CO transformation will be less efficient than the OLS estimator. Chipman [1979] considers a model with an intercept and a time trend and finds analytically that for $0 \leq \rho < 1$, the OLS estimator is always at least 75% as efficient as the GLS estimator. Similar findings have been indicated for a general X matrix by Krämer [1981]. Hoque [1980] has considered the model with a trended variable as in Maeshiro but without an intercept. His results, in contrast to Maeshiro, show CO to be more efficient than the OLS estimator. These analytical results indicate that the effect of dropping the first row of B in CO depends crucially on the nature of the X matrix. We should note however that these analytical results assume that the value of ρ is known. For the more practical case of estimated ρ no analytical study appears to have been done.

4.6 "EXACT" DW TEST WITHOUT TABLES

Recall the DW statistic from (1a) as

$$d = \frac{\hat{u}'A\hat{u}}{\hat{u}\hat{u}} = \frac{u'M'AMu}{u'M'Mu} = \frac{u'M'AMu}{u'Mu}, \tag{37}$$

where A is from (1b) and $M = [I - X(X'X)^{-1}X']$ is an idempotent matrix. The sampling distribution of d is such that its moments are functions of the eigenvalues of (the matrix product) MA denoted by ν_i. Let n^* denote $T-p-1$. Now Durbin-Watson [1950, p. 419] give

$$E(d) = (1/n^*)\sum_{i=1}^{n^*} \nu_i = \bar{\nu} = (1/n^*)Tr\ MA, \tag{38}$$

$$Var(d) = [2/n^*(n^*+1)]\sum_{i=1}^{n^*}(\nu_i - \bar{\nu})^2$$

$$= [2/n^*(n^*+1)][Tr(MA)^2 - (TrMA)^2/n^*]. \tag{39}$$

These are known for any given X. Equating them to the mean and variance of a beta random variable will give us the shape parameters of beta as follows.

The sampling distribution of $(d/4)$ may be assumed to have the beta density

$$\frac{1}{B(p,q)}\left(\frac{d}{4}\right)^{p-1}\left(1-\frac{d}{4}\right)^{q-1}, \tag{40}$$

with shape parameters p and q. The mean of such a beta variable is $4p/(p+q)$ and variance is $16pq(p+q)^{-2}(p+q+1)^{-1}$. Equating these with known $E(d)$ and $var(d)$, we have

$$p+q = \frac{E(d)[4-E(d)]}{var\ (d)} - 1$$

$$p = \frac{1}{4}(p+q)E(d) \tag{41}$$

$$q = (p+q) - p$$

which yields the shape parameters. Since there is some approximation involved in using the beta density we have placed quotes around "exact" in the title above.

Computer programs (e.g., BDTR by IBM [1968]) may be used to compute the area under the beta density. To test the null hypothesis that the errors u follow iid normal distribution, against the alternative hypothesis that they follow an AR(1) process, we require the critical value of $(d/4)$ at the lower tail of the beta distribution. If the critical value is less than the observed $d/4$ we reject the null hypothesis. In fact the "exact" probability that the null hypothesis is accepted for observed $d/4$ may be computed as the area under beta density from $d/4$ to 1. For example, rather than always considering a 5% or 1% level and consulting the tables provided by Durbin and Watson which have "indeterminate regions", a computer program can routinely provide the probability that there is no AR(1) serial correlation among regression errors. One does not need to compute eigenvalues ν_i of MA themselves. All we need are the Traces (sum of all diagonal elements) of two matrices, MA and $MAMA$. If the so-called hat matrix $\hat{H} = X(X'X)^{-1}X'$ is stored in the computer, note that $Tr\ MA = Tr\ A - Tr\ \hat{H}A$, $Tr\ MAMA = Tr[A^2 - 2\hat{H}A^2$

$+ \hat{H}A^2\hat{H}]$, which involve matrix multiplications, which are relatively inexpensive. Durbin and Watson [1971] and Vinod [1973] discuss methods which may be used if higher accuracy than what is implicit in the beta approximation of (40) is desired, despite increased computational costs.

4.7 COMMENTS ON MA ERRORS

In the model $y = X\beta + u$, $u \sim N(0,\sigma^2\Omega)$ let the errors follow a first order moving average process MA(1): $u_1 = \epsilon_t + \theta\epsilon_{t-1}$, $|\theta| < 1$. Consider the expectation of the $T \times T$ matrix based on errors u_t, that is $E(uu')$, to yield the expression for Ω to be

$$\Omega = \begin{vmatrix} 1+\theta^2 & \theta & 0 & .. & 0 & 0 \\ \theta & 1+\theta^2 & \theta & .. & 0 & 0 \\ 0 & \theta & 1+\theta^2 & .. & 0 & 0 \\ . & . & . & .. & . & . \\ . & . & . & .. & . & . \\ 0 & 0 & 0 & .. & 1+\theta^2 & \theta \\ 0 & 0 & 0 & .. & \theta & 1+\theta^2 \end{vmatrix} . \tag{42}$$

If θ is known or estimated one can use the GLS estimator: $b_{GLS} = (X'\Omega^{-1}X)^{-1}X'\Omega^{-1}y$, which would require inversion of this tri-diagonal matrix Ω on a computer. There is some discussion in the literature on approximating the inverse. It is doubtful whether these approximations can be justified, since there are many low cost computational algorithms available for inverting such matrices. Nicholls et al. [1975] have recently argued that MA(1) errors are more plausible alternative to the null hypothesis of iid errors in some applications. A study of the power of the Durbin-Watson (DW) type test in simulation and in theoretical work by Durbin and Watson [1971] suggests that the DW type test has a good power against MA(1) alternatives, even if it is not explicitly designed to test MA errors.

4.8 COMMENTS ON GENERAL AR AND/OR MA PROCESSES

Godfrey [1978a] has applied Silvey's Lagrange multiplier approach to the problem of testing AR(p), MA(q) order tests. These tests are asymptotically equivalent to likelihood ratio tests. There are two advantages of these tests: (i) they are applicable even when X contains lagged values of the dependent variable. (ii) they are computationally simple because they use OLS residuals: $\hat{u}_t$, $t = 1,\ldots,T$. Denote

$$\hat{u} = y - Xb$$

$$(u_i^*)' = (0,\ldots,0,\, \hat{u}_1,\ldots,\hat{u}_{T-i})$$

$$U_q^* = [u_1^* : u_2^* :\ldots: u_q^*]$$

$$V = (X'X)^{-1}$$

$$\hat{\sigma}^2 = T^{-1}\sum_{t=1}^{T}\hat{u}_t^2.$$

Godfrey's likelihood ratio type test statistic is

$$\ell = \hat{u}'U_q^*[U_q^{*\prime}U_q^* - U_q^{*\prime}XVX'U_q^*]U^{*\prime}\hat{u}/\hat{\sigma}^2 \tag{43}$$

to be compared to the critical value of a χ^2 with q degrees of freedom. Large values of ℓ indicate rejection of the null hypothesis of iid errors, in favor of AR(q) or MA(q) errors. One feature of this test which may be puzzling to the practitioner, is that the test statistic is the same for both AR and MA processes as alternatives. This is due to the asymptotic (approximate) nature of the test. The power of the test (43) in small samples is unknown. If we reject the null hypothesis, it is not clear whether we should correct for AR(q) or MA(q) as the alternative. Godfrey's [1978b] statistics are useful for testing the null hypothesis of AR(1) errors against AR(2) errors.

When the model contains lagged values of the dependent variable (y_{t-1}) as one of the regressors (say the first), the DW test is strictly speaking inapplicable. If we denote by d the normal test statistic for the DW test against AR(1) alternative, Durbin's [1970] h statistic is given by

$$h = (1 - \frac{d}{2})(\frac{T}{1-T(SE_1)^2})^{1/2}, \tag{44}$$

where SE_1 is the standard error for the regression coefficient associated with y_{t-1} in the standard OLS computations. Since h is shown to be an asymptotically normal variable with zero mean and unit variance, $N(0,1)$ its tables are widely available. Hence this is a rather simple test to implement. One problem is that the denominator in (44) may become negative. In that case Durbin suggests regressing $\hat{u}_t$ on $\hat{u}_{t-1}$ and regressors including y_{t-1}, and then testing the significance of the regression coefficient of $\hat{u}_{t-1}$ by the usual methods. Some Monte Carlo evidence suggests that the usual DW test may often provide the correct conclusion for these problems, even if it is (strictly speaking) not applicable here.

The tests discussed above for ARMA errors are concerned with the time domain. Certain frequency domain methods are attractive when one's ideas regarding the alternatives to iid errors are less specific. Durbin [1969] constructs cumulative periodogram tests for small samples, where the periodogram statistics are p_i, $i=1,...,m$, and $m=(T-1)/2$, or $T/2$ depending on whether T is odd or even. We define

$$a_i = (2/T)^{½} \sum_{t=1}^{T} \hat{u}_t \cos \frac{2\pi it}{T},$$

$$b_i = (2/T)^{½} \sum_{t=1}^{T} \hat{u}_t \sin \frac{2\pi it}{T},$$

$$p_i = a_i^2 + b_i^2,$$

$$cum_i = \sum_{j=1}^{i} p_j / \sum_{j=1}^{m} p_j, \tag{45}$$

where cum_i is the scaled cumulative sum up to i of periodogram statistics plotted with (i/m) on the horizontal axis, and cum_i on the vertical axis. Durbin [1969] provides tables of c_o values (intercepts) which are used to plot two boundary lines: $c_o + (i/m)$ and $- c_o + (i/m)$. If the observed cum_i points lie outside the band we reject the null hypothesis d serial independence. Some recent computer packages produce such cumulative periodograms.

4.9 THE HETEROSCEDASTICITY PROBLEM

If the regression errors u_t do not have a common variance σ^2, the variance changes with t, i.e., it equals σ_{tt}, we have the so-called heteroscedasticity problem. This problem is often faced by a researcher dealing with cross section data on households, firms, individuals or regions. Therefore, in this section we represent by t = 1, ..., T certain cross sectional data points rather than time units.

When heteroscedasticity is present, the covariance matrix Euu′ = Ω becomes

$$\Omega = \begin{bmatrix} \sigma_{11} & \cdots & 0 \\ \cdot & & \cdot \\ \cdot & & \cdot \\ 0 & \cdots & \sigma_{TT} \end{bmatrix}. \tag{46}$$

Since two cross sectional units can be assumed to be independent, the Ω matrix has zeros for its off diagonal elements. The nature of this Ω is therefore different from the Ω for the AR(1) case in (27). It is of course possible to face a situation where one has both heteroscedasticity and AR(1). This will however involve a large number of unknown parameters which may not be estimable jointly.

Since Ω is a diagonal matrix, the matrix B in (32) becomes diagonal, say

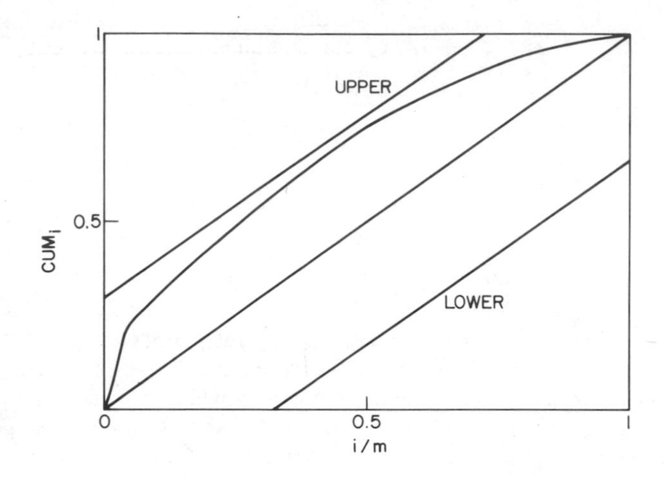

Figure 4.1 Cumulative Periodogram

$B = \Omega^{-1/2}$ is $\sigma_{tt}^{-1/2}$. Thus the GLS estimator of β similar to (2a) is given by

$$\hat{\beta} = (X'\Omega^{-1/2}\Omega^{-1/2}X)^{-1}X'\Omega^{-1/2}\Omega^{-1/2}y$$

$$= (X'\Omega^{-1}X)^{-1}X'\Omega^{-1}y. \tag{2b}$$

This estimator is unbiased. Further we know that

$$MtxMSE(\hat{\beta}) = V(\hat{\beta}) = (X'\Omega^{-1}X)^{-1}. \tag{47}$$

The OLS estimator of β which amounts to ignoring heteroscedasticity, $\Omega = I$, is

$$b = (X'X)^{-1}X'y. \tag{2c}$$

This is also unbiased and its covariance matrix is

$$MtxMSE(b) = V(b) = (X'X)^{-1}X'\Omega X(X'X)^{-1}. \tag{47a}$$

If Ω is known, the GLS estimator of β is the best linear unbiased estimator (BLUE). Recall that the GLS estimator is essentially the OLS estimator on: $\Omega^{-1/2}y = \Omega^{-1/2}X\beta + \Omega^{-1/2}u$. Therefore the proof of BLUE property of the GLS follows directly from Chapter 2.

4.9.1 Sathe-Vinod Bounds

If Ω is unknown, as is generally true in practical situations, $\hat{\beta}$ of (2b) is not operational. Furthermore, the dispersion matrices in (47) and (47a) of β and b are also not operational, because they contain the unknown Ω. Suppose that the model is made "operational" by using some estimate Ω.

Following Sathe and Vinod [1974] consider the diagonal elements of the ratio of dispersion matrices

$$P_{tt} = [(X'X)^{-1}X'\Omega X(X'X)^{-1}]_{tt}/[E\,\hat{\sigma}^2(X'X)^{-1}]_{tt}, \tag{48}$$

and prove that following bounds hold true as an algebraic fact:

$$P_{tt}^{LO} = \frac{(T-p)\min_t(\omega_t)}{\sum_{t=p+1}^{T}\omega_t} \le P_{tt} \le \frac{(T-p)\max_t(\omega_t)}{\sum_{t=1}^{T-p}\omega_t} = P_{tt}^{UP}, \tag{49}$$

where ω_t = Diag $(\Omega\,\hat{\Omega}^{-1})$. In order to use (49) we need to guess a pessimistic Ω associated with $\hat{\Omega}$ to obtain P_{tt}^{UP}. Now observed Student's $|t|$ times $(1/P_{tt}^{UP})^{1/2}$ gives us how large the $|t|$ can become. This is used to check whether the conclusion regarding statistical significance of regression coefficients might be reversed.

Now we discuss estimation of Ω for the purpose of obtaining better estimates of both β and its dispersion matrix.

4.9.2 Alternative Estimates of Diagonals of Ω

Let us write the OLS residual vector as

$$\hat{u} = Mu, \tag{50}$$

where $M = I - X(X'X)^{-1}X' = M^2$ is an idempotent matrix of rank T-p. Then $E\hat{u}=0$ and $E\hat{u}\hat{u}' = M\Omega M$. Therefore, if we write the diagonal terms of the matrix $E\hat{u}\hat{u}'$ and correspondingly of the matrix $M\Omega M$ we shall get (in a column vector)

$$E\hat{\dot{u}} = \dot{M}\sigma, \tag{51}$$

where $\hat{\dot{u}} = \hat{u}*\hat{u}$, $\dot{M} = M*M$, σ is a TX1 vector $\sigma = [\sigma_{11}, \ldots, \sigma_{TT}]'$ which estimates Diag Ω, and where the operator * represents the Hadamard product (A*B implies the multiplication of each element of A with the corresponding element of B). In fact, the TX1 vector $\hat{\dot{u}}$ contains $(\hat{u}_t)^2$, whereas the TXT matrix $\dot{M}$ contains a square of the corresponding element of M.

If we now write $\eta = \hat{\dot{u}} - E\hat{\dot{u}}$, we have

$$\hat{\dot{u}} = \dot{M}\sigma + \eta. \tag{52}$$

This appears as a regression of squared OLS residuals on the matrix $\dot{M}$.

Applying OLS to (52) we then have the estimator of σ as

$$\hat{\sigma} = (\dot{M}'\dot{M})^{-1}\dot{M}'\hat{\dot{u}} = \dot{M}^{-1}\hat{\dot{u}}. \tag{53}$$

It is interesting to note that this estimator $\hat{\sigma}$ is Rao's [1970] minimum norm quadratic unbiased estimator (MINQUE) of σ. For a survey of this topic see P.S.R.S. Rao [1977] and Kleffe [1977]. Note that $\hat{\sigma}$ is defined when $\dot{M}$ is nonsingular. A necessary and sufficient condition for its nonsingularity is given by Mallela [1972]. If $\dot{M}$ is singular one could write the estimator $\hat{\sigma}$ as $\hat{\sigma} = \dot{M}^{+}\hat{\dot{u}}$, where $\dot{M}^{+}$ is any generalized inverse of $\dot{M}$. Alternatively one can use the following generalized ridge estimator

$$\hat{\sigma}_K = (\dot{M}'\dot{M}+K)^{-1}\dot{M}'\hat{\dot{u}}, \tag{54}$$

where K is the matrix of ridge biasing parameters. When K = kI; k > 0, the estimator in (54) is known as the "ordinary" ridge estimator. The family of ridge estimators will be discussed in detail in Chapters 7 and 8.

Notice that $M = I - \hat{H}$ where $\hat{H} = X(X'X)^{-1}X'$ is the so called "hat matrix", with elements of order T^{-1} because $X'X$ is of order T (assuming that $\lim \frac{X'X}{T}$ is a positive definite matrix as $T \to \infty$). Therefore we can use the expansion $\dot{M} = I + \hat{H}_1 + \hat{H}_2$ where $\hat{H}_1$ is the matrix of order T^{-1}, and $\hat{H}_2$ is the matrix of order T^{-2}. Thus, for sufficiently large T we can approximate $\dot{M}$ by I, i.e., $\dot{M} \simeq I$. Using this in (52) or (53) we can get an approximate estimator of σ as

$$\hat{\sigma}_a = \hat{\dot{u}}. \tag{55}$$

This estimator simply means that σ_{tt}; t = 1, ..., T, is estimated by the squared OLS residuals $\hat{u}_t^2$. Intuitively, this estimator should be less efficient than $\hat{\sigma}$ for small samples.

To obtain an estimator which is more efficient than $\hat{\sigma}$ and $\hat{\sigma}_a$ let us consider $\eta = \hat{\dot{u}} - \dot{M}\sigma$ in (52) and let us assume the normality of the disturbance vector u. It can then be verified that E $\eta\eta' = 2\dot{\psi}$, where $\psi = (M\,\Omega\,M)*(M\,\Omega\,M)$ is the matrix of the squared elements of $M\,\Omega\,M$. Thus the GLS estimator of σ form (52) is

$$\hat{\sigma}_{GLS} = (\dot{M}'\dot{\psi}^{-1}\dot{M})^{-1}\dot{M}'\dot{\psi}^{-1}\hat{\dot{u}}, \tag{56}$$

provided $\dot{\psi}$ and $\dot{M}\dot{\psi}^{-1}\dot{M}$ are nonsingular. Since GLS is BLUE, it is more efficient than both $\hat{\sigma}$ and $\hat{\sigma}_a$. Notice that $\dot{\psi}$ depends on Ω. Therefore in practice, one could first estimate Ω by using $\hat{\sigma}$ or $\hat{\sigma}_a$, then substitute this estimate of Ω in $\dot{\psi}$, and then obtain an "operational version" of $\hat{\sigma}_{GLS}$. The efficiency of this operational estimator compared to $\hat{\sigma}$ and $\hat{\sigma}_a$ is not clear.

A problem which one may face while using the above mentioned estimators of σ is that some of the elements of the estimated σ vector may turn out to be negative. Certainly this is not interesting, since the true variances are always nonnegative. There are two solutions to this problem. The first solution is the replacement of negative estimates with zeros. This will however inject certain bias in the estimate of σ. Alternatively one can use a quadratic programming estimator, obtained by minimizing $(\hat{u}-\dot{M}\sigma)'(\hat{u}-\dot{M}\sigma)$, subject to $\sigma \geq 0$ (see, e. g., Liew [1976]).

4.9.3 Estimation of β and V(b)

Recall that our problem was to make $\hat{\beta}$ of (2b) and V(b) of (47a), both of which depend on Ω, operational. We can do so by substituting its diagonal by the T elements of the estimators $\hat{\sigma}, \hat{\sigma}_a, \hat{\sigma}_k, \hat{\sigma}_{GLS}$ in Ω. Thus

$$\tilde{\beta} = (X'\hat{\Omega}^{-1}X)^{-1}X'\hat{\Omega}^{-1}y, \tag{57}$$

and

$$V(\hat{b}) = (X'X)^{-1}X'\hat{\Omega}X(X'X)^{-1}, \tag{58}$$

are the operational estimators of β and V(b), respectively.

4.9.4 Properties of the Estimators

If the disturbances in the model $y = X\beta + u$ are heteroscedastic then the OLS estimator b is unbiased but inefficient. The correct V(b) is given in (47a) and its estimate is in (58). Using b and its standard errors given form (58), one can form Students' t-ratios for testing the null hypothesis about the individual regression coefficients. But for small samples it is not clear whether these t ratios will actually follow the t-distribution. For large samples, $\hat{\Omega}$ based on any estimator of σ will not be consistent. However, the estimator $X'\hat{\Omega}X/T$ of $X'\Omega X/T$ will be consistent. Intuitively, this is so because there is more information introduced through the matrix X. Analytically, White [1980] has proved the consistency of $X'\hat{\Omega}X/_T$ where $\hat{\Omega}$ is obtained by $\hat{\sigma}_a$ in (55). The implication of all this is that even though the OLS is inefficient one can use it for estimation and testing of regression coefficients. This is especially useful, because for a large sample the GLS estimator $\tilde{\beta}$ may not be consistent, and for a small sample variance and the sampling distribution of $\tilde{\beta}$ are not known.

4.9.5 Some Further Results, Extensions and Applications

J. N. K. Rao [1971] and Ullah and Singh [1973] have developed MINQUE type estimators $\hat{\sigma}, \hat{\sigma}_a, \hat{\sigma}_{GLS}$ in multiple sets of regression (Chapter 10) models with heteroscedasticity. Hartley and Jayatillake [1973], Horn, Horn and Duncan [1975] and Horn and Horn [1975] have given alternative estimators of σ that overcome the problem of negative estimates, but again the properties of the resulting $\hat{\beta}$'s have not been analysed. Kmenta [1971], Fuller and Rao [1978],

Drèze [1979] and Taylor [1977, 1978] have considered the estimation problems when the sample is divided into say n groups, and heteroscedasticity is present between the groups but not within the groups.

There is a considerable amount of literature dealing with the problem of heteroscedasticity by imposing various restrictions on the behavior of σ_{tt} for t=1, .., T. Essentially the aim of these restrictions is to reduce the number of T unknown variances. A popular heteroscedastic error model in this context assumes that σ_{tt} is a linear function of a function of a set of L < T known exogenous variables, $z_t' = [z_{1t},...,z_{Lt}]$, and α is an LX1 vector of unknowns. For example, if population or income is one of the regressors our z_t may be simply that regressor, or some known function of one or more such regressors. We specify

$$\sigma_{tt} = f(z_t'\alpha) + \text{error} . \tag{59}$$

Now the number of unknown parameters T is reduced to L parameters α. If Z is a T × L matrix the estimators of α corresponding to $\hat{\sigma}, \hat{\sigma}_K$ and $\hat{\sigma}_{GLS}$ discussed above can be written by substituting $\dot{M}Z$ for M in (53) (54) and (56). Singh and Ullah [1974], Raj and Ullah [1981], Harvey [1974], Amemiya [1977] and others have considered these types of models.

In Hildreth and Houck's random coefficients model, one uses $Z = \dot{X}$, where X is simply the matrix of squared elements of X. This model has been used extensively in applied work, see Raj and Ullah [1981]. For models with the information $\sigma_{tt} = z_t'\alpha$, the estimators of α and hence of σ_{tt} would be consistent. Vinod and Raj [1978] used Hildreth and Houck's model with ridge type estimators for σ in (54), along with ridge type modification of β, and analyzed a production function by using certain Bell system data for 1947-76. They report that ridge-type estimators provide more sensible estimates of β, and also reduce the number of negative estimates in σ.

4.9.6 Tests For Heteroscedasticity

It is useful to consider the problem of detecting heteroscedasticity before estimating the model. Apart from some *ad hoc* methods like plotting the absolute values of the OLS residuals against an exogenous variable, there are several statistical tests which are used for detecting heteroscedasticity. These are briefly mentioned below.

Let us assume that the T observations can be grouped into n subgroups. If heteroscedasticity is present, "within group" variances will be different from each other. Bartlett's [1937] likelihood ratio test is designed for this. If the observations can be rank ordered according to increasing variance we can use Goldfeld and Quandt's [1965] test. Harrison and McCabe [1979] and Szroeter [1978] suggest a test based on Durbin-Watson type bounds. It should be possible to develop more exact tests similar to Section 4.6 above, which avoids using tabulated bounds. It is conjectured that such tests would be more powerful than the ones based on BLUS residuals suggested by Theil [1971], Harvey and Phillips [1974] and others.

In Glejser's test for heteroscedasticity we start with some specification of Z and $f(z_t'\alpha)$ in (59). We obtain α's by the linear regression

$$|\hat{u}_t| = \alpha_1 + \sum_{i=2}^{L} \alpha_i z_{it}. \tag{59a}$$

The null hypothesis $H_0{:}\alpha_2 = \alpha_3 = \ldots = \alpha_L = 0$, is tested by the normal methods. If the test rejects this null hypothesis, heteroscedasticity is not a serious problem.

Breusch and Pagan [1979] regress squared scaled residuals $\hat{u}_t^2/\hat{\sigma}^2$ on the specified function of Z_i and jointly test the null hypothesis that all α coefficients are zero. The one-half "regression sum of squares" in (59a) is shown to be distributed as a Chi-square variable with L degrees of freedom.

White [1980] develops a large sample test which does not require a specification of the correct form of heteroscedasticity as in (59). Specifically, he forms a regression of squared residuals on the sum of squares and products of the p regressors as

$$\hat{u}_t^2 = \alpha_0 + \sum_{i=1}^{p} \sum_{j=1}^{p} \alpha_{ij} z_{it} z_{jt} + \epsilon_t, \tag{59b}$$

where $E\epsilon_t{=}0$. He then defines the standard R^2 statistic for testing the joint hypothesis $H_0{:}\alpha_{ij} = 0$ as the statistic for testing heteroscedasticity. This is

$$TR^2 \sim \chi^2_{p(p+1)/2} \tag{60}$$

for large T, where R^2 is the (constant-adjusted) multiple correlation coefficient from the regression (59b). White has shown that this is identical with testing the insignificant difference $\Delta = (X'X)^{-1}X'\hat{\Omega}X(X'X)^{-1} - \hat{\sigma}^2(X'X)^{-1}$, where the first term is in (58) and $\hat{\sigma}^2(X'X)^{-1}$ is the usual estimator of the OLS variance covariance matrix $\sigma^2(X'X)^{-1}$ under the homoscedasticity of variances; $\hat{\sigma}^2 = \frac{1}{T}(y-Xb)'(y-Xb)$. Thus the acceptance of H_0 above implies the acceptance of the insignificant difference Δ. Now noting that $\hat{\sigma}^2 (X'X)^{-1}$ becomes inconsistent under heteroscedasticity (Goldfeld and Quandt, [1972, Chapter 3]) but $(X'X)^{-1}X'\hat{\Omega}X\ (X'X)^{-1}$ is consistent, the insignificant Δ implies that, in fact $\hat{\sigma}^2(X'X)^{-1}$ is consistent for the data set under consideration, and that the disturbances are homoscedastic. The small sample power property of White's test has not been investigated.

In the first eight sections of this chapter, we have indicated the Durbin-Watson type tests and their generalizations, and emphasized the effect of ARMA errors on significance testing for regression coefficients. For parameter estimation, "iterative Prais-Winston" estimator may be recommended following the simulation by Park and Mitchell [1980]. If the problem of autocorrelated errors occurs in conjunction with multicollinearity, a ridge-type correction can be applied to each of the Prais-Winston iterations. This is quite straight forward because once the transformation matrix B of (32) is specified, we end up with OLS estimation of (33). Incorporating ridge parameters k at this point presents no difficulties. One example in conjunction with ridge regression is given in Vinod [1974], where a trans-log production function is fitted to 41 observations.

In the ninth section of this chapter we have outlined some of the recent developments in regard to the heteroscedasticity problem, including some new Hildreth and Houck type or MINQUE-type estimators of variances, large sample tests by White [1980] and some ridge-type modifications.

EXERCISES

4.1 Consider the quarterly time series data from Vinod [1973, p. 125] given in Table 4.3, and regress Investment on GNP and Time.

(1) Compute the residuals $\hat{u}_t$. Hence compute the DW type statistics d_1 to d_4.

(2) Compute the exact confidence levels for d_1 to d_4 along the lines suggested in Section 4.6 (Hint: for d_2 the A of (1b) becomes the Kronecker product of A and the 2×2 identity matrix).

(3) What is the difference between testing directly (and only) for the 4th order as by Wallis versus testing sequentially for d_1 to d_4?

(4) Since the test for d_2 is conditionally applied, how will you modify the significance levels in the sequence of tests for d_1 to d_4?

4.2 Compute Durbin's cumulative periodogram for the above data.

4.3 What are some of the hazards in using estimated GLS estimator? If the main interest is in testing the significance of regression coefficients, how can we avoid GLS estimation and be certain that our conclusions will not be reversed?

4.4 Consider an arbitrary set of true u and $M = I - X(X'X)^{-1}X'$ for the first exercise above. Show that serially correlated $\hat{u} = Mu$ do not necessarily imply serially correlated u and vice versa. What does this say about residual plots in general?

Table 4.3

Quartery Data On US Manufacturing Investment

Time	Investment	GNP	Time	Investment	GNP
01	2.96	410.6	23	3.34	524.2
02	3.73	416.2	24	3.88	537.7
03	3.83	420.6	25	3.14	547.8
04	4.43	429.5	26	3.69	557.2
05	3.50	436.9	27	3.72	564.4
06	4.18	439.9	28	4.13	572.0
07	4.01	446.3	29	3.27	577.4
08	4.26	441.5	30	3.92	584.2
09	2.90	434.7	31	3.95	594.7
10	2.94	438.3	32	4.56	605.8
11	2.66	451.4	33	3.79	617.7
12	2.93	464.4	34	4.53	628.0
13	2.46	474.0	35	4.67	638.9
14	3.02	486.9	36	5.59	645.1
15	3.02	484.0	37	4.54	662.7
16	3.57	490.5	38	5.47	675.4
17	3.09	503.0	39	5.73	690.0
18	3.76	504.7	40	6.72	708.4
19	3.62	504.2	41	5.61	725.9
20	4.01	503.3	42	6.78	736.7
21	3.00	503.6	43	6.84	748.8
22	3.46	514.9	44	7.75	762.1

5

Muticollinearity and Stability of Regression Coefficients

The purpose of this chapter is to discuss multicollinearity related concepts such as ill-conditioning, their causes, and consequences, including effects on the stability of estimated regression coefficients.

5.1 WHAT IS MULTICOLLINEARITY?

Consider the usual regression model of Chapter 1

$$y = X\beta + u = H\Lambda^{1/2}G'\beta + u = X^{*}\gamma + u \ , \tag{1}$$

where X is a matrix of regressors which is replaced by its "singular value decomposition" of equation (14) of Chapter 1, and where $X^{*} = H\Lambda^{1/2}$. We make the following basic assumptions: (i) the matrix X of regressors is fixed (ii) the parameters β are of primary interest and (iii) the specification (1) is the true model.

5.1.1 Nonorthogonality

The matrix X is said to have orthogonal regressors when it is such that $X'X = I$. For example, when $\Lambda = I$ all eigenvalues of $X'X$ are equal to unity, and X consists of orthogonal regressors. Hence, nonorthogonality of regressors implies that $\Lambda \neq I$.

Multicollinearity (or collinearity) mean linear dependence among the columns of X. A distinction is often made between exact multicollinearity and near multicollinearity defined as follows.

5.1.2 Exact Multicollinearity

Exact multicollinearity exists when the rank of X is less than p, i.e., the columns of X denoted by $x_1,x_2,...,x_p$ are linearly dependent. In other words, there exist nonzero constants a_i, $(i=1,...,p)$ such that

$$\sum_{i=1}^{p} a_i x_i = 0 . \tag{2}$$

5.1.3 Near Multicollinearity

Near multicollinearity or multicollinearity exists when there are nonzero constants a_i such that

$$\sum_{i=1}^{p} a_i x_i \approx 0 ,$$

where $\approx$ denotes approximate equality.

When the matrix product $X'X = G\Lambda G'$ is singular, its inverse does not exist. Exact multicollinearity implies that $X'X$ is singular, i.e. the smallest eigenvalue $\lambda_p = 0$. Near multicollinearity means that $X'X$ is nearly singular and the smallest eigenvalue is close to zero. When eigenvalues are different from unity, we have nonorthogonality. In particular, the smallest eigenvalue $\lambda_p \neq 1$ does not imply that $\lambda_p = 0$ or $\lambda_p \approx 0$. Thus nonorthogonality does not imply singularity or multicollinearity. However, multicollinearity $(\lambda_p=0)$ does imply nonorthogonality $(\lambda_p \neq 1)$.

The first Nobel memorial prize in Economics was awarded to Professor Ragnar Frisch from Oslo, who was the first researcher to seriously study the multicollinearity problem in 1934. He recognized the relevance of the pattern of eigenvalues λ_i. Since Frisch worked on the problem before the modern era of high speed computers, it was not practical to advocate a numerical study of the pattern of eigenvalues of the $X'X$ matrix, as well as its submatrices. Frisch had suggested certain cumbersome graphical methods which have not been popular among economists or others. Since it is a simple matter to compute the eigenvalues on computers, we recommend that all serious practitioners of regression methods should compute such eigenvalues.

5.1.4 Ill-Conditioned Matrix

In addition to statisticians and economists, numerical analysts have been concerned with near singularity of a matrix and have derived the so-called "condition number," $K^{\#}$, to index the extent of ill-conditioning of a matrix. They define

$$K^{\#} = \lambda_1^{1/2}\lambda_p^{-1/2} \,, \tag{4}$$

which is a ratio of the "singular" values of X (i.e. square roots of the eigenvalues of $X'X$). The phrase "singular" in singular values is not to be confused with the singularity of the matrix itself. The presence of multicollinearity can be checked by merely looking at λ_p without worrying about λ_1. For a singular $X'X$ matrix $\lambda_p = 0$, there is exact multicollinearity, and $K^{\#} \to \infty$. For orthogonal data having $X'X = I$, we have $K^{\#} = 1$. Thus the condition number lies in the half-open interval $[1, \infty)$. For any two matrices, the larger the $K^{\#}$ the worse the "conditioning." An ill-conditioned matrix of available data is defined by a large condition number. There is some ambiguity in defining "large" similar to an ambiguity in defining "closeness" to zero. It is not advisable to resolve the ambiguity by choosing fixed benchmarks for all problems. However, benchmarks such as 0.01 is small and 100 is large may be assigned for classes of problems in particular areas of scientific investigation.

Remark 1 In terms of the above definition, we note that nonorthogonal does not imply multicollinear, which in turn does not imply ill-conditioned.

For convenience, ridge regression literature often ignores the distinction between multicollinearity, non-orthogonality and ill-conditioning. We shall follow this tradition unless there is a serious possibility of a misunderstanding.

5.1.5 Predictive vs. Nonpredictive Multicollinearity

The above discussion of multicollinearity considers only the matrix of regressors X. The entire data consist of vector y and the matrix X. Certain information about multicollinearity is contained in an augmented correlation matrix.

Webster, Gunst and Mason [1974] suggest that we should ignore the theoretical distinction between the stochastic y variables and nonstochastic X variables, augment the correlation matrix $X'X$ by incorporating one additional column and one row having correlation r_{yi}, and then consider the "eigenvalue eigenvector decomposition" of the augmented matrix. Denote the $(p+1)$ augmented eigenvalues by λ_i^A and the corresponding $(p+1)$ augmented eigenvectors by g_i^A with elements g_{ij}^A $(i,j=0,1,\ldots,p)$. From the geometric interpretation of the eigenvectors Webster, Gunst and Mason (*inter alia*) show that g_{0j}^A (the first element of each eigenvector) measures the "predictive value" of that eigenvector. If λ_j^A and g_{0j}^A are small, the jth eigenvector reveals a "nonpredictive multicollinearity." The removal of these eigenvectors is then suggested. Hawkins [1975] suggests the largeness of the following ratio as a good diagnostic indicator.

$$K_{HAW}^{\#} = \max_j (g_{0j}^A/\lambda_j^A) \,, \tag{5}$$

where the superscript refers to augmentation by including y discussed above.

5.2 MULTICOLLINEARITY AND INESTIMABILITY

Professor R. C. Bose defined the concept of estimability in 1944 in his paper entitled "The Fundamental Theorem of Linear Estimation." More accessible references are Morrison [1967], and Rao [1973]. It is convenient to discuss the concept in the context of a simple case where $p = 2$. Denote by r_{12} the correlation coefficient between two regressors and r_{yi} the correlation coefficient between y and x_i. The ordinary least squares estimator $b = (X'X)^{-1}X'y$ requires the computation of the inverse

$$(X'X)^{-1} = \begin{bmatrix} 1 & -r_{12} \\ -r_{12} & 1 \end{bmatrix} [\det(X'X)]^{-1} ,$$

where $\det(X'X) = 1 - r_{12}^2$, is the determinant. Note that $X'X$ has been standardized as discussed in Chapter 1. Also, we have

$$b = \begin{bmatrix} r_{y1} - r_{12}r_{y2} \\ r_{y2} - r_{12}y_{y1} \end{bmatrix} [\det(X'X)]^{-1} . \tag{6}$$

Note that if $r_{12} = 1$, $\det(X'X) = 0$; there is exact multicollinearity, and b is not defined. However, note that

$$\begin{aligned} b_1 + b_2 &= [r_{y1}(1-r_{12}) + r_{y2}(1-r_{12})] \cdot [\det(X'X)]^{-1} \\ &= (r_{y1}+r_{y2})/(1+r_{12}) , \end{aligned} \tag{7}$$

which remains well defined even with $r_{12} = 1$. By contrast,

$$b_2 - b_1 = (r_{y2}-r_{y1})/(1-r_{12}) \tag{8}$$

is not defined (e.g., tends to ∞). It is said that $\pm(\beta_1+\beta_2)$ is estimable, whereas $\pm(\beta_2-\beta_1)$ is inestimable. Denote by $L'\beta$ a linear combination of the parameter vector β. When $p = 2$ we have noted that $L'\beta$ with $L' = (1,1)$ is estimable but $L' = (-1,1)$ is inestimable.

In the presence of exact multicollinearity, there exist at least some linear combinations of β vector which are inestimable. In particular, it can be shown that all individual regression coefficients β_i cannot be estimated for singular $X'X$ when $\lambda_p = 1 - r_{12} = 0$. Exact multicollinearity leads to many "observationally equivalent" estimates of the expectations of y as explained by Theil [1971, p 148]. Econometricians define a concept of "identifiability" for simultaneous equation systems (see Section 11.3 of Chapter 11) where misleading "observationally equivalent" systems may be observed if the system is not properly "identified." B. M. S. Lee's [1979] doctoral dissertation related "identifiability" to multicollinearity by showing that both can be thought in terms of estimability. Electrical control engineers define a concept of "observability" which is also related to identifiability and estimability (see Graupe, [1972]). The main point of these concepts is to warn the user about the danger of being misled by nonsensical estimates. The danger is particularly present when we have near multicollinearity, which may not be immediately detected.

5.3 EFFECTS OF MULTICOLLINEARITY

It is well-known that the variance covariance matrix of the OLS estimator is given by $\sigma^2(X'X)^{-1}$, and its trace (sum of diagonals) represents its unweighted mean squared error

$$\mathrm{MSE}(b) = \sigma^2 Tr\,(X'X)^{-1} \,. \tag{9}$$

In terms of the eigenvalues of $X'X$ we have

$$\mathrm{MSE}(b) = \sigma^2 \sum_{i=1}^{p} \lambda_i^{-1} > \sigma^2 \lambda_p^{-1} \,. \tag{10}$$

5.3.1 High Variance of b_i

For near multicollinearity, $\lambda_p \rightarrow 0$ and (10) implies that $\mathrm{MSE}(b) \rightarrow \infty$, that is, b is subject to very large variance. Often, this is revealed by the low values of the usual t ratios whose denominator has the square root of the diagonals of $(X'X)^{-1}$, which are called variance inflation factors (VIF) by Marquardt [1970].

For the standardized data $X'X$ is the correlation matrix and the $(i,j)^{th}$ element of the inverse $(X'X)^{-1}$ may be denoted by a superscript notation as r^{ij}. Farrar and Glauber [1967] were the first to suggest looking at the values of r^{ii} to diagnose multicollinearity. Marquardt [1970] suggests a rule of thumb that

$$VIF(i) = r^{ii} > 5 \tag{11}$$

indicates harmful multicollinearity.

Theil [1971, p. 166] shows that

$$r^{ii} = \frac{1}{(1-R_i^2)\|x_i\|^2} \,, \tag{12}$$

where $\|x_i\|^2 = x_i' x_i$, and where R_i^2 represents the squared multiple correlation coefficient when x_i (ith column of X) is regressed on the remaining $(p-1)$ regressors. When multicollinearity is present, there is a linear relation among the regressors as in (2). Since multiple correlation measures the explanatory strength of such a linear relationship, R_i^2 will be large (close to 1), the denominator of r^{ii} will be close to zero, and hence r^{ii} will be large. In other words, multicollinearity leads to high variance of regression coefficients, b_i.

High variance of regression coefficients would generally mean that the null hypothesis $H_0{:}\beta_i = 0$ will be more likely to be accepted. This is a healthy feature of the traditional significance tests.

A practical difficulty arises when the conclusions from significance tests for essentially similar models are in a hopeless conflict with each other. For example, one can observe situations similar to the following hypothetical example. Let $p = 5$, and superscript * denote a significant regression coefficient. We can observe the following kinds of sets from multicollinear data:

(b_1^*, b_2^*), (b_1^*, b_3^*), (b_1^*, b_4^*), (b_1^*, b_5^*), (b_2^*, b_3^*), (b_2^*, b_4^*),
(b_2^*, b_5^*), (b_3^*, b_4), (b_3, b_5^*), (b_4^*, b_5), (b_1, b_2, b_5^*), (b_1, b_2^*, b_3),
(b_1, b_2^*, b_3, b_4^*), $(b_1^*, b_2, b_3^*, b_4, b_5^*)$ etc.

It is not clear which variables should be retained in the final specification.

Unfortunately, this very real and commonplace difficulty cannot even be expressed in a theoretically rigorous manner, without first specifying the true model for comparison. If we knew the true specification of the model, there is no reason to try any other model, and there is no conflict. Therefore, our hypothetical practitioner gets scorn from the purist statistician for working with bad data. A part of the blame is attributable to high variance emanating from multicollinearity. For a geometrically appealing discussion of the fact that nonorthogonality leads to wide confidence intervals for regression coefficients the reader is referred to an Econometrics textbook by Wonnacott & Wonnacott [1970, p. 302].

Another practical difficulty with the estimated students' t values based on multicollinear data is that these values are highly unstable, and often change their signs and relative magnitudes with minor perturbations in data.

5.3.1.1 *Analysis of Specific High Variances*

The eigenvalue-eigenvector decomposition of $X'X = G\Lambda G'$ (see Chapter 1) allows us to study in the further detail the specific culprit terms which are responsible for inflating a specific $r^{ii} = (G\Lambda^{-1}G')_{ii}$. For standardized data, we know that

$$VIF(i) = r^{ii} = \sum_{j=1}^{p} g_{ij}^2 \lambda_j^{-1}, \tag{13}$$

where g_{ij} denotes the $(i,j)^{th}$ term of G matrix of eigenvectors. Since λ_p is the smallest, it is usually the case that $j = p$ is the culprit. However, sometimes g_{ip} is also small and the pth term in the summation on the right-hand side of (13) is small compared to $(p-1)^{th}$ or $(p-2)^{th}$ term. Then λ_{p-1} or λ_{p-2} can be the culprits. Ragnar Frisch [1934] has hinted a study along these lines. Unfortunately, he was ahead of his time in suggesting this before computers were available.

5.3.2 Wrong Sign for Regression Coefficients

Multicollinearity can result in b_i *appearing* to "have the wrong sign" (Farrar and Glauber [1967], Mullet [1976]), i.e. opposite to the *a priori* expectations of the researcher. Of course, the "wrong" sign may be due to incorrect specification, outliers, serial correlation, etc.

The exact causes of wrong signs may be many, and what appears to be a wrong sign may not even be wrong. Therefore extreme caution is needed in relying heavily on the appearance of wrong signs. Most regression practitioners know about this problem, even though it is not a well-defined problem, in a puristic sense. Of course, in a Bayesian framework where the "correct" sign can be assumed to be known from a prior distribution, the "wrong" sign is well-defined.

Obenchain and Vinod [1974] point out that when the estimated regression coefficients b_i are interpreted as partial derivatives $\partial y/\partial x_i$ the wrong sign problem is particularly serious. From a puristic viewpoint one can argue that such an interpretation of regression coefficients amounts to an "abuse" of regression analysis (Box [1966]). However, if causality is assumed and not inferred from the nonexperimental data the partial derivative interpretation is appropriate for "price elasticities" in Houthakker and Taylor [1966], "marginal elasticities" in Vinod [1972] or "multipliers" in Goldberger [1959]. For such applications the wrong signs problem cannot be ignored.

Returning to the $p = 2$ case, we shall now illustrate how "near inestimability" arising from near multicollinearity may lead to "wrong" signs. Consider an example where $r_{12} = 0.99$, $r_{y1} = 0.4$, $r_{y2} = 0.5$, $\det(X'X) = 0.0199$, $b_1 = -4.77$, $b_2 = 5.22$, $b_1 + b_2 = 0.45$, and $b_2 - b_1 = 9.99$. The correlation coefficient between b_1 and b_2 is -0.99 which may be expressed by an elongated confidence ellipse similar to the one in Malinvaud [1966], p. 190. Hence there is a trade-off between magnitudes of b_1 and b_2. The "near inestimability" of $\beta_2 - \beta_1$ seems to have made $b_2 - b_1$ large, whereas the trade-off has made the signs of b_1 and b_2 opposites. We shall see later that a RIDGE TRACE is designed to reveal such trade-offs graphically, especially when p is large (See Figure 7.1 of Chapter 7).

Sastry [1970] argued that when x_1 is proportional to x_2 for the multicollinear model, it is reasonable to assume that $r_{y1} = r_{y2}$. He derives the following limits for $\tilde{b}_1$ and $\tilde{b}_2$ the regression coefficients in original units (before standardization) by using L'Hopital's rule for the case when (8) is a ratio of two zeroes

$$\lim_{r_{12}\to 1} \tilde{b}_1 = \frac{r_{y1}}{2} \frac{\sigma_y}{\sigma_1} \tag{14}$$

and

$$\lim_{r_{12}\to 1} \tilde{b}_2 = \frac{r_{y1}}{2} \frac{\sigma_y}{\sigma_2}, \tag{15}$$

where $\sigma_y^2 = \sum_{t=1}^{T} (y_t - \bar{y})^2/(T-1)$, and $\sigma_j^2 = \sum_{t=1}^{T} (x_{jt} - \bar{x}_j)^2/(T-1)$. Note that equations (14) and (15) imply that standardized $|b_1| \to |b_2|$ which is what should be expected since $r_{y1} \to r_{y2}$. On the other hand, the two OLS coefficients can be quite different from each other and have opposite signs, suggesting that there is something wrong with OLS.

Visco [1978] and Leamer [1975] have shown by algebraic methods that omitting a variable with relatively low t value cannot correct the "wrong sign" of a regressor having a higher t value. However, replacing a less significant regressor by another can indeed remove the "wrong sign" problem in certain cases. Also, omission of two collinear regressors at the same time can change the signs of significant regressors, and even make them insignificant.

5.3.3 Unstable Regression Coefficients

When important practical decisions are based on the results of a regression analysis, the researcher needs a stable or robust result. The concept of stability of b_1 values can be rigorously defined (without having to resort to Bayesian methods) by using some classical concepts in perturbation theory developed by Householder, Truing, Varga, Von Neumann, Wilkinson and others. Wilkinson [1965] is regarded by numerical analysts as a modern-day classic reference for these developments.

We consider perturbations of the X matrix and y vectors which arise due to "measurement errors," data revisions, and modifications made from time to time by the data producing agencies (governmental or private) in addition to the rounding errors. In October 1979, the Federal Reserve Board admitted a staggering error of over 3 billion dollars in the (M_1) money supply figures, which made newspaper headlines.

Beaton, Rubin and Barone [1976] report an interesting experiment with Longley's [1967] multicollinear data on GNP, unemployment, etc. They perturb the available data *beyond the last published digit* by adding uniform random numbers from -0.5 to $+0.499$. These minuscule variations drastically change most b_i (e.g., from -232 to $+237$). By contrast, $b_k = (X'X+kI)^{-1}X'y$ are stable with respect to such perturbations. The conventional tests of significance do not provide adequate information about this kind of stability. In fact, Vinod [1981] shows that Student's t values are themselves quite unstable.

Recall the definition of the "condition number" $K^{\#} = \lambda_1^{1/2}\lambda_p^{-1/2}$. Denote the norm of a vector y by $\|y\| = \left[\sum_{t=1}^{T} y_t^2\right]^{1/2} = (y'y)^{1/2}$. Also, denote the "2-norm" of a matrix, $\|X\|_2 = \max_i(\lambda_i^{1/2})$, its largest "singular value," i.e., the square root of the largest eigenvalue of $X'X$.

5.3.3.1 *Effect of Perturbations in y*

Denote the perturbed y by y^p, and perturbed $\hat{y} = Xb = X(X'X)^{-1}X'y$ by $\hat{y}^p = X(X'X)^{-1}X'y^p$. Stewart [1973, p. 194] explains that the following upper bound originally obtained by Golub [1969] is reachable

$$\frac{\|b-X^{+}y^{p}\|}{\|b\|} \le K^{\#}\frac{\|\hat{y}-\hat{y}^{p}\|}{\|\hat{y}\|}, \tag{16}$$

where $X^{+} = (X'X)^{-1}X'$, $\hat{y}$ and $\hat{y}^{p}$ are projections of y and y^{p} respectively on the column space of X. Thus, the effect of perturbations in y on b may be amplified by $K^{\#}$ which is usually greater than unity. In the presence of multicollinearity, $K^{\#} >> 1$ and the effect can be severe. Also, if $\|\hat{y}\|$ is small (poor fit) relative to $\|\hat{y}-\hat{y}^{p}\|$ the effect is severe.

5.3.3.2 Effect of Perturbation in X

Denote the perturbed matrix by X^{p} and write $X^{p} = (X+E^{p})$ where E^{p} is a matrix of perturbations. Now, Golub decomposes $E^{p} = E_{1}^{p} + E_{2}^{p}$, where E_{1}^{p} is the component of E^{p} lying in the column space of X, and E_{2}^{p} is the component orthogonal to the column space of X. We have the approximate reachable bound involving a squared and cubed condition number:

$$\frac{\|b-(X+E^{p})^{+}y\|}{\|b\|} \le K^{\#}\|E_{1}^{p}\|_{2} + K^{\#\,2}\|E_{2}^{p}\|_{2}\frac{\|y-\hat{y}\|}{\|\hat{y}\|} + K^{\#\,3}\|E_{2}^{p}\|_{2}^{2}. \tag{17}$$

For further explanation and exact expressions see Stewart [1969]. The main point is that the upper bound can be large when $K^{\#}$ is large. Data perturbations in X and y may change $\|b\|$ anywhere in the interval between 0 and the right-hand-sides of (16) and (17), respectively. Assuming that the practitioner is "averse" to the risk of large perturbation effects, he would prefer models with low $K^{\#}$. Regarding the bounds in (16) and (17) the reader is warned that (i) they are often too large compared to what might be expected in random perturbations, (ii) they are sensitive to changes in units of measurements. The readers who are interested in numerical methods and algorithms (e.g. Gram Schmidt, Householder) are referred to Chambers [1977].

5.3.4 Additional Effects of Multicollinearity

It is normally assumed by practical workers that forecasting accuracy is reasonably measured by squared multiple correlation coefficient R^2 or some transformation of R^2. It can be shown that the forecasting error is large when λ_p is small. Any misspecification of the model can lead to erroneous conclusions when multicollinearity is present. Also any violations of the basic assumptions of least squares procedure discussed in Chapter 1 tend to be serious for multicollinear data.

5.4 AN EXAMPLE OF THE EFFECT OF PERTURBATION

In this section we use the illustrative example of Chapter 1 having $T = 9$, two regressors and some apparent multicollinearity. Tables 5.1a, 5.1b and 5.1c report the original data and the data associated with our ten perturbations, for y, x_1 and x_2 respectively. This perturbation is beyond the published digit, and primarily includes the rounding errors. Aggregation or other causes can also create such perturbations. Published tables of random numbers were used to add a digit beyond the published digit in an appropriate manner. Table 5.2 reports the regression coefficients obtained when the OLS method is applied to the ten sets of perturbed data along ten rows. We also report certain summary statistics, such as means, standard deviations, medians, quartiles, etc. associated with the ten solutions. It is clear that our mild perturbations lead to considerable changes in regression coefficients. Table 5.3 reports the summary statistics for certain ridge solutions, (to be discussed in Chapters 7 and 8) where the instability is seen to be considerably reduced.

For example, the standard deviation (SD) of the 10 perturbed values of b_1 in Table 5.2 for OLS is 0.0074, which is reduced to 0.0028 in Table 5.3 for the ridge regression estimator $b_k = (X'X+kI)^{-1}X'y$, of equation (1) Chapter 7 below, with $k = 0.1667$. The range of b_1 values (Maximum — Minimum) from Table 5.2 is 0.0207, which is reduced in Table 5.3 to 0.0104 for the ridge estimator having $k = .1667$, which is nearly half of the OLS range, in addition to being positive.

5.5 FINAL REMARKS

For large examples having more regressors and greater degree of multicollinearity the instability of OLS regression coefficients is more dramatic. In Beaton, Rubin and Barone [1976] one regression coefficient changes radically from a large negative number −232 to a large positive number +237. A practitioner faced with such instability cannot possibly rely on the OLS solution, despite all the theoretically appealing properties including BLUE, maximum likelihood, etc. By contrast, ridge-type solutions do not suffer from such instability and may be more useful to the practitioner, even though they may suffer from theoretical problems associated with an unknown bias. Various theoretical problems associated with ridge-type methods discussed below in Chapters 7 and 8 suggest that these methods need to be used with caution.

5.5.1 Common-Sense Type Remedies for Multicollinearity

We need not discuss some well-known common-sense type remedies for multicollinearity, such as obtaining better data, better specification, removal of

TABLE 5.1a
Perturbed Raw Data for y

	Perturbation No.									
orig. data	1	2	3	4	5	6	7	8	9	10
41.9	41.91	41.86	41.92	41.90	41.87	41.85	41.94	41.90	41.92	41.87
45.0	44.96	44.96	44.95	45.02	45.00	45.02	44.99	44.97	45.03	44.99
49.2	49.18	49.21	49.24	49.19	49.16	49.21	49.19	49.20	49.21	49.20
50.6	50.56	50.60	50.62	50.56	50.64	50.55	50.62	50.59	50.62	50.59
52.6	52.56	52.56	52.58	52.56	52.59	52.56	52.57	52.56	52.64	52.55
55.1	55.11	55.05	55.08	55.06	55.06	55.12	55.06	55.07	55.14	55.11
56.2	56.15	56.23	56.16	56.17	56.23	56.15	56.23	56.16	56.19	56.17
57.3	57.31	57.25	57.31	57.34	57.27	57.28	57.33	57.29	57.29	57.31
57.8	57.84	57.76	57.84	57.82	57.75	57.84	57.79	57.76	57.79	57.83

many of the other causes for failure of basic assumptions in regression analysis (e.g. serially correlated or non-normal errors, outliers, simultaneous equations bias, etc.). We assume that the researcher knows enough to try these first, and does not treat ridge methods mechanically without thought.

5.5.2 Non-Bayesian Flavor of the Instability Problem

Since the lack of stability of OLS regression coefficients associated with multicollinearity is generally outside the scope of Bayesian regression, the Bayesian view of multicollinearity expounded by Leamer [1973] appears to be incomplete, although it does contain valuable insights. Vinod [1981] develops a Bayesian formulation of measurement (e.g. rounding) errors to discuss instability.

To summarize, we have seen that multicollinearity can seriously affect the signs and relative magnitudes of individual b_i. Even though the OLS estimator b has several attractive statistical properties (BLUE, maximum likelihood, etc.), from (10) we note that $\text{MSE}(b) \to \infty$ as $\lambda_p \to 0$. Thus b may be too far distant from the true β. Also the solution b may be too unstable with respect to small perturbations in the X, y data, especially because the condition number $K^{\#} = (\lambda_1 / \lambda_p)^{1/2}$ may be too large. We shall see in Chapters 7 and 8 whether ridge regression yields alternatives to b which have a lower MSE and/or which are more stable with respect to data perturbations. We have assumed that the basic model specification (1) is correct. The reader is warned that certain effects similar to the effect of multicollinearity can arise from misspecification, such as omission of relevant or inclusion of irrelevant variables. See e.g., Kmenta [1971, Chapter 10].

TABLE 5.1b
Perturbed Raw Data for x_1

orig. data	Perturbation No. 1	2	3	4	5	6	7	8	9	10
12.4	12.40	12.35	12.36	12.36	12.44	12.42	12.36	12.37	12.44	12.42
16.9	16.86	16.85	16.87	16.92	16.93	16.93	16.85	16.88	16.92	16.92
18.4	18.40	18.37	18.37	18.37	18.37	18.36	18.41	18.39	18.40	18.41
19.4	19.40	19.36	19.42	19.41	19.38	19.39	19.35	19.41	19.36	19.43
20.1	20.06	20.07	20.12	20.06	20.07	20.05	20.07	20.14	20.05	20.11
19.6	19.63	19.61	19.61	19.63	19.56	19.59	19.57	19.58	19.59	19.60
19.8	19.75	19.84	19.81	19.83	19.78	19.84	19.77	19.76	19.81	19.83
21.1	21.12	21.12	21.07	21.11	21.13	21.08	21.05	21.10	21.11	21.11
21.7	21.69	21.70	21.68	21.71	21.74	21.66	21.71	21.66	21.67	21.70

TABLE 5.1c
Perturbed Raw Data for x_2

orig. data	Perturbation No. 1	2	3	4	5	6	7	8	9	10
28.2	28.18	28.15	28.21	28.22	28.18	28.22	28.21	28.16	28.15	28.17
32.2	32.21	32.23	32.15	32.23	32.18	32.23	32.17	32.20	32.24	32.15
37.0	36.98	37.02	36.99	37.02	36.99	37.00	36.97	36.95	36.99	36.99
37.0	36.99	37.01	37.02	37.01	37.01	36.99	37.04	37.02	37.03	36.96
38.6	38.63	38.55	38.60	38.64	38.61	38.64	38.57	38.63	38.63	38.61
40.7	40.66	40.74	40.74	40.68	40.66	40.68	40.72	40.69	40.73	40.72
41.5	41.45	41.47	41.50	41.52	41.53	41.46	41.47	41.46	41.48	41.46
42.9	42.90	42.88	42.86	42.92	42.90	42.92	42.92	42.88	42.91	42.88
45.3	45.31	45.27	45.26	45.29	45.25	45.31	45.30	45.29	45.31	45.33

EXERCISES

5.1 Familiarize yourself with the inner workings of the computer programs available to you for least squares regression, and check whether numerical accuracy is adequate by using Longley [1967] data, and its perturbations.

5.2 Consider a perturbation in the tth row and i^{th} column of X matrix. Write the partial derivative of b_i with respect to this perturbation. Use Taylor series expansion to simplify the partial derivative. Consider similar expansion for

TABLE 5.2
Perturbed Regression Coefficients (k = 0, OLS)

Perturbation No.	b_1	b_2
1	-.1045	1.0861
2	-.0867	1.0692
3	-.0852	1.0685
4	-.0925	1.0750
5	-.0933	1.0755
6	-.0838	1.0670
7	-.1014	1.0830
8	-.0960	1.0779
9	-.0971	1.0790
10	-.0811	1.0640
Summary Statistics:		
Mean	-.0934	1.0757
SD	.0074	.0069
Median	-.0933	1.0755
Minimum	-.1045	1.0670
Maximum	-.0838	1.0861
Quartile 25	-.0985	1.0690
Quartile 75	-.0862	1.0803
Unperturbed OLS	-0.17	1.12

standard errors and t values and show that the sensitivity is large when multicollinearity is present.

5.3 For the example of Chapter 1 having T=9 compute the augmented X matrix, $X^A = [y, X]$, "non-predictive multicollinearity" from its first eigenvector, and Hawkins' diagnostic indicator (5).

5.4 Find the limits for variances and t values using L'Hopital's rule. Show that the variances approach infinity and t values approach zero. (Reference: Crocker [1971]).

Table 5.3

PERTURBED REGRESSION COEFFICIENTS

SUMMARY STATISTICS FOR RIDGE

	$k = .0417$		$k = .1667$	
	b_1	b_2	b_1	b_2
Mean	.1460	.8158	.3055	.5992
SD	.0046	.0043	.0028	.0026
Median	.1461	.8159	.3055	.5995
Minimum	.1393	.8070	.3008	.5938
Maximum	.1549	.8220	.3112	.6035
Quartile 25	.1411	.8128	.3034	.5974
Quartile 75	.1489	.8199	.3077	.6015

Note: Ridge solution refers to the estimator $b_k = (X'X+kI)^{-1}X'y$

5.5 Let U denote a $T \times p$ matrix of rounding errors perturbations in X generated by a uniform random variable. If $W=X+qU$, compare $b^*=W^+y$ with X^+y. (Recall that $X^+=(X'X)^{-1}X'$). Letting q vary over (-1 to +1) integrate out q from the residual sum of squares and develop an "enduring" estimator for β (Reference: Vinod [1981]).

5.6 Show that the diagonal elements of a "hat matrix" given by $\hat{H}=XX^+$ can be used to check whether any single observation has an undue effect on the estimates of y. Show that the diagonals of XX^+ can be obtained by subtracting the predicted y values from similar predicted y value for another comparable regression run where we add unity to all y values.

6

Stein-Rule Shrinkage Estimators

6.1 MOTIVATION FOR SHRINKAGE: BASEBALL EXAMPLE

Baseball is a popular American game played with a bat and a ball between two teams of nine players each. In order to score, the player at bat must hit the ball according to certain rules. In professional baseball the balls are thrown with such precision and speed that it is considered good when a player hits successfully more than a third of the times he is at bat. There are a large number of games played in a professional season so that players get a chance to be at bat over 300 times. Considerable statistics are available for each baseball season in newspapers like the New York Times and elsewhere. Professors Efron and Morris [1973] and later in an article in the May 1977 issue of *Scientific American* explain the concepts of Stein-Rule estimators with an example about batting averages of 18 baseball players in the 1970 season. The purpose of this exercise is to estimate batting ability. Denote

θ_i = true batting ability of the ith player $(i = 1, \ldots, p)$, $p = 18$

y_i = batting average calculated after each player had 45 times at bat in the early part of the 1970 season

$= \sum_{t=1}^{T} y_{it}/T$, *where* $T = 45$.

$\bar{y}$ = the grand mean of all 18 baseball players

$= (\sum_{i=1}^{p} y_i)/p$, *where* $p = 18$.

Since averaging is one of the most common statistical operations it is not surprising that y_i may be thought to be a natural estimator of θ_i. Of course, we do not expect the observed values to be always equal to the average.

Let the underlying statistical model be given by including an error term:

$$y_i = \theta_i + u_i, \; i = 1, ..., p \; , \tag{1}$$

where the u_i's are normally and independently distributed, $u_i \sim NID(0, \sigma^2)$ with zero mean and common finite variance σ^2. Here we are assuming a simple error structure. We will write (1) in the context of linear regression in (14a) below. In this simple model the matrix X of regressors is the identity matrix.

Thus $E(y_i) = \theta_i$ implies that the arithmetic mean is an unbiased estimator. To prove that it is a maximum likelihood (ML) estimator we write the likelihood function

$$L(\theta_i, \sigma^2 | y_i) = \frac{1}{\sigma\sqrt{2\pi}} \, exp[-\frac{1}{2\sigma^2}(y_i - \theta_i)^2]. \tag{2}$$

To maximize this, the first-order condition is

$$\partial L / \partial \theta_i = 0 = (positive \; number)\,[-2\,(y_i - \theta_i)]. \tag{3}$$

This implies that y_i is a ML or least squares estimator of θ_i. The mean squared error (MSE) of y_i is given by the expectation of squared error loss:

$$\begin{aligned} MSE(y_i) &= E\,(y_i - \theta_i)^2 \\ &= E\,u_i^2 \\ &= \sigma^2. \end{aligned} \tag{4}$$

Thus, the ML estimator y_i of θ_i has constant MSE(or risk).

If we write the model (1) in terms of $p \times 1$ vectors as $y = \theta + u$, the ML estimator of θ is y, and its MSE is the sum of the $MSE(y_i)$ of (4), and remains constant. We have $MSE(y) = E(y-\theta)'(y-\theta) = E\sum_{i=1}^{i=p} (y_i-\theta_i)^2 = p\,\sigma^2$

In 1955, Stein proposed the following nonlinear and biased estimator of θ_i for the case when σ^2 is assumed to be known for convenience. The estimator is

$$\hat{\theta}_{iS} = (1 - \frac{c\,\sigma^2}{\sum_{i=1}^{p} y_i^2})\, y_i = y_i + f(y_i), \tag{5}$$

where S represents Stein, c is a positive constant, and $f(y_i) = (-c\,\sigma^2 y_i)/\sum_{i=1}^{p} y_i^2$. In terms of $p \times 1$ vectors θ vector is estimated by $\hat{\theta}_S = y + f(y)$.

Now the MSE of $\hat{\theta}_S$ is given by James and Stein [1961] as follows

$$\begin{aligned} MSE(\hat{\theta}_S) &= E \sum_{i=1}^{p} (\hat{\theta}_{iS} - \theta_i)^2 \\ &= E \sum_{i=1}^{p} (y_i + f(y_i) - \theta_i)^2 \\ &= E \sum_{i=1}^{p} \left[(y_i - \theta_i)^2 + 2(y_i - \theta_i) f(y_i) + f^2(y_i)\right] \\ &= E \sum_{i=1}^{p} (y_i - \theta_i)^2 + E \sum_{i=1}^{p} \left[f^2(y_i) + 2E\, f'(y_i)\right]. \end{aligned} \tag{6}$$

The details of the argument which leads to the writing of the expectation of $(y_i - \theta_i)f(y_i)$ as the expectation of the derivative of $f(y_i)$ denoted by $f'(y_i)$ are given at the end of this chapter in Explanatory Note 6.1. We can write

$$MSE(\hat{\theta}_S) = p\,\sigma^2 - c\,\sigma^2[2(p-2) - c]\; E[1/\chi^2(p, \lambda_N)], \tag{7}$$

where $\chi^2(p, \lambda_N)$ is a non-central chi-square with p degrees of freedom and the non-centrality parameter $\lambda_N = \sum_{i=1}^{p} \theta_i^2/2$.

The first term in this expression is $MSE(y)$ for the ML estimator and the second term must be positive for $p > 2$ and $c < 2(p-2)$. It is obvious that under these circumstances $MSE(\hat{\theta}_S) < MSE(y)$. It can be shown that when θ_i is close to zero (i.e., $\lambda_N = 0$) $MSE(\hat{\theta}_S) = 2\sigma^2$ which is smaller than $MSE(y) = p\,\sigma^2$, for $p > 2$, and the MSE reduction is appreciable for large p. For large $\theta_i\,(\lambda_N \to \infty)$ we have $MSE(\hat{\theta}_S) \to \mathrm{MSE}(y)$ from below. Thus the Stein-rule estimator achieves a lower MSE for all possible values of λ_N. This result implies that for $p > 2$ the OLS estimate is inadmissible.

At first sight, this seems to be surprising because it suggests that $\hat{\theta}_S$ may be better than y which has been known to be the best available estimator since the time of Gauss and Legendre in the early 19th Century. We should note, however, that the ML estimator is best in the class of unbiased estimators. The Stein result above suggests that if we are willing to come out of the class of linear and unbiased estimators we can obtain an estimator which is better than the ML estimator in the lower MSE (risk) sense, provided there are at least three parameters to be estimated. The estimator $\hat{\theta}_S$ in (5) is biased, non-linear, and sensitive to changes in the units of measurement. It is pointed out by Efron and Morris that there was no great rush to use the Stein-rule estimator in the 1960's and early 1970's by the practitioners of statistics, despite its theoretical superiority. This may be due to a lack of faith in hard-to-understand mathematical results or uncertainty as to their relevance in the real world where the normal distribution, independence of errors, quadratic loss based on MSE may not be realistic.

The baseball example suggests that $\hat{\theta}_{iS}$ does have practical implications for certain kinds of problems where MSE as the objective function is relevant and meaningful. Of course, there are applications where $\hat{\theta}_{iS}$ should not be preferred, which will be noted later.

In expression (5) for $\hat{\theta}_{iS}$, the constant c is an arbitrary positive number. We can choose it from (7) to minimize the MSE to yield the optimum choice of c. We may differentiate MSE($\hat{\theta}_{iS}$) with respect to c, and set the derivative equal to zero as follows:

$$MSE(\hat{\theta}_{iS}) = \delta_1 + c^2\delta_2 - 2c(p-2)\delta_2 ,$$

where δ_1 and δ_2 are some positive numbers. Now,

$$\partial MSE/\partial c = 0 = 2c\,\delta_2 - 2(p-2)\delta_2$$

implies that

$$c = (p-2).$$

With this optimum choice of the constant c in (5) we get

$$\hat{\theta}_{iS}^* = [1 - \frac{(p-2)\sigma^2}{\sum_{i=1}^{p} y_i^2}]\, y_i , \tag{8}$$

defined for p strictly larger than 2 only ($p > 2$). This shrinks the ML estimate y_i towards zero. A so-called "positive part" version of this estimator was conjectured to be superior to $\hat{\theta}_{iS}^*$ by Stein [1960]. The conjecture was later proved by Baranchik [1964] in an unpublished Technical Report. The "positive part" Stein-rule estimator is given by

$$\hat{\theta}_{iS}^+ = \begin{cases} [1-(p-2)\sigma^2 W^{-1}]y_i, & \text{when } 1 > (p-2)\sigma^2 W^{-1} > 0 \\ 0, & \text{when } 1 < (p-2)\sigma^2 W^{-1} , \end{cases} \tag{9}$$

where $W = \sum_{i=1}^{p} y_i^2$. The superscript + is sometimes used on $[1-(p-2)\sigma^2 W^{-1}]$ to suggest that it is a "positive part" in the sense it is defined by (9).

It is stated by Sclove [1968] that using $\hat{\theta}_{iS}^+$ amounts to using a preliminary test of significance of y_i with respect to the null hypothesis $\theta_i = 0$. If the test is accepted at a specified level of significance, we use $\theta_i = 0$ instead of y_i, and when the test is rejected we use the optimum solution $\hat{\theta}_{iS}^*$. Sclove notes that one implicitly obtains a "very large significance level," especially in small samples. In large samples ($T-p \rightarrow \infty$) it is found to be the 50% level.

In the Stein-rule estimators given in (5) and (8), σ^2 is assumed to be known. When σ^2 has an unknown value, James and Stein (JS), consider the following estimator

$$\hat{\theta}_{iJS} = \left(1 - \frac{cz}{\sum_{i=1}^{p} y_i^2}\right) y_i , \tag{9a}$$

where z is a random variable independent of y, distributed as σ^2 times a χ^2 variable with n degrees of freedom. This estimator will be taken up in the following sections.

The procedure of shrinkage toward zero in the Stein-rule estimations was thought to be unsatisfactory by Efron and Morris for the baseball problem. They use Lindley's (in Stein [1962] pp. 285-287) modified Stein-rule estimator given by

$$\hat{\theta}_{iL}^{+} = \bar{y} + [1 - \frac{(p-3)\sigma^2}{\sum_{i=1}^{p}(y_i - \bar{y})^2}]^{+}(y_i - \bar{y}), \quad p > 3, \tag{10}$$

where the superscript + suggests "positive part" as before, and L represents Lindley's estimator. Note that this shrinks the ML estimator y_i toward the grand mean $\bar{y}$ of batting averages, rather than towards zero. It is convenient to derive Lindley's Bayesian model in terms of its own notation separately (see Chapter 3 Section 6 and Section 6.2.2 of this chapter).

For the baseball problem Table 6.1 gives the batting averages y_i in column (1). The true unknown batting ability may be thought to be the batting average for the remainder of the baseball season. At the end of 1970 this unknown ability became known. It is given in column (2), entitled θ_i. In this example the baseball fans and managers would like to predict θ_i from their knowledge of the batting averages in the initial part of the season (after the players have been 45 times at bat, say). Column (3) of Table 6.1 gives the $\hat{\theta}_{iS}^{+}$ or "positive part" Stein-rule estimator. For this example $\hat{\theta}_{iS} \equiv \hat{\theta}_{iS}^{+}$ because the shrinkage factor $[1 - (p-2)\sigma^2 W^{-1}]$ is non-negative. For σ^2 we use the minimum MSE estimate $\hat{\sigma}^2 = \sum_{i=1}^{p}(y_i - \bar{y})^2/(p+1) = 0.004405 = z$, where we have the grand mean $\bar{y} = 0.277714$. Efron and Morris consider the binomial model and use a variance stabilizing transformation to ensure that $\sigma = 1$. A more appropriate estimate of σ^2 is

$$\tilde{\sigma}^2 = \sum_{i=1}^{p}\sum_{t=1}^{T}(y_{it} - \bar{y})^2/(T - p + 1) \tag{11}$$

which requires knowledge of individual y_{it} values.

TABLE 6.1
1970 Batting Averages for 14 Major League Baseball Players

Name	y_i = Batting Average for First 45 at Bats	θ_i = Batting Average for Remainder of Season	$\hat{\theta}_{iS}^+$	$\hat{\theta}_{iL}^+$
	(1)	(2)	(3)	(4)
Clemente (Pitts, NL)	0.400	0.346	0.381547	0.310324
F. Robinson (Balt, AL)	0.378	0.298	0.360562	0.304457
F. Howard (Wash, AL)	0.356	0.276	0.339577	0.298590
Johnstone (Cal, AL)	0.333	0.221	0.317638	0.292457
Berry (Chi, AL)	0.311	0.273	0.296653	0.286590
Spencer (Cal, AL)	0.311	0.270	0.296653	0.286590
Kessinger (Chi, NL)	0.289	0.263	0.275668	0.280724
Santo (Chi, NL)	0.244	0.269	0.232744	0.268724
Unser (Wash, AL)	0.222	0.264	0.211759	0.262857
Williams (Chi, AL)	0.222	0.256	0.211759	0.262857
Scott (Bos, AL)	0.222	0.304	0.211759	0.262857
Petrocelli (Bos, AL)	0.222	0.264	0.211759	0.262857
Companeris (Oak, AL)	0.200	0.285	0.190774	0.256990
Munson (NY, AL)	0.178	0.319	0.169788	0.251124
Sum, SE	3.888	3.908	3.708637	3.888000
Average, MSE	0.277714	0.279143	0.264903	0.277714

Name	$(y_i - \theta_i)^2$	$(\hat{\theta}_{iS}^+ - \theta_i)^2$	$(\hat{\theta}_{iL}^+ - \theta_i)^2$
	(5)	(6)	(7)
Clemente	0.002916	0.001264	0.001273
F. Robinson	0.006400	0.003914	0.000042
F. Howard	0.006400	0.004042	0.000510
Johnstone	0.012544	0.009339	0.005106
Berry	0.001444	0.000559	0.000185
Spencer	0.001681	0.000710	0.000275
Kessinger	0.000676	0.000160	0.000314
Santo	0.000625	0.001315	0.000000
Unser	0.001764	0.002729	0.000001
Williams	0.001156	0.001957	0.000047
Scott	0.006724	0.008508	0.001693
Petrocelli	0.001764	0.002729	0.000001
Companeris	0.007225	0.008879	0.000785
Munson	0.019881	0.022264	0.004607
Sum, SE	0.071200	0.68370	0.014839
Average, MSE	0.005086	0.004884	0.001060

Since these were unavailable, the above estimate $\hat{\sigma}^2 = 0.004405$ is a convenient short cut. For the theoretical discussion σ^2 is simply assumed to be known. Column (4) of Table 6.1 gives Lindley's version of the Stein rule estimator

$$\hat{\theta}_{iL} = 0.278 + [1 - \frac{(p-3)\ (0.0044)}{\sum_{i=1}^{p} (y_i - \bar{y})^2}]\ (y_i - 0.278). \tag{12}$$

Columns (5), (6) and (7) give the squared errors $(y_i - \theta_i)^2$, $(\hat{\theta}_{iS}^{+} - \theta_i)^2$ and $(\hat{\theta}_{iL}^{+} - \theta_i)^2$, respectively; and the sum of squared errors (sum of SE's) and their average (i.e., MSE) are given in the two rows at the bottom of the table. Note that θ_i are known here, and $MSE(y) = p^{-1} \sum_{i=1}^{p} (y_i - \theta_i)^2$. This is a mean of SE's and it is an appropriate estimate of the mathematical expectation of the sum of SE's. Similarly, the column averages for col.s (6) and (7) give the MSE's for Stein and Lindley estimators respectively.

From this example, it is clear that the Stein-rule estimator which shrinks towards zero reduces the $MSE(y) = 0.0050857$ to $MSE(\hat{\theta}_S) = 0.0048836$. Shrinking toward the grand mean leads to an even smaller $MSE(\hat{\theta}_L) = 0.0010599$. It can be verified that $\hat{\theta}_L$ or $\hat{\theta}_S$ are "closer" to the true parameter vector θ. This shows that we would be better off estimating the θ_i by these methods.

6.1.1 Effect of Changes in Origin and Scale

If the batting averages are measured in percentages, the maximum likelihood estimates are modified in an "equivariant" way in which y_i is replaced by $100y_i$. Now the $\bar{y}$ would be replaced by 100 $\bar{y}$ and both $\sum_{i=1}^{p} \sum_{t=1}^{T} (y_{it} - \bar{y})^2$ and $\sum_{i=1}^{p} (y_i - \bar{y})^2$ would become $(100)^2$ times the old estimates. Thus $\hat{\sigma}^2$ is replaced by $(100)^2$ times the old value. Hence $\hat{\sigma}^2 / \sum_{i=1}^{p} y_i^2$ would remain unchanged with the cancellation of $(100)^2$ from the numerator and denominator. Thus the shrinkage fraction $[1-(p-2)\hat{\sigma}^2 W^{-1}]^+$ for the "positive part" estimator remains unchanged, and therefore $\hat{\theta}_{iS}$ is merely replaced by 100 $\hat{\theta}_{iS}$ in an equivariant way (i.e., multiplying y_i by a constant c changes $\hat{\theta}_{iS}$ to $c\ \hat{\theta}_{iS}$).

When the "origin" is changed, the scores y_{it} are replaced by $y_{it}^* = y_{it} + d$ where d is some constant. The transformation from y_{it} to y_{it}^* is sometimes called a "translation". An unfortunate property of the Stein-rule estimator is that $\hat{\theta}_{iS}$ is sensitive to change of origin, i.e., it is not translation equivariant: $\hat{\theta}_{iS}^* \neq \hat{\theta}_{iS} + d$ when $y_i^* = y_i + d$. It can be verified that the new shrinkage factor will be different because $\sum y_i^2$ will become $\sum (y_i + d)^2$.

6.1.2 Equivariance of Lindley-Stein Estimator

If we replace y_{it} by $y_{it}^* = c\, y_{it} + d$ we have $y_i^* = c\, y_i + d$, $\bar{y}^* = c\,\bar{y} + d$, $y_i^* - \bar{y}^* = c\, y_i + d - (c\,\bar{y} + d) = c(y_i - \bar{y})$, and $\sum_{i=1}^{p} (y_i^* - \bar{y}^*)^2 = c^2 \sum_{i=1}^{p} (y_i - \bar{y})^2$.

Denoting the estimates $\hat{\sigma}^{*2}$ based on y_{it}^* data

$$\hat{\sigma}^{*2} = \sum_{i=1}^{p} \sum_{t=1}^{T} (y_{it}^* - \bar{y}^*)^2/(T - p + 1)$$

$$= c^2 \hat{\sigma}^2 ,$$

where $\hat{\sigma}^2$ is from the y_{it} data defined in (11). Now

$$\hat{\theta}_{iL}^* = \bar{y}^* + [1 - \frac{(p-3)\hat{\sigma}^{*2}}{\sum_{i=1}^{p} (y_i^* - \bar{y}^*)^2}]^+ (y_i^* - \bar{y}^*)$$

$$= c\,\bar{y} + d + [1 - \frac{(p-3)c^2\hat{\sigma}^2}{c^2 \sum_{i=1}^{p} (y_i - \bar{y})^2}]^+ (c\, y_i + d - c\bar{y} - d)$$

$$= c\,\bar{y} + [1 - \frac{(p-3)\hat{\sigma}^2}{\sum_{i=1}^{p} (y_i - \bar{y})^2}]^+ c(y_i - \bar{y}) + d$$

$$= c\,\hat{\theta}_{iL} + d$$

This shows that the Lindley-Stein estimator $\hat{\theta}_{iL}$ is "equivariant" with respect to a change of origin and scale (i.e., with respect to linear transformations).

In our example the $MSE(\hat{\theta}_S) > MSE(\hat{\theta}_L)$, and we may be tempted to conclude that $\hat{\theta}_{iL}$ is always preferable to $\hat{\theta}_{iS}$. However, when $p = 3$ the $\hat{\theta}_{iL}$ estimator is not defined. There are applications where $\hat{\theta}_{iL}$ may not be preferred over $\hat{\theta}_{iS}$. In the baseball application, the change of origin does not make much sense for batting averages. Hence we need not insist on equivariance with respect to change of origin, and $\hat{\theta}_{iS}$ may not be unreasonable. In those applications where the origin change is meaningful, a practitioner should be cautious when using $\hat{\theta}_{iS}$ because the estimator can be arbitrarily changed when the origin changes.

TABLE 6.2
Absolute Errors for Batting Average Estimates

Player No.	$\lvert y_i - \theta_i \rvert$	$\lvert \hat{\theta}_{iS} - \theta_i \rvert$	$\lvert \hat{\theta}_{iL} - \theta_i \rvert$
	(1)	(2)	(3)
1	0.054000	0.035547	0.035676
2	0.080000	0.062562	0.006457
3	0.080000	0.063577	0.022590
4	0.112000	0.096638	0.071457
5	0.038000	0.023653	0.013590
6	0.041000	0.026653	0.016590
7	0.026000	0.012668	0.017724
8	0.025000	0.036256	0.000276
9	0.042000	0.052241	0.001143
10	0.034000	0.044241	0.006857
11	0.082000	0.092241	0.041143
12	0.042000	0.052241	0.001143
13	0.085000	0.094226	0.028010
14	0.141000	0.149212	0.067876
Sum	0.882000	0.841957	0.330533
Average	0.063000	0.060140	0.023610

6.1.3 Closeness to Parameters

Using squared errors instead of absolute errors may be criticized because it tends to give higher weight to large errors. The intuitive idea of "closeness" is more in terms of the absolute errors $|y_i - \theta_i|$, $|\hat{\theta}_{iS} - \theta_i|$, and $|\hat{\theta}_{iL} - \theta_i|$. For our example, Table 6.2 gives these absolute errors. It is interesting to note that the basic conclusion regarding the choice between estimators does not change under this criterion.

6.2 STEIN-RULE IN THE REGRESSION CONTEXT

We now consider the Stein-rule estimator for the regression model

$$y = X\beta + u \ , \tag{13}$$

where y is a $T \times 1$ vector, X is a $T \times p$ nonstochastic matrix with rank p, and $u \sim N(0, \sigma^2 I)$ is a $T \times 1$ disturbance vector as defined in the earlier chapters.

To see how the model in (1) is related with the above regression model, we rewrite (13) as

$$y = H\Lambda^{1/2}G'\beta + u \tag{14}$$
$$= H\theta + u \ ,$$

where $\theta = \Lambda^{1/2}G'\beta$ and we have used the singular value decomposition $X = H\Lambda^{1/2}G'$ such that $X'X = G\Lambda G'$ as discussed in Chapter 1. The $T \times p$ matrix H is such that $H'H = I$. Also, G is a $p \times p$ matrix of eigenvectors of $X'X$ and $\Lambda^{1/2}$ is a $p \times p$ diagonal matrix of square roots of the eigenvalues of $X'X$. Premultiplying by H' we can get

$$w = \theta + v \ , \tag{14a}$$

where $w = H'y$ and $v = H'u \sim N(0, \sigma^2 I)$. The model (14a) is then a generatization of the model considered in (1) for $i = 1, ..., p$. Note that equation (1) for the baseball example is a special case of (14a) where $H = I$. Therefore, since $\beta = G\ \Lambda^{-1/2}\theta$, using the expression for Stein-rule estimators of θ in the earlier section we can write the Stein-rule estimators for β in (13). For example, the Stein estimator for β is

$$\hat{\beta}_S = G\ \Lambda^{-1/2}\hat{\theta}_S = G\ \Lambda^{-1/2}[1 - \frac{c\,\sigma^2}{w'w}]w \tag{15}$$
$$= [1 - \frac{c\,\sigma^2}{w'w}]\ G\ \Lambda^{-1/2}w$$
$$= [1 - \frac{c\,\sigma^2}{b'X'Xb}]\ b\ ,$$

where $c > 0$ is any constant, $b = G\ \Lambda^{-1/2}w = (X'X)^{-1}X'y$ is the OLS estimator of β (see Section 1.3) and $w = \Lambda^{1/2}G'b$ such that $w'w = b'G\Lambda G'b = b'X'Xb$. The James and Stein (JS) estimator of β corresponding to (9a) is

$$\hat{\beta}_{JS} = G\ \Lambda^{-1/2}\hat{\theta}_{JS} \tag{16}$$
$$= [1 - (\frac{c}{n})\ \frac{\hat{u}'\hat{u}}{b'X'Xb}]\ b\ ,\ n = T - p,$$

where c is an arbitrary constant to be determined later on in this chapter, and where σ^2 has been substituted by its unbiased estimator

$$s^2 = \frac{\hat{u}'\hat{u}}{n} = \frac{(y - Xb)'(y - Xb)}{n}. \tag{17}$$

The $\hat{u}'\hat{u}$ term is independent of b and it is distributed as σ^2 times a chi-square with n degrees of freedom. In Section 6.2.2 we write Lindley's Stein-like estimator (see equation (24)) in a similar fashion.

We now look at the alternative interpretation and derivations of the Stein-rule estimator.

6.2.1 Bayesian Interpretation of the Stein-Rule

Consider the sample information $y \sim N(X\beta, \sigma^2 I)$, and the prior information for β as multivariate normal with the mean vector β_0 and known variance covariance matrix $\sigma_\beta^2 \Omega_0$, that is, $\beta \sim N(\beta_0, \sigma_\beta^2 \Omega_0)$. The posterior density of β given σ^2 is then multivariate normal with the mean vector (Bayes estimator)

$$\beta^* = (\frac{X'X}{\sigma^2} + \frac{\Omega_0^{-1}}{\sigma_\beta^2})^{-1} (\frac{X'Xb}{\sigma^2} + \frac{\Omega_0^{-1}\beta_0}{\sigma_\beta^2}) \tag{18}$$

$$= \beta_0 + [I - (\Omega_0^{-1} + \frac{\sigma_\beta^2}{\sigma^2} (X'X)^{-1} \Omega_0^{-1}](b - \beta_0),$$

and the variance covariance matrix

$$\Omega^* = (\frac{X'X}{\sigma^2} + \frac{\Omega_0^{-1}}{\sigma_\beta^2})^{-1}, \tag{19}$$

(for detail see Section 3.6.1). The posterior mean β^* is the Bayes estimator of β, which is optimal relative to a quadratic (squared error) loss function.

If we now consider the prior distribution of $\beta \sim N(0, \sigma_\beta^2 (X'X)^{-1})$, then the Bayes estimator simplifies to

$$\beta^* = (\frac{X'X}{\sigma^2} + \frac{X'X}{\sigma_\beta^2})^{-1} \frac{X'X}{\sigma^2} b \tag{20}$$

$$= (\frac{1}{\sigma^2} + \frac{1}{\sigma_\beta^2})^{-1} \frac{b}{\sigma^2}$$

$$= \frac{\sigma_\beta^2}{\sigma^2 + \sigma_\beta^2} b$$

$$= [1 - \frac{\sigma^2}{\sigma^2 + \sigma_\beta^2}]\ b.$$

The estimator β^* shrinks the OLS estimator b by the factor $0 \le \sigma_\beta^2[\sigma^2 + \sigma_\beta^2]^{-1} \le 1$. However, this estimator is not operational, since it depends on unknowns σ^2 and σ_β^2. But we note from $y = X\beta + u$ that $b = \beta + (X'X)^{-1}X'u$. Thus

$$E(\frac{b'X'Xb}{p}) = p^{-1} E[\beta'X'X\beta + u'X(X'X)^{-1}X'u + 2\beta'X'u] \tag{21}$$

$$= p^{-1} E(\beta'X'X\beta) + E(u'X(X'X)^{-1}X'u) + 2E(\beta'X'u)$$

$$= p^{-1} [p\sigma_\beta^2 + \sigma^2 p]$$

$$= \sigma_\beta^2 + \sigma^2 ,$$

where $E(u'X(X'X)^{-1}X'u) = p\sigma^2$, $E(\beta'X'X\beta) = \sigma_\beta^2 p$, and $E(\beta'X'u) = 0$ by assuming β and u to be independent. Note that in the Bayesian context β is a random variable. The independence of β and u imply that the prior and sample information are independent.

Now if we replace $\sigma_\beta^2 + \sigma^2$ in β^* of (20) by its unbiased estimator $b'X'Xb/p$, then we can write

$$\beta^* = [1 - \frac{p\sigma^2}{b'X'Xb}]b. \tag{22}$$

This estimator is Stein's estimator $\hat{\beta}_S$ for $c = p$. Further, replacing σ^2 in β^* by its unbiased estimator $s^2 = \hat{u}'\hat{u}/n$, we get the form of James and Stein's estimator, $\hat{\beta}_{JS}$, given in (16).

6.2.1.1 *Lindley's Form of Stein's Estimator of* β

Let us consider the model $y = X\beta + u$ such that $y \sim N(X\beta, \sigma^2 I)$ and the prior $\beta \sim N(\mu\iota, \sigma_\beta^2(X'X)^{-1})$, where our notation ι is a $p \times l$ vector all of whose elements are unity, and μ is a scalar.

Recall that it has been shown in Chapter 3 that for given σ_β^2 and σ^2, the posterior density of β is multivariate normal. The mean of this posterior density leads to Lindley's Bayesian estimator indicated by the subscript L as follows.

$$\beta_L^* = (\frac{X'X}{\sigma^2} + \frac{X'X}{\sigma_\beta^2})^{-1} (\frac{X'Xb}{\sigma^2} + \frac{(X'X)\mu\iota}{\sigma_\beta^2}) . \tag{23}$$

$$= (\frac{1}{\sigma^2} + \frac{1}{\sigma_\beta^2})^{-1} (\frac{b}{\sigma^2} + \frac{\mu\iota}{\sigma_\beta^2})$$

$$= \frac{\sigma_\beta^2}{\sigma^2 + \sigma_\beta^2} (b + \frac{\sigma^2}{\sigma_\beta^2}\mu\iota)$$

$$= \mu\iota + (1 - \frac{\sigma^2}{\sigma^2 + \sigma_\beta^2}) (b - \mu\iota).$$

This result follows from β^* of (18) above by substituting $\Omega_0 = (X'X)^{-1}$ and

$\beta_0 = \mu\iota$.

Substituting the unbiased estimators for σ^2 and $\sigma^2 + \sigma_\beta^2$, respectively, we can write

$$\beta_L^* = \mu\iota + (1 - \frac{p}{n} \frac{\hat{u}'\hat{u}}{b'X'Xb - \mu^2\iota'X'X\iota}) (b - \mu\iota) . \tag{24}$$

which is Lindley's [1962] estimator in the regression context. Note that $p^{-1} E(b'X'Xb - \mu^2\iota'X'X\iota)$ can be shown to be equal to $\sigma^2 + \sigma_\beta^2$. The main idea behind Lindley's estimator is therefore to use Stein's method for the deviations $(b - \mu\iota)$ rather than b. This would avoid the difficulty of arbitrary origin that appears in Stein's estimator. In effect, Lindley's estimator shrinks b towards the prior mean $\mu\iota$, whereas in Stein's case it shrinks toward the prior mean zero. Although Stein's prior may be thought to be a special case of Lindley's prior, Stein's prior is consistent with the practical workings of classical inference in Statistics. Roughly speaking, the effect of one variable on another is assumed to be zero in the absence of specific evidence to the contrary.

For some models (see Chapter 9) μ can easily be defined. However, in many applications the unknowns μ remains a problem for the usage of β_L^*. One can consider $\mu = \iota'b/p$, the simple average of $b_1, \ldots, b_p$, as was done in (10) for the model (15). This would be meaningful provided $b_1, \ldots, b_p$ are in the same unit of measurement. Alternatively, one could consider $\mu = \hat{\mu} = \iota'X'Xb/\iota'X'X\iota$. This particular choice is explained in the next section. Substituting, $\hat{\mu}$ in (24) one can write β_L^* as

$$\hat{\beta}_L^* = \hat{\mu}\iota + [1 - \frac{p}{n} \frac{\hat{u}'\hat{u}}{(b - \hat{\mu}\iota)'X'X(b - \hat{\mu}\iota)}] (b - \hat{\mu}\iota) , \tag{24a}$$

where $(b - \hat{\mu}\iota)'X'X(b - \hat{\mu}\iota) = b'X'Xb - \hat{\mu}^2\iota'X'X\iota$. Note that the sampling properties of $\hat{\beta}_L^*$ would be different from those of James and Stein's estimator, where $\hat{\mu} = 0$. The necessary condition for the MSE of this estimator to be lower than the MSE of OLS becomes $p \geq 4$ when we use Lindley's mean correction. The reader should see Exercise 6.2 at the end of this chapter.

6.2.2 The Minimum Weighted Mean Squared Error (MWMSE) Interpretation of the Stein Rule

Consider a class of estimators

$$\tilde{\beta} = q_o\iota + q_1 b , \tag{25}$$

where q_0 and q_1 are arbitrary scalars, and ι is a $p \times 1$ vector of unit elements. For $q_0 = 0$, $\tilde{\beta} = q_1 b$. Now write the weighted MSE of $\tilde{\beta}$ as

$$
\begin{aligned}
WMSE(\tilde{\beta}) &= E(\tilde{\beta}-\beta)'W(\tilde{\beta}-\beta) \qquad (26)\\
&= E(q_0\iota + q_1 b - \beta)'W(q_0\iota + q_1 b - \beta)\\
&= E[q_1(b-\beta) + (q_1-1)\beta + q_0\iota]'W[q_1(b-\beta) + (q_1-1)\beta + q_0\iota]\\
&= q_1^2\, E(b-\beta)'W(b-\beta) + (q_1-1)^2\beta'W\beta + 2q_0(q_1-1)\iota'W\beta\\
&\quad + q_0^2\,\iota'W\,\iota,
\end{aligned}
$$

where W is a given positive definite matrix, and we have used $E(b-\beta)=0$. Since $(b-\beta) \sim N(0, \sigma^2(X'X)^{-1})$, we can verify that $E(b-\beta)'W\ (b-\beta) = \sigma^2 h$, where $h = Tr[(X'X)^{-1}W]$. Thus

$$WMSE(\tilde{\beta}) = q_1^2\,\sigma^2 h + (q_1-1)^2\beta'W\beta + 2q_0\,(q_1-1)\iota'W\beta + q_0^2\,\iota'W\iota. \qquad (27)$$

The partial differentiation of this expression with respect to q_0 and q_1 provides the minimizing conditions as

$$
\begin{aligned}
\frac{\partial WMSE(\tilde{\beta})}{\partial q_0} &= 2(q_1-1)\iota'W\beta + 2q_0\iota'W\iota = 0\,, \qquad (28)\\
\frac{\partial WMSE(\tilde{\beta})}{\partial q_1} &= 2q_1\sigma^2 h + 2(q_1-1)\beta'W\beta + 2q_0\iota'W\beta = 0\,.
\end{aligned}
$$

The optimal values of q_0 and q_1 that minimize $WMSE\ (\tilde{\beta})$ can be obtained by solving these equations. The solutions are:

$$
\begin{aligned}
q_0 &= \frac{\sigma^2 h\,\mu}{\sigma^2 h\,\mu + (\beta-\mu\iota)'W(\beta-\mu\iota)} \qquad (29)\\
q_1 &= \frac{(\beta-\mu\iota)'W(\beta-\mu\iota)}{\sigma^2 h\,\mu + (\beta-\mu\iota)'W(\beta-\mu\iota)}
\end{aligned}
$$

where $\mu = \iota'W\beta/\iota'W\iota$.

Substituting (29) in (25) we get the *MWMSE* estimator as

$$\tilde{\beta} = \mu\iota + \left[1 - \frac{\sigma^2 h}{\sigma^2 h + (\beta-\mu\iota)'W(\beta-\mu\iota)}\right](b-\mu\iota). \qquad (30)$$

This estimator involves unknown parameters and thus it cannot be used in practice. However, since

$$
\begin{aligned}
E(b-\mu\iota)'W(b-\mu\iota) &= E[(b-\beta) + (\beta-\mu\iota)]'W[(b-\beta)\\
&\quad + (\beta-\mu\iota)] \qquad (31)\\
&= E(b-\beta)'W(b-\beta) + (\beta-\mu\iota)'W(\beta-\mu\iota)\\
&\quad + 2E(b-\beta)'W(\beta-\mu\iota)\\
&= \sigma^2 h + (\beta-\mu\iota)'W(\beta-\mu\iota)\,,
\end{aligned}
$$

we can replace $\sigma^2 h + (\beta-\mu\iota)'W(\beta-\mu\iota)$ in (30) by its unbiased estimator

$(b - \mu\iota)'W(b - \mu\iota)$. We then get the operational (Op) form of $\tilde{\beta}$ as

$$\tilde{\beta}_{Op} = \mu\iota + [1 - \frac{h}{n} \frac{\hat{u}'\hat{u}}{(b - \mu\iota)'W(b - \mu\iota)}] (b - \mu\iota) \tag{32}$$

provided μ is known. If $\mu = \iota'W\beta/\iota'W\iota$, we can replace it by its unbiased estimator $\hat{\mu} = \iota'Wb/\iota'W\iota$. In this case $\tilde{\beta}_{Op}$ becomes

$$\tilde{\beta}_{Op} = \hat{\mu}\iota + [1 - \frac{h}{n} \frac{\hat{u}'\hat{u}}{(b - \hat{\mu}\iota)'W(b - \hat{\mu}\iota)}] (b - \hat{\mu}\iota). \tag{33}$$

We note that $\tilde{\beta}_{Op}$, both in (32) and (33), may no longer be an *MWMSE* estimator.

In the special case when $W = X'X$ the estimator $\tilde{\beta}_{Op}$ in (33) becomes identical with the operational form of Lindley's estimator in (24a). Further, if $W = X'X$ and $\mu = 0$ in (32), then $\tilde{\beta}_{Op}$ becomes James and Stein's (JS) estimator given in equation (16) above.

6.2.3 Stein-Rule Estimator as a Combination of Restricted and Unrestricted Least Squares Estimators

Let us consider the set of m linear restrictions on the parameter β as $r = R\beta$, where r and R contain $m \times 1$ and $m \times p$ matrices of known constants respectively. Then as derived in Chapter 3, the restricted least squares estimator of β is

$$b_R = b - (X'X)^{-1}R'[R(X'X)^{-1}R']^{-1}(Rb - r) , \tag{34}$$

where b is the OLS estimator. The test statistic for the null hypothesis $r = R\beta$ is the same as in equation (19a) of Chapter 3, namely,

$$w = \frac{(Rb - r)'[R(X'X)^{-1}R']^{-1}(Rb - r)/m}{\hat{u}'\hat{u}/(T-p)}, \tag{35}$$

which is distributed as Snedecor's central $F(m, T - p)$ when $r = R\beta$ is true. If w is greater than the critical value c at a given significance level we reject the null hypothesis, otherwise we accept it. Recall that the preliminary test estimator, which depends on the test of significance, was defined in equation (75) of Chapter 3 as

$$b_{PT} = (1 - I(w))b + I(w)b_R ,$$

where $I(w)$ is the indicator function such that $I(w) = 1$ if $w \leq c$, and $I(w) = 0$ if $w > c$.

If we now define $I(w) = c^*/w$, then we can obtain an alternative estimator, proposed by Bock [1975],

$$\hat{\beta}_{BK} = (1 - \frac{c^*}{w})b + \frac{c^*}{w}b_R \qquad (36)$$
$$= b_R + (1 - \frac{c^*}{w})(b - b_R) ,$$

where c^* is a suitably chosen positive number, subscript BK represents Bock, and w is from (35). Note that $\hat{\beta}_{BK}$ like b_{PT}, is a combination of the OLS (unrestricted) and the restricted estimator. However, unlike b_{PT}, $\hat{\beta}_{BK}$ involves a more definite form in place of $I(w)$. The determination of c^* will be discussed in the following sections.

The estimator $\hat{\beta}_{BK}$ in (36) is a more general form of Stein-rule estimator compared to Lindley's estimator in (24a). While in $\hat{\beta}_{BK}$ we shrink the OLS estimator b towards b_R, in (24a) it is shrunken towards its overall average. The Bayesian interpretation of (36) follows from 6.2.1. In a special case when $R = I$, *we have* $b_R = r$ from (34). Then

$$\hat{\beta}_{BK} = r + [1 - \frac{c^{**}\hat{u}'\hat{u}}{(b - r)'X'X(b - r)}](b - r), \qquad (37)$$

where $c^{**} = c^* m/ (T-p)$. It is then clear that if $r = \hat{\mu}\iota$ we get Lindley's estimator in (24a), and if $r = 0$ we get $\hat{\beta}_{JS}$, James and Stein's estimator in (16) as a special case of Bock's estimator.

6.3 PROPERTIES OF THE STEIN-RULE ESTIMATOR

First, we note from the discussion in the previous section that the Stein-rule estimator can be written in the following form:

$$\hat{\beta}_{SR} = [1 - \frac{k_1\hat{u}'\hat{u}}{b'X'Xb}]b, \qquad (38)$$

where $k_1 > 0$ is an arbitrary scalar constant. It is proper to consider k_1 since various forms of Stein-type estimators differ with respect to this constant. The reason for subscript 1 in k will become clear from the discussion of estimators in Section 6.4.

The following approximate results are proved in the Explanatory Note 6.2 at the end. The approximations are true in "small sigma asymptotic" sense discussed by Kadane [1970, 1971].

$$Bias\,(\hat{\beta}_{SR}) = E\,(\hat{\beta}_{SR} - \beta) = -\frac{nk_1\sigma^2}{\beta'X'X\beta}\,\beta. \qquad (39)$$

$$MtxMSE(\hat{\beta}_{SR}) = E(\hat{\beta}_{SR} - \beta)(\hat{\beta}_{SR} - \beta)' = \sigma^2(X'X)^{-1} + \frac{nk_1\sigma^4}{(\beta'X'X\beta)^2}[\beta\beta'\{4I + k_1(n+2)\} - 2(\beta'X'X\beta)(X'X)^{-1}] \quad (40)$$

$$MSE(\hat{\beta}_{SR}) = E(\hat{\beta}_{SR} - \beta)'(\hat{\beta}_{SR} - \beta) = Tr\ MtxMSE(\hat{\beta}_{SR}) = \sigma^2 Tr(X'X)^{-1} + \frac{nk_1\sigma^4}{(\beta'X'X\beta)^2}[\beta'\beta\{4 + k_1(n+2)\} - 2(\beta'X'X\beta)Tr(X'X)^{-1}]. \quad (41)$$

In the above results the Bias is of order σ^2 and both *MtxMSE* and MSE are of order σ^4 *a la* Kadane. The direction of the bias is opposite to the sign of β. As $\sigma \to 0$, the bias term vanishes. Further, the relative bias of a component of $\hat{\beta}_{SR}$ is a decreasing function of σ.

Let $\lambda_1 \geq \lambda_2 \geq \cdots \geq \lambda_p$ be the eigenvalues of the matrix $X'X$. Then according to a result in Rao [1965, p. 59], $\lambda_{min} \leq \beta'X'X\beta/\beta'\beta \leq \lambda_{max}$, where $\lambda_{min} = \lambda_p$ and $\lambda_{max} = \lambda_1$. It can then be verified from (41) that for $p \geq 3$, $MSE(\hat{\beta}_{SR}) - MSE(b) < 0$, when

$$0 < k_1 \leq \frac{2}{n+2}(d-2); \; d = \lambda_p \sum_{i=1}^{p} \lambda_i^{-1} > 2, \quad (42)$$

where $\sum_{i=1}^{p} \lambda_i^{-1} = Tr(X'X)^{-1}$. This implies that the Stein-rule estimator does better, for all values of β, than the OLS estimator in the small-σ approximate MSE sense. Thus, for small σ it is a minimax estimator with respect to β. In fact it has been shown by various authors that even the exact $MSE(\hat{\beta}_{SR}) < MSE(b)$ for the above range of k_1 (also see Explanatory Note 6.3 at the end of this chapter). When $X'X = I$, that is when we are dealing with the orthogonal regression model, or when the weight is $X'X$ for $WMSE(\hat{\beta}_{SR}) = E(\hat{\beta}_{SR} - \beta)'X'X(\hat{\beta}_{SR} - \beta)$ then the range of k_1 becomes $0 < k_1 \leq \frac{2}{n+2}(p-2); p > 2$. These results imply that b (OLS) is inadmissible for $p \geq 3$. A sufficient condition for inadmissibility is given in Section 2.7.2 above by the imprecise descriptive phase: "someone better exists."

We note that under the MSE criterion the necessary condition for *SR* estimator to be better than OLS requires the date matrix X to be such that $d = \lambda_p \sum_{i=1}^{p} \lambda_i^{-1} > 2$. In many applications this condition may not be satisfied. First, for $d > 2$ we need the models to have three or more regressors. This is because for $p = 2$, $d = \lambda_2[\frac{1}{\lambda_1} + \frac{1}{\lambda_2}] = \frac{\lambda_2}{\lambda_1} + 1 < 2$. Also, $p \geq 3$ is required because the $MSE(\hat{\beta}_{SR})$ becomes infinity for $p \leq 2$, see Ullah and

Ullah [1978]. Further, if the data have severe multicollinearity λ_p could be very small, making $d < 2$. All these comments imply that one gains by using the Stein-rule estimator, under the MSE criterion, when there are at least three regressors in the model and the data do not show strong multicollinearity.

To see the behavior of $\hat{\beta}_{SR}$ under the *MtxMSE* criterion, it would be convenient to consider the ith component of $MtxMSE(\hat{\beta}_{SR})$:

$$E(\hat{\beta}_{SR} - \beta)_i^2 - E(b - \beta)_i^2 = \frac{nk_1\sigma^4}{(\beta'X'X\beta)^2}\,[\beta_i^2\{4 + k_1(n+2)\}$$

$$- 2(\beta'X'X\beta)\,(X'X)_{ii}^{-1}], \tag{43}$$

where $(X'X)_{ii}^{-1}$ is the ith diagonal element of $X'X$. It is now clear that for $k_1 > 0$, the right-hand side is negative when

$$0 < k_1 < \frac{2}{n+2}\,(\frac{\beta'X'X\beta}{\beta_i^2}(X'X)_{ii}^{-1} - 2);\ \frac{\beta'X'X\beta}{\beta_i^2}(X'X)_{ii}^{-1} > 2. \tag{44}$$

This result shows that there is no single positive value of k_1, for all elements of β, for which *SR* does better than OLS. Instead, k_1 changes with each element, and in fact should be k_{1i}, $i = 1, ..., p$. Even for a particular element of β, *SR* does better than OLS only in the parameter space which satisfies $\frac{\beta'X'X\beta}{\beta_i^2}(X'X)_{ii}^{-1} > 2$. This is consistent with the well-known result that OLS is admissible for $p \leq 2$.

6.3.1 Sampling Distribution of the Stein-rule Estimator

Let us write the sampling error of the ith component of the *SR* estimator as

$$(\hat{\beta}_{SR} - \beta)_i = (b - \beta)_i - k_1\,\frac{\hat{u}'\hat{u}}{b'X'Xb}\,b_i,$$

where $0 < k_1 \leq \frac{2}{n+2}\,(d-2)$. Alternatively, we can write $k_1 = \frac{a}{n+2}$, $n = T - p$, $0 < a < 2(d-2)$, such that $\lim Tk_1 = a$ as $T \to \infty$. Thus k_1 is, at most, of the order 1/T in magnitude. Also, we assume that $\lim \frac{X'X}{T}$ is a positive definite matrix as $T \to \infty$, i.e., $X'X$ is of order T in magnitude. Recall from Chapter 2 that $\sqrt{T}(b - \beta)_i \sim N(0, \sigma^2(\frac{X'X}{T})_{ii}^{-1})$ as $T \to \infty$. Thus

$$Asy\cdot\sqrt{T}\ (\hat{\beta}_{SR} - \beta)_i = Asy\cdot\sqrt{T}\ (b - \beta)_i \sim N(0, \sigma^2(\frac{X'X}{T})_{ii}^{-1}).$$

This is because $s^2 \to \sigma^2$ and $b_i \to \beta_i$ as $T \to \infty$, and thus $k_1 ns^2 b_i / b'X'Xb$ tends to zero, $(s^2 = \hat{u}'\hat{u}/n)$ in probability, as $T \to \infty$.

We now present the asymptotic expansion, up to the order $1/T$, of the distribution of the Stein-rule (SR) estimator. The result is a direct outcome of a Theorem in Sargan [1975], which gives

$$Pr[\sqrt{T}\ (\hat{\beta}_{SR} - \beta)_i \leq x] = F(\frac{x}{\sigma_i^*}) + f(\frac{x}{\sigma_i^*})(\lambda_{0i} + \frac{x}{\sigma_i^*}\lambda_{1i}), \qquad (45)$$

where $F(\ )$ and $f(\ )$ represent the standard normal distribution function and density function, respectively, $\sigma_i^* = \sigma(\frac{X'X}{T})_{ii}^{-1/2}$ is the asymptotic standard error of $\hat{\beta}_{SR}$, $\lambda_{0i} = nk_1\sigma^2\beta_i/2(\beta'X'X\beta)\sigma_i^*$ and $\lambda_{1i} = [1 - \frac{2\beta_i(q_i'\beta)\sigma^2 T}{\beta'X'X\beta\ \sigma^{*2}}]\frac{nk_1\sigma^2}{2(\beta'X'X\beta)}$; q_i is the ith column of the matrix $\frac{X'X}{T}$. The elements λ_{0i} and λ_{1i} are of order $1/T$ in magnitude. (The notation q_i here should not be confused with similar notation in Section 6.2.3 above.)

An alternative representation of the above distribution function is

$$Pr[\sqrt{T}\ (\hat{\beta}_{SR} - \beta)_i \leq x] = F(\lambda_{0i} + \frac{x}{\sigma^*_i}(1 + \lambda_{1i})) = F(\frac{x}{\sigma^*_i} + \lambda_{0i} + \frac{x}{\sigma^*_i}\lambda_{1i}).$$

This expression can be easily evaluated on a desk calculator by using a standard normal table. Note that the Taylor series expansion of the right-hand side, up to order $1/T$, would give the expression in (45).

If we now normalize $\sqrt{T}\,(\hat{\beta}_{SR} - \beta)_i$ as

$$Z_i = \frac{\sqrt{T}\ (\hat{\beta}_{SR} - \beta)_i}{\sigma^*_i}$$

and write $x/\sigma_i^* = r_i$ then, up to order $1/T$,

$$P[Z_i \leq r_i] = F(r_i) + f(r_i)\ (\lambda_{0i} + r_i\lambda_{1i}).$$

6.3.2 Comparison with the OLS estimator

In the case of the OLS estimator

$$Pr[\sqrt{T}\ (b - \beta)_i \leq x] = F(\frac{x}{\sigma_i^*}).$$

Now we compare the performance of the SR estimator with the OLS estimator under the criterion of smaller absolute distance from the true unknown parameter (see Chapter 2 for this criterion). The result in this section is true for any positive k_1 provided it is of order T^{-1}.

From (45), write

$$Pr[\sqrt{T}\ (\hat{\beta}_{SR} - \beta)_i \leq x] - Pr[\sqrt{T}\ (b - \beta)_i \leq x]$$

$$= f\left(\frac{x}{\sigma_i^*}\right)\left(\lambda_{0i} + \frac{x}{\sigma_i^*}\lambda_{1i}\right).$$

Therefore

$$Pr[|\sqrt{T}\ (\hat{\beta}_{SR} - \beta)_i| \leq x] - Pr[|\sqrt{T}\ (b - \beta)_i| \leq x]$$

$$= Pr[\sqrt{T}\ (\hat{\beta}_{SR} - \beta)_i \leq x] - Pr[\sqrt{T}\ (b - \beta)_i \leq x]$$

$$- Pr[\sqrt{T}\ (\hat{\beta}_{SR} - \beta)_i \leq -x] + Pr[\sqrt{T}\ (b - \beta)_i \leq -x]$$

$$= 2f\left(\frac{x}{\sigma_i^*}\right)\frac{x}{\sigma_i^*}\lambda_{1i}\ ,$$

where λ_{1i} is from equation (45). The right hand side is negative if $\lambda_{1i} < 0$, that is, if k_1 is positive and

$$1 - \frac{2\beta_i(q_i'\beta)\sigma^2 T}{(\beta'X'X\beta)\sigma_i^{*2}} < 0.$$

Thus for some regions of the parameter space, the Stein-rule estimator may not be "closer" than OLS to the true parameter. This result, like the *MtxMSE* criterion, indicates that thc *SR* cstimator will not do better than OLS in the entire parameter space of β.

6.4 SOME EXTENSIONS OF THE STEIN-RULE ESTIMATOR

6.4.1 A Minimum MtxMSE Estimator

Consider a class of linear estimators

$$\hat{\beta} = Cy,$$

where C is the arbitrary $p \times T$ matrix. The matrix mean squared error (*MtxMSE*) of $\hat{\beta}$ can be written as

$$MtxMSE(\hat{\beta}) = E(\hat{\beta} - \beta)(\hat{\beta} - \beta)' \tag{46}$$

$$= E(CX\beta + Cu - \beta)(CX\beta + Cu - \beta)'$$

$$= (CX - I)\beta\beta'(CX - I)' + \sigma^2 CC',$$

where we have used $y = X\beta + u$, and $u \sim N(0, \sigma^2 I)$. The matrix C for which (46) is a minimum is obtained by solving

$$\frac{\partial MtxMSE\ (\hat{\beta})}{\partial C} = 2C(X\beta\beta'X' + \sigma^2 I) - 2\beta\beta'X' = 0 \tag{47}$$

for C. This gives $C = \beta\beta'X'(X\beta\beta'X' + \sigma^2 I)^{-1}$. Substituting this in $\hat{\beta} = Cy$ gives the minimum *MtxMSE* estimator as

$$\hat{\beta}_{TH} = \beta\beta'X'(X\beta\beta'X' + \sigma^2 I)^{-1}y. \tag{48}$$

This estimator was derived by Theil [1971, p. 125], and the subscript *TH* refers to him.

An alternative form of (48) is

$$\begin{aligned}\hat{\beta}_{TH} &= \frac{\beta\beta'X'}{\sigma^2}[I - X\beta(\sigma^2 + \beta'X'X\beta)^{-1}\beta'X']y \\ &= \frac{1}{\sigma^2}[\beta'X'y - \frac{\beta'X'X\beta\beta'X'y}{\sigma^2 + \beta'X'X\beta}]\beta \\ &= \frac{\beta'X'y}{\sigma^2 + \beta'X'X\beta}\beta ,\end{aligned} \tag{49}$$

where $(X\beta\beta'X' + \sigma^2 I)^{-1} = [I - X\beta(\sigma^2 + \beta'X'X\beta)^{-1}\beta'X']/\sigma^2$ by using matrix inversion results in Rao [1973, p. 33]. (If A and D are square non-singular matrices of order m and n respectively, and if B is an $m \times n$ matrix, we have $(A+BDB')^{-1} = A^{-1} - A^{-1}B\ (B'A^{-1}B+D^{-1})^{-1}B'A^{-1}$.

Note that $\hat{\beta}_{TH}$ depends on the unknown β and σ^2, and thus it cannot be used in practice in its present form. In view of this, Farebrother [1976] suggested using the unbiased OLS estimators b and s^2 for β and σ^2, respectively, in (49). This gives

$$\hat{\beta}_{FA} = \frac{b'X'y}{s^2 + b'X'Xb}b , \tag{50}$$

where $s^2 = \hat{u}'\hat{u}/(T-p) = (y - Xb)'(y - Xb)/(T-p)$. Note that $\hat{\beta}_{FA}$ does not have any "minimum" MSE or *MtxMSE* property, because of the substitution of OLS *estimates* of β and σ^2.

6.4.2 An Iterative Estimator

The estimator $\hat{\beta}_{FA}$ was an outcome of substituting, at the first stage, the OLS estimators of β and σ^2 in (49). Let us denote $\hat{\beta}_{FA} = \hat{\beta}_{FA}^{(1)}$, where (1) represents the first stage or iteration. We can now obtain the second iteration of $\hat{\beta}_{FA}$ as

$$\hat{\beta}_{FA}^{(2)} = \frac{\hat{\beta}_{F}^{(1)}X'y}{(s^2)^{(1)} + \hat{\beta}_{FA}^{(1)}X'X\hat{\beta}_{FA}^{(1)}}\hat{\beta}_{FA}^{(1)} , \tag{51}$$

where $(s^2)^{(1)}$ has been obtained by using $\hat{\beta}_{FA}^{(1)}$. In general, $\hat{\beta}_{FA}^{(2)} \neq \hat{\beta}_{FA}^{(1)}$, and it is tempting to iterate until a stationary (or fixed point) value is obtained. Vinod

[1976t] gives an analytic solution to the following iteration:

$$\hat{\beta}_{FA}^{(j+1)} = \frac{\hat{\beta}_F^{(j)\prime}X'y}{(s^2)^{(j)} + \hat{\beta}_F^{(j)\prime}X'X\hat{\beta}_F^{(j)}}\hat{\beta}_{FA}^{(j)}, \tag{52}$$

where j denotes the iteration number, and where

$$(s^2)^{(j)} = (y - X\hat{\beta}_{FA}^{(j)})'(y - X\hat{\beta}_{FA})/(T-p). \tag{53}$$

A stationary solution, i.e., one which will not change from one iteration to the next is found by setting $\hat{\beta}_{FA}^{(j)} = \hat{\beta}_{FA}^{(j+1)} = \hat{\beta}_{VN}$ as $j \to \infty$, VN represents Vinod. It is possible to write the equations (52) and (53) as a mapping f, having β_{VN} as its fixed point, i.e., satisfying $f(\beta_{VN}) = \beta_{VN}$. This leads to the following constraint

$$\hat{\beta}_{VN}'X'y = (T-p)^{-1}(y-X\hat{\beta}_{VN})'(y-X\hat{\beta}_{VN}) + \hat{\beta}_{VN}'X'X\hat{\beta}_{VN}. \tag{54}$$

In general, there are many points $\hat{\beta}_{VN}$ satisfying (54) depending on the starting point b of the iteration. The fixed point solution can be analytically derived if we use minimization of the residual sum of squares as the objective function, and the fixed point relation as a constraint. This avoids the need to specify starting points of the iteration.

Vinod [1976t] proves that the solution to the minimization problem can be written as $\beta_{VN}-b\,\delta$, where b is the OLS solution and δ is a shrinkage factor. The geometrical interpretation of the minimizing solution is that the solution must lie along the ray through the origin and b. Substituting $b\,\delta$ for β_{VN} in (54) we have

$$\delta b'X'y = \frac{1}{T-p}(y'y - 2\delta b'X'y + \delta^2 b'X'Xb) + \delta^2 b'X'Xb, \tag{55}$$

which is a quadratic in δ and has two solutions. We use the following equations

$$b'X'y = y'X(X'X)^{-1}X'y = b'X'Xb = R^2(y'y), \tag{56}$$

where R^2 denotes the usual squared multiple correlation coefficient. Thus the two solutions for δ are

$$\delta = \frac{1}{2(T-p+1)}\left\{(T-p+2) \pm [(T-p+2)^2 - 4(T-p+1)/R^2]^{1/2}\right\} \tag{57}$$

These solutions for δ are real provided the radicand in (57) is nonnegative, i.e.,

$$(T-p+2)^2R^2 \geq 4(T-p+1). \tag{58}$$

This condition is always satisfied asymptotically ($T \to \infty$). Typically, when (58) fails for finite T, R^2 is near zero, and a preliminary F test of the null hypothesis $\beta = 0$ is not rejected. Thus we may perform such a test, and if $\beta = 0$ is accepted then shrinking all the way to zero ($\delta = 0$) is appropriate. After all, $\hat{\beta}_{VN} = 0$ is a solution to (54). On the other hand, (58) may have failed due to numerical reasons, and the hypothesis $\beta = 0$ cannot be accepted.

In this situation, rather than shrinking to zero, making the radicand zero and $\delta = (T-p)/(2T-2p+2)$ makes intuitive sense. This intuition was supported by a simulation experiment in Vinod [1976t] and criticized by Draper and Van Nostrand [1979] on the grounds that a stable solution should have $\delta = 0$. These authors considered the stability properties of Hemmerle's [1975] independently derived analytic solution to Hoerl and Kennard's iterative estimator (see Chapter 7) and incorrectly attributed those properties to the above iteration.

The stability (convergence) of the iteration (52) may be studied by considering whether the difference $\hat{\beta}_{FA}^{(j+1)} - \hat{\beta}_{FA}^{(j)}$ is monotonically decreasing. This involves computing the derivative of the difference and checking whether it is negative, as follows.

Substituting b for $\hat{\beta}_{FA}^{(j)}$ and using (56) we write

$$\hat{\beta}_{FA}^{(j+1)} - \hat{\beta}_{FA}^{(j)} = \frac{\delta(y'y)R^2 b}{(1-\delta)^2(T-p)^{-1}R^2 y'y + \delta^2 R^2 y'y} - \delta b \tag{59}$$

$$= \frac{(\delta^2 - \delta)b}{pos.\ const.},$$

where *pos. const.* represents a positive constant in the denominator, which is not a function of b. Now differentiating this function with respect to b yields the following condition for stability

$$\delta^2 - \delta < 0, \tag{60}$$

which is true as long as $\delta < 1$. This shows that both solutions in (57) are stable. Our reason for choosing the positive solution is that it will be closer to the least squares value $\delta = 1$, and therefore may have a lower bias. However the negative solution is also stable and may be useful in certain situations (e.g., when shrinkage $\delta < 1/2$ is desirable).

Note that in the above iteration Vinod used the usual "unbiased" estimator of σ^2 having the denominator $T-p$. It is well-known, Rao [1973, p. 316], that the denominator $(T-p+2)$ yields an estimator with a lower MSE than the unbiased estimator. Hence using $(T-p+2)$ in (53) we find that the two solutions analogous to (57) and (58) are respectively:

$$\delta = 1/2 + \frac{1}{2(T-p+3)}\{1 \pm [(T-p+1)^2 - 4(T-p+3)R^{-2}]^{1/2}\}. \tag{61}$$

Now, the condition for a non-negative radicand similar to (58) is given by

$$(T-p+4)^2 R^2 \geq 4(T-p+3). \tag{62}$$

6.4.3 Double k-Class Estimator

Let us write the estimator $\hat{\beta}_{TH}$ in (49) as

$$\hat{\beta}_{TH} = \frac{\beta'X'y}{\sigma^2 + \beta'X'X\beta}\beta = \frac{(y-u)'y}{\sigma^2 + (y-u)'(y-u)}\beta , \tag{63}$$

where we use $X\beta = y - u$. Suppose that instead of replacing $X\beta$ by $y - u$ we replace it by $y - ku$, where k is any arbitrary number. This amounts to saying that we have the regression model as $y = X\beta + ku$. Then a natural extension of $\hat{\beta}_{TH}$ is

$$\hat{\beta}_{TH(k)} = \frac{(y-ku)'y}{\sigma^2 + (y-ku)'(y-ku)}\beta. \tag{64}$$

Now substituting $\hat{u}$, b and s_k^2 for u, β and σ^2 we can write an operational version:

$$\tilde{\beta}_{ko} = \frac{(y-k\hat{u})'y}{s_k^2 + (y-k\hat{u})'(y-k\hat{u})}b , \tag{65}$$

where $k\hat{u} = y - Xb = y - X(X'X)^{-1}X'y = My$; $M = I - X(X'X)^{-1}X'$, and $s_k^2 = k^2\hat{u}'\hat{u}/n$; $n = T - p$. Noting that $k\hat{u}'y = y'My$ and $k^2\hat{u}'\hat{u} = y'My$ we can write an alternative form of $\tilde{\beta}_{ko}$ as

$$\begin{aligned}\tilde{\beta}_{ko} &= \frac{y'y - k^2\hat{u}'\hat{u}}{y'y - k^2(1-1/n)\hat{u}'\hat{u}}b , \\ &= [1 - \frac{k^2\hat{u}'\hat{u}/n}{y'y - k^2(1-1/n)\hat{u}'\hat{u}}]b. \end{aligned} \tag{66}$$

For $k = 1$, the reader can verify that $\tilde{\beta}_{ko}$ is equal to Farebrother's estimator $\hat{\beta}_{FA}$. For $k = 0$, $\tilde{\beta}_0 = b$.

A further generalization of $\tilde{\beta}_{ko}$ can be written as

$$\tilde{\beta}_{k_1,k_2} = [1 - \frac{k_1\hat{u}'\hat{u}}{y'y - k_2\hat{u}'\hat{u}}]b , \tag{67}$$

where $k_1 > 0$ and k_2 are any arbitrary stochastic or nonstochastic scalars. This estimator called "double k-class" estimator has been proposed by Ullah and Ullah [1978]. It is interesting to note that for $k_2 = 1$, $y'y - \hat{u}'\hat{u} = b'X'Xb$, and in this case $\tilde{\beta}_{k_1,k_2}$ is the James and Stein estimator defined in equation (16) above.

It has been shown by Ullah and Ullah that the *MtxMSE* of $\tilde{\beta}_{k_1,k_2}$ exists for any nonstochastic k_1 and $0 < k_2 < 1$, and also for $k_2 = 1$ provided $p \geq 3$. However, if one allows for negative values of k_2, then it follows from their proof that the *Mtx*MSE will exist for $-\infty < k_2 < 1$, and for $k_2=1$ provided $p \geq 3$. It is proved in the Explanatory Note 6.3 that the *exact* MSE of $\tilde{\beta}_{k_1,k_2}$ dominates OLS in the sense that:

$$E(\tilde{\beta}_{k_1,k_2} - \beta)'(\tilde{\beta}_{k_1,k_2} - \beta) - E(b-\beta)'(b-\beta) < 0$$

when

$$0 < k_1 < \frac{2}{n+2}(d-2) \text{ and } -\infty < k_2 \leq 1, \tag{68}$$

where $d=\lambda_p \sum_{i=1}^{p} \lambda_i^{-1} > 2$ is as given in (42). This result is interesting in three respects. First, it shows that the Stein-rule range in (42) is also valid for any $k_2 < 1$. Secondly, it has been established that the condition for dominance of the Stein-rule estimator (k_2=1) in (42), based on the approximate MSE expression, is identical with the exact condition in (68). This shows that Kadane's small-σ asymptotics is a very simple and useful tool to derive the conditions for dominance of estimators of the type in (67). In fact this observation is also true for a more general class of estimators considered in chapters 8, 9 and 13. Thirdly, we note that the MSE of OLS under the "scaled" loss function $(b-\beta)'(b-\beta)/\sigma^2$ is $Tr(X'X)^{-1}$, which is a constant. Thus (68) is also a condition for the minimaxity of the double k-class estimator $\tilde{\beta}_{k_1,k_2}$ under the "scaled" loss $(\tilde{\beta}_{k_1,k_2}-\beta)'(\tilde{\beta}_{k_1,k_2}-\beta)/\sigma^2$. This implies that $\tilde{\beta}_{k_1,k_2}$ is a family of minimax estimators of the parameters β.

Vinod [1980] studied the choice of $k_2 \leq 1$ which will reduce the MSE of $\tilde{\beta}_{k_1,k_2}$ for the range of k_1 in (68). His simulation finding suggested that $k_2 = 1-k_1$ and large negative k_2 do better with respect to MSE.

6.5 FINAL REMARKS

Stein-rule (SR) estimator in the regression context is useful especially for situations where the signs and relative magnitudes of the OLS regression coefficients are already appropriate on *a priori* grounds, and the number of regressors p is strictly larger than 2. It should be pointed out that SR estimators shrink all regression coefficients toward zero, and are not guaranteed to reduce the MSE of OLS. Such a guarantee is available only for the problem of finding *the mean of a multivariate normal variable.*

Lindley's form of SR estimator shrinks the regression coefficients toward their own average, and is useful for $p > 3$. Its attractiveness over SR estimator arises from its equivariance property discussed in Section 6.1.2 above. We have discussed some extensions of SR estimator including (i) an operational version of the minimum weighted MSE (MWMSE) estimator given in (33), (ii) M. E. Bock's estimator in (37), (iii) Minimum MSE matrix estimator of Theil in (48), (iv) Farebrother's estimators in (50), and Vinod's analytic solution (57) to Farebrother's iterative procedure, and (v) the Double k-class estimator by Ullah and Ullah in (67). These extensions have attractive theoretical properties and dominate OLS in certain regions of the parameter space. Simulation studies in Vinod [1976t, 1980] and his unpublished work indicate that these extensions can be useful in reducing the MSE of OLS in practical situations.

Formulation of regression problem in terms of the mean of a multivariate normal variable is given in Section 1.3.2 of Chapter 1.

$$\text{MSE}(a) = E(a - \alpha)'(a - \alpha),$$

where $a = Hy'$ in the OLS estimator of $\alpha = \wedge^{1/2}G'\beta$. We can write $(a-\alpha) = H'X - \alpha = H'H\wedge^{1/2}G'b - \wedge^{1/2}G'\beta = \wedge^{1/2}G'(b-\beta)$. Hence the MSE is related to the "predictive MSE" (see equation (12) of Chapter 2) as follows:

$$\begin{aligned}\text{MSE}(a) &= (b-\beta)'G\wedge^{1/2}\wedge^{1/2}G'(b-\beta)\\ &= (b-\beta)'(X'X)(b-\beta)\\ &= PMSE(b),\end{aligned}$$

Thus, we note that accepting the *PMSE* criterion for β is equivalent to the MSE criterion for α. It is well known that the Stein-rule (SR) estimator is guaranteed to reduce the MSE(a). Hence SR estimator is also guaranteed to reduce PMSE(b), and therefore it may be recommended for certain forecasting applications. We have noted in the previous section that the generalization of the SR estimator by introducing k_1 and k_2 yields a family of minimax estimators under "scaled" loss after appropriate choices of k_1 and k_2.

The sampling properties of a more general class of Stein-type estimators are discussed in Chapter 13 under normal as well as non-normal distribution of errors, and for a more general quadratic loss function. Chapter 13 also discusses the stability of Stein-rule estimators.

Explanatory Notes

Explanatory Note 6.1

Integration by Parts and MSE of a Biased Estimator

Let $y \sim N(\eta, \sigma^2)$ be a normally distributed real random variable with mean η and variance σ^2. Now, let $x = y/\sigma$, $\xi = \eta/\sigma$ and temporarily assume that σ is known. Thus we have

$$x \sim N(\xi, 1).$$

Consider a function $f(x)$ such that $E[|f'(x)|] < \infty$, where f' denotes the derivative with respect to x. Now, x is the maximum likelihood (ML) estimator of ξ, and by definition

$$E[f'(x)] = (2\pi)^{-1/2} \int_{-\infty}^{\infty} f'(x) \exp[-\frac{1}{2}(x - \xi)^2]\, dx. \tag{68}$$

Integration by parts yields the right hand side to be a sum of two terms F_1 and

F_2 defined by

$$F_1 = (2\pi)^{-1/2}[f(x)\exp[-\frac{1}{2}(x-\xi)]^2]_{-\infty}^{+\infty} \quad (69)$$

$$F_2 = (2\pi)^{-1/2}\int_{-\infty}^{\infty} f(x)(x-\xi)\exp[-\frac{1}{2}(x-\xi)]^2\,dx \quad (70)$$
$$= E[(x-\xi)f(x)].$$

For a finite well-behaved $f(x)$ it is clear that $F_1 = 0$, because the density vanishes at the end points. Hence, we have the following identity

$$E[f'(x)] = E[(x-\xi)f(x)]. \quad (71)$$

This identity can be exploited to obtain the MSE of an almost arbitrary biased estimator $x + f(x)$ of ξ

$$MSE[x+f(x)] = E[x+f(x)-\xi]^2 \quad (72)$$
$$= E[x^2+f^2+\xi^2+2xf-2x\xi-2f\xi]$$
$$= E[(x-\xi)^2+f^2+2(x-\xi)f],$$

where we have written f for $f(x)$, and f^2 for $[f(x)]^2$. Note that $E(x-\xi)^2 = 1$ is the variance of x. Upon substitution of (71) into (72) we have

$$MSE[x+f(x)] = 1 + E[f(x)]^2 + 2E[f'(x)], \quad (73)$$

which involves the expectation of observable x, and ξ is absent. Since both sides of (73) have the same expectation the unbiased estimate of

$$\text{MSE}[x+f(x)]$$

is given by

$$UMSE[x+f(x)] = 1 + [f(x)]^2 + 2f'(x). \quad (74)$$

Stein has proved a general result for the case when x is a $p \times l$ vector.

Now let the Stein-rule estimator for the ith dimension of a p-dimensional problem similar to equation (8) be

$$x_{iS} = (1 - \frac{p-2}{\sum_{i=1}^{p} x_i^2})\,x_i \quad (75)$$

$$= x_i + \frac{(2-p)x_i}{\sum_{i=1}^{p} x_i^2},$$

which defines $f(x_i) = (2-p)x_i/\sum_{i=1}^{p} x_i^2$. Differentiating this with respect to x_i

we have

$$f'(x_i) = \frac{\sum_{i=1}^{p} x_i^2(2-p) - (2-p)x_i(2x_i)}{(\sum_{i=1}^{p} x_i^2)^2}$$

Thus from (72) we have

$$MSE[x_i + f(x_i)] = 1 + E[\frac{2-p}{\sum_{i=1}^{p} x_i^2} x_i]^2 + 2E[\frac{2-p}{\sum_{i=1}^{p} x_i^2} + \frac{2(p-2)x_i^2}{(\sum_{i=1}^{p} x_i^2)^2}].$$

Since x_i is independent of x_j $(i \neq j = 1, ..., p)$ we can simply add $MSE(x_i)$ to obtain $MSE(x)$, where x is a p-vector of x_i values;

$$MSE[x + f(x)] = p + E\frac{(p-2)^2 \sum_{i=1}^{p} x_i^2}{(\sum_{i=1}^{p} x_i^2)^2} + E\left[\frac{2(2-p)p}{\sum_{i=1}^{p} x_i^2} + \frac{4(p-2) \sum_{i=1}^{p} x_i^2}{(\sum_{i=1}^{p} x_i^2)^2}\right].$$

Denoting $W = \sum_{i=1}^{p} x_i^2$, we have

$$\begin{aligned} MSE[x + f(x)] &= p + E\frac{(p-2)^2}{W} + 2(2-p)p\ E(\frac{1}{W}) + 4(p-2)E(\frac{1}{W}) \\ &= p + [(p-2)^2 + 2(2-p)p + 4(p-2)]E(W^{-1}) \\ &= p + [p^2 - 4p + 4 + 4p - 2p^2 + 4p - 8]E(W^{-1}) \\ &= p + [-p^2 + 4p - 4]E(W^{-1}) \\ &= p - (p-2)^2 E(W^{-1}). \end{aligned}$$

James and Stein [1961] write $E(W^{-1}) = E(\frac{1}{p-2+n})$ where n is a Poisson variable with mean $\lambda = \sum_{i=1}^{p} \theta_i^2/2$ (the non-centrality). The density of n is $(e^{-\lambda}\lambda^n/n!)$. Hence

$$E(W^{-1}) = \sum_{n=0}^{\infty} \frac{e^{-\lambda}\lambda^n}{n!}(p-2+n)^{-1} \geq 0. \quad (76)$$

This expression is equivalent to an alternative expression based on confluent hypergeometric expression in Ullah [1974]. It is now clear that the $MSE(x_S) - \text{MSE}(x) = -(p-2)^2 E(W^{-1})$; which is clearly negative when $p > 2$.

Unknown σ^2

Now we relax the assumption that σ^2 is known in $x \sim N(\xi, \sigma^2)$. We assume that in addition to x we can also observe another real random variable (residual sum of squares) S distributed as a "Central Chisquare" variable with n^* degrees of freedom (d.fr.) (e.g., $n^* = T - p$). Note that S is distributed independently of x, and S/n^* is the usual unbiased estimator of σ^2. We consider the biased estimate $S(n^* + 2)^{-1} = \hat{\sigma}^2$ whose MSE is known to be less than the MSE of S/n^* as an estimate of σ^2, as spelled out in Rao [1973, p. 316]. From the properties of a χ^2 variable we know that $E(S/\sigma^2) = n^*$ and $E(S^2/\sigma^4) = n^*(n^* + 2)$. Hence,

$$E(\hat{\sigma}^2) = \sigma^2 n^*(n^* + 2)^{-1} \text{ and } E(\hat{\sigma}^4) = \sigma^4 n^*(n^* + 2)^{-1} \quad (77)$$

Now, consider the $MSE[x + \hat{\sigma}^2 f(x/\hat{\sigma})]$ which equals

$$E(x + \hat{\sigma}^2 f - \xi)^2 = E[(x - \xi)^2 + \hat{\sigma}^4 f^2 + 2\hat{\sigma}^2(x - \xi)f]$$

$$= \sigma^2 + n^*(n^* + 2)^{-1}[\sigma^4 f^2 + 2\sigma^2(\partial f/\partial x)],$$

where we use (71).

Thus, the presence of unknown σ^2 implies that we use $n^*(n^* + 2)^{-1}$ instead of unity, and $\hat{\sigma}^2 f(x/\hat{\sigma})$ for $f(x)$ in (72). In other words, we lose only the proportion $2/(n^* + 2)$ of the reduction in MSE attained under the assumption of known σ^2. Thus, there is no great harm in assuming $\hat{\sigma}^2 = \sigma^2$ in this context unless n^* is very small.

Explanatory Note 6.2

Kadane's Small-σ Expansion

Here we shall analyze Kadane's [1970, 1971] expansion to the Stein-rule estimator. Let us write the model as $y = X\beta + \sigma u$ where $\sigma u \sim N(0, \sigma^2 I)$ or $u \sim N(0, I)$. In the text we had considered the distribution of u as $N(0, \sigma^2 I)$. However, now we explicitly take into account the effect of variation in u, and that is why it will be advantageous to write u as σu. If σ is small then the model is good, whereas if it is large then the model is not well explained by X. Kadane's algebraic results involve a reinterpretation of Nagar's [1959] expansion for large T.

Kadane's principle of small-σ expansion for the moments of an estimator involves (i) assuming σ to approach zero; (ii) expanding the sampling error of the estimator under consideration in higher orders of σ; and (iii) taking term by term expectations of each term in the expansion of the sampling error and its function.

Let us consider the Stein-rule estimator and write its sampling error after replacing $\hat{u}$ and b by expressions involving u and β:

$$\hat{\beta}_{SR} - \beta = b - \beta - \frac{k_1 \hat{u}'\hat{u}}{b'X'Xb} b \tag{78}$$

$$= \sigma (X'X)^{-1} X'u - k_1 \sigma^2 \frac{u'Mu}{g} (\beta + \sigma (X'X)^{-1} X'u),$$

where $M = I - X(X'X)^{-1}X'$, $g = \beta'X'X\beta + 2\sigma\beta'X'u + \sigma^2 u'M^*u$ and $M^* = X(X'X)^{-1}X'$.

Now, for sufficiently small σ we write g^{-1} and g^{-2} in terms of the following Binomial expansions:

$$(1 + z)^{-1} = 1 - z + z^2 \cdots, \text{ and } (1 + z)^{-2} = 1 - 2z + 3z^2 \cdots .$$

Thus we have

$$\frac{1}{g} = \frac{1}{\beta'X'X\beta} \left[1 + \frac{2\sigma\beta'X'u + \sigma^2 u'M^*u}{\beta'X'X\beta} \right]^{-1} \tag{79}$$

$$= \frac{1}{\beta'X'X\beta} [1 - (\frac{2\sigma\beta'X'u + \sigma^2 u'M^*u}{\beta'X'X\beta}) + ...]$$

and similarly

$$\frac{1}{g^2} = \frac{1}{(\beta'X'X\beta)^2} [1 - 2(\frac{2\sigma\beta'X'u + \sigma^2 u'M^*u}{\beta'X'X\beta}) + ...]. \tag{80}$$

Now using (79) in the expression for $\hat{\beta}_{SR} - \beta$, and collecting terms up to order σ^2 we obtain

$$\hat{\beta}_{SR} - \beta = \sigma (X'X)^{-1} X'u - k_1 \sigma^2 \frac{u'Mu}{\beta'X'X\beta} \beta.$$

Taking expectations on both sides and noting that $Eu'Mu = T - p = n$ we get the bias as stated in (39).

Likewise, using (79) and (80) in the expression for $(\hat{\beta}_{SR} - \beta)(\hat{\beta}_{SR} - \beta)'$ and collecting only those terms which contribute up to order σ^4, we obtain

$$(\hat{\beta}_{SR} - \beta)(\hat{\beta}_{SR} - \beta)' = \sigma^2 Auu'A' + k_1^2 \sigma^4 (u'Mu)^2 \frac{\beta\beta'}{(\beta'X'X\beta)^2}$$

$$- 2k_1 \sigma^4 (u'Mu) \frac{Auu'A'}{\beta'X'X\beta} + 2\sigma^4 \frac{k_1(u'Mu)}{(\beta'X'X\beta)^2} (Auu'X\beta\beta'$$

$$+ \beta\beta'uu'A'),$$

where $A = (X'X)^{-1}X'$ is such that $MA' = 0$. Thus $u'Mu$ and Au are independently distributed. Finally, taking expectation on both sides,

$$E(\hat{\beta}_{SR} - \beta)(\hat{\beta}_{SR} - \beta)' = \sigma^2(X'X)^{-1} + k_1^2\,\sigma^4\, n(n+2)\,\frac{\beta\beta'}{(\beta'X'X\beta)^2}$$

$$- 2k_1\sigma^4\, n\,\frac{(X'X)^{-1}}{\beta'X'X\beta} + 2\sigma^4\,\frac{k_1 n}{(\beta'X'X\beta)^2}\,(2\beta\beta'),$$

where we have used $Euu' = I$, $E(u'Mu) = n$ and $E(u'Mu)^2 = n(n+2)$. Note that $u'Mu$ is a central chi-square with $n = T - p$ degrees of freedom. This expression for the MSE matrix of the Stein-rule estimator is also given in equation (40).

Explanatory Note 6.3

Exact Condition for dominance of Double k-class Estimator

This explanatory note refers to the double k-class estimator of equation (67). Let us write the linear regression model as

$$y = X\beta + u = H\wedge^{1/2}G'\beta + u = Z\alpha + u,$$

where X is a matrix of p regressors which is replaced by its "singular value decomposition" of equation (14) of Chapter 1, and where $Z = H\wedge^{1/2}$ and $\alpha = G'\beta$. Notice that $Z'Z = \wedge$ and $X'X = G\wedge G'$, because $H'H = I$ and $G'G = I$. Thus we can write

$$b = (X'X)^{-1}X'y = Ga;\ a = \wedge^{-1/2}H'y = \wedge^{-1}Z'y\ , \qquad (81)$$

and

$$y'y = \hat{u}'\hat{u} + b'X'Xb = s + a'\wedge a\ ;\ s = \hat{u}'\hat{u}. \qquad (82)$$

Further, from (67) we have

$$\tilde{\beta}_{k_1,k_2} = \tilde{\beta} = \left[1 - \frac{k_1 s}{a'\wedge a + (1-k_2)s}\right]Ga$$

$$= G\tilde{\alpha}, \qquad (83)$$

where, denoting $W_2 = a'\wedge a + (1-k_2)s$,

$$\tilde{\alpha} = \left[1 - \frac{k_1 s}{W_2}\right]a. \qquad (84)$$

The subscript 2 in W reminds us that W depends on k_2.

Observe that $E(\tilde{\beta}-\beta)'(\tilde{\beta}-\beta) = E(\tilde{\alpha}-\alpha)'G'G(\tilde{\alpha}-\alpha) = E(\tilde{\alpha}-\alpha)'(\tilde{\alpha}-\alpha)$. Now let $\tilde{\alpha}_i$ be the i-th component of $\tilde{\alpha}$ and write it as

$$\tilde{\alpha}_i = \left[1 - \frac{k_1 s}{W_2}\right] a_i \ .$$

We can then write its sampling error as

$$\tilde{\alpha}_i - \alpha_i = (a_i - \alpha_i) - k_1 \frac{s a_i}{W_2} \ . \tag{85}$$

Hence we can write

$$E(\tilde{\alpha}_i - \alpha_i)^2 = E(a_i - \alpha_i)^2 + k_1^2 E(\frac{s^2 a_i^2}{W_2^2}) - 2k_1 E(\frac{s}{W_2} a_i (a_i - \alpha_i))$$

$$= \frac{\sigma^2}{\lambda_i} + k_1^2 E(\frac{s^2 a_i^2}{W_2^2}) - 2k_1 \frac{\sigma^2}{\lambda_i} E[\frac{s}{W_2} - 2\lambda_i a_i^2 \frac{s}{W_2^2}]$$

$$= \frac{\sigma^2}{\lambda_i} + k_1 E\left[\frac{s}{W_2}\left\{\frac{(k_1 s + 4\sigma^2) a_i^2}{W_2} - 2\frac{\sigma^2}{\lambda_i}\right\}\right];$$

where the third term in the second equality is obtained through integration by parts with respect to a_i. Further summing over $i = 1, \ldots, p$, we write

$$E(\tilde{\alpha} - \alpha)'(\tilde{\alpha} - \alpha) = \sum_{i=1}^{p} E(\tilde{\alpha}_i - \alpha_i)^2$$

$$= \sigma^2 \sum_{i=1}^{p} \lambda_i^{-1} + k_1 E\left[\frac{s}{W_2}\left\{\frac{(k_1 s + 4\sigma^2) \sum_{i=1}^{p} a_i^2}{W_2} - 2\sigma^2 \sum_{i=1}^{p} \lambda_i^{-1}\right\}\right] \tag{86}$$

In obtaining the above result we use the result that $V(a) = \sigma^2 \Lambda^{-1}$.

Now note that, for $k_2 \leq 1$, we have

$$\frac{\sum_{i=1}^{p} a_i^2}{W_2} = \frac{a'a}{a'\Lambda a + (1 - k_2)s} \leq \frac{a'a}{a'\Lambda a} \leq \frac{1}{\lambda_p}, \tag{87}$$

where λ_p is the minimum eigenvalue of $X'X$. Therefore, noting that a and s are independently distributed, we have

$$E(\tilde{\alpha} - \alpha)'(\tilde{\alpha} - \alpha) - E(a - \alpha)'(a - \alpha) \leq \frac{k_1}{\lambda_p} E\left[\frac{1}{W_2}\left\{(k_1 s + 4\sigma^2)s - 2\sigma s^2 \lambda_p \sum_{i=1}^{p} \lambda_i^{-1}\right\}\right]$$

$$\leq \frac{k_1}{\lambda_p} E(\frac{1}{W_2}) E\left[k_1 s^2 + 4\sigma^2 s - 2\sigma^2 ds\right],$$

where $d = \lambda_p \sum_{i=1}^{p} \lambda_i^{-1}$ is as given in (42). Thus we have

$$E(\tilde{\alpha}-\alpha)'(\tilde{\alpha}-\alpha)-E(a-\alpha)'(a-\alpha) \leq 0, \tag{88}$$

when either we have

$$k_1\left[k_1 n(n+2)\sigma^4+4\sigma^4 n-2\sigma^4 nd\right] \leq 0,$$

or that

$$k_1\left[k_1-\frac{2(d-2)}{n+2}\right] \leq 0.$$

The latter inequality can be written as the following condition used in equation (68) of the text:

$$0 < k_1 \leq \frac{2}{n+2}(d-2)\,. \tag{89}$$

In deriving the above result we use $Es = E\hat{u}'\hat{u} = n\sigma^2$, and $E\, s^2 = n(n+2)\,\sigma^4$. The conditions $0 < k_1 \leq \frac{2}{n+2}(d-2)$ and $k_2 \leq 1$ are the sufficient conditions for the dominance of $\tilde{\beta}_{k_1,k_2}$ over the OLS estimator.

Now we obtain the necessary conditions for dominance. Following Alam and Hawkes [1978, p.171], let $\alpha_i \to \infty (i=1,\ldots,p)$. Then substituting α_i^2 for a_i^2 and $\alpha'\Lambda\alpha$ for $a'\Lambda a$ on the right hand side of $E(\tilde{\alpha}_i-\alpha_i)^2$, we obtain

$$E(\tilde{\alpha}_i-\alpha_i)^2 - \frac{\sigma^2}{\lambda_i} = k_1\, E\left[\frac{s(k_1 s+4\sigma^2)\alpha_i^2}{(\alpha'\Lambda\alpha)^2} - \frac{2\sigma^2 s}{\alpha'\Lambda\alpha}\cdot\frac{1}{\lambda_i}\right] \tag{90}$$

$$= \frac{k_1}{\alpha'\Lambda\alpha}\left[(k_1\, n(n+2)\sigma^4+4\sigma^2 n)\frac{\alpha_i^2}{\alpha'\Lambda\alpha} - 2\sigma^4 n\frac{1}{\lambda_i}\right]$$

$$= \frac{n\sigma^4 k_1}{\alpha'\Lambda\alpha}\left[(k_1(n+2)+4)\frac{\alpha_i^2}{\alpha'\Lambda\alpha} - \frac{2}{\lambda_i}\right].$$

Therefore, summing over $i=1,\ldots p$, on both sides of (90) we write

$$E(\tilde{\alpha}-\alpha)'(\tilde{\alpha}-\alpha)-E(a-\alpha)'(a-\alpha) = \frac{n\sigma^4 k_1}{\alpha'\Lambda\alpha}\left[(k_1(n+2)+4)\frac{\alpha'\alpha}{\alpha'\Lambda\alpha} - 2\sum_{i=1}^{p}\lambda_i^{-1}\right] \tag{91}$$

$$\geq \frac{n\sigma^4 k_1}{\alpha'\Lambda\alpha}\left[(k_1(n+2)+4)\,\frac{1}{\lambda_1} - 2\sum_{i=1}^{p}\lambda_i^{-1}\right],$$

where λ_1 is the maximum eigenvalue. Now, if $E(\tilde{\alpha}-\alpha)'(\tilde{\alpha}-\alpha) - E(a-\alpha)'(a-\alpha)$ is negative for all α, then

$$0 < k_1 \leq \frac{2(d_{\max}-2)}{n+2} \quad ; d_{\max} = \lambda_1 \sum_{i=1}^{p} \lambda_i^{-1}\,. \tag{92}$$

Inequality (92) provides a necessary condition for dominance of $\tilde{\alpha}$ over a.

For the special case $k_2=1$, it is clear from above that James and Stein's [1961] Stein-rule estimator in (38) also dominates OLS for the necessary constraint on k_1 in (92) below and for the sufficient constraint given in (89). If $X'X = I$, the ranges of k_1 in (89) and (92) reduce to

$$o < k_1 \leq \frac{2(p-2)}{n+2} \;;\; p > 2 \tag{93}$$

which is then both necessary and sufficient for $\tilde{\alpha}$ to be better than a provided $-\infty < k_2 \leq 1$.

EXERCISES

6.1 Consider the problem of estimating the conditional mean of the forecast, $Ey_o = X_o\beta = \mu_o$, in the model $y_o = X_o\beta+\mu_o$, where y_o and u_o are $\tau\times 1$ vectors of τ future periods, and X_o is $\tau\times p$ matrix of future regressor values. Let $\hat{\mu}_o = X_o b$ be the ordinary least squares estimator of the conditional mean forecast. Then show that in the class of estimators $\tilde{\mu}_o = g\hat{\mu}_o$;$g$ is an arbitrary constant, the optimal estimator for the conditional mean, under the predictive MSE criterion $(\tilde{\mu}_o - \mu_o)'(\tilde{\mu}_o - \mu_o)$, is

$$\tilde{\mu}_o^* = \left[\beta'X'_oX_o\beta \;/\; (\beta'X'_oX_o\beta + \sigma^2 h)\right]X_o b \;,$$

where $h=Tr[(X'X)^{-1}X'_oX_o]$. Suggest an operational version of this estimator $\tilde{\mu}_o^*$ [Hint: substitute the least square estimators of β and σ^2].

6.2 Consider the linear model with T observations and p regressors as, $y = X\beta+u$, and the problem of estimating the parameter β by

(i) Stein-Rule: $\hat{\beta}_1 = [1-\dfrac{c\hat{u}'\hat{u}}{b'X'Xb}]b \quad ; \hat{u} = y-Xb$

(ii) Lindley's type: $\hat{\beta}_2 = \hat{\mu} + [1-\dfrac{c\hat{u}'\hat{u}}{(b-\hat{\mu})'X'X(b-\hat{\mu})}](b-\hat{\mu})$ where $\hat{\mu} = \iota'X'Xb \;/\; \iota'X'X\iota$ and ι is a column of ones.

Show that the estimator $\hat{\beta}_2$ can also be written as

$$\hat{\beta}_2 = (I-J)b + \left[1-\frac{c\hat{u}'\hat{u}}{b'Ab}\right]b$$

where $J = I-\iota(\iota'X'X\iota)^{-1}\iota'X'X = J^2$ and $A = J'X'XJ = X'XJ$. Further obtain the WMSE from $(\hat{\beta}-\beta)'X'X(\hat{\beta}-\beta)$, up to order σ^4, for $\hat{\beta}_1$ and $\hat{\beta}_2$ and show that $0 < c < \dfrac{2}{n+2}(p-2)$, $p>2$ and $0 < c < \dfrac{2}{n+2}(p-3)$,$p>3$ are the dominance conditions for the estimators $\hat{\beta}_1$ and $\hat{\beta}_2$, respectively; $n = T-p$.

6.3 Consider the Porcelain Cement data, (Table 6.3), where y the heat evolved in drying is explained by four chemical variables including tricalcium aluminate, tricalcium silicate, etc. x_1 to x_4. These data are from Hald [1952, p.647].

TABLE 6.3
Cement Data

i	x_1	x_2	x_3	x_4	y
1	7.0	26.0	6.0	60.0	78.5
2	1.0	29.0	15.0	52.0	74.3
3	11.0	56.0	8.0	20.0	104.3
4	11.0	31.0	8.0	47.0	87.6
5	7.0	52.0	6.0	33.0	95.9
6	11.0	55.0	9.0	22.0	109.2
7	3.0	71.0	17.0	6.0	102.7
8	1.0	31.0	22.0	44.0	72.5
9	2.0	54.0	18.0	22.0	93.1
10	21.0	47.0	4.0	26.0	115.9
11	1.0	40.0	23.0	34.0	83.8
12	11.0	66.0	9.0	12.0	113.3
13	10.0	68.0	8.0	12.0	109.4

Use these data to compute the estimators of the previous Exercise and their approximate standard errors. Choose an arbitrary set of β and σ^2 values and simulate the performance of $\hat{\beta}_1$ and $\hat{\beta}_2$ from the previous exercise.

6.4 Consider the model $y = X\beta + u$, with no intercept. Show that the double k-class estimator of β, $\hat{\beta} = \left[1 - \frac{k_1 s}{W}\right] b$, $s = \hat{u}'\hat{u}$ and $W = y'y - k_2\hat{u}'\hat{u} = b'X'Xb + (1-k_2)s$, can be written as

$$\hat{\beta} = \left[1 - \frac{k_1(1-R^2)}{1-k_2(1-R^2)}\right] b,$$

where R^2 is the squared multiple correlation coefficient. Also show that

$$\frac{s}{W} = \sum_{j=o}^{\infty} (k_2)^j \left(\frac{x_1^2}{x_1^2+x_2^2}\right)^{j+1}, \quad when \ -1 < k_2 \leq 1$$

$$\frac{s}{W} = \frac{1}{1-k_2} \sum_{j=o}^{\infty} \left(1 - \frac{1}{1-k_2}\right)^j \frac{x_1^2 (x_2^2)^j}{(x_1^2+x_2^2)^{j+1}}, \quad when \ -\infty < k_2 < -1$$

where $s/\sigma^2 = x_1^2$ and $b'X'Xb/\sigma^2 = x_2^2$. Using this or by other methods obtain the expectation of $\hat{\beta}$. (References: See Dwivedi, Srivastava and Hall [1980] and Ullah and Ullah [1978]).

7

Ridge Regression

7.1 INTRODUCTION AND HISTORICAL COMMENTS

Ridge regression (RR) due to Hoerl and Kennard [1970a] amounts to adding a biasing constant $k(0 \leq k \leq \infty)$ to the diagonal of the correlation matrix among regressors before inverting it for least squares estimation. In general, one considers $X'X$ matrix instead of the correlation matrix. This is the most common ridge estimator, called the "ordinary" ridge estimator (ORE), which was originally designed to tackle the the multicollinearity problem discussed in Chapter 5. Since the constant k is arbitrary, we have the following "family" or class of estimators of β in the model $y = X\beta + u$, rather than a unique estimator.

Hoerl and Kennard's ORE of the $p \times 1$ vector of the parameters β is given by

$$b_k = (X'X + kI)^{-1}X'y \ , \tag{1}$$

where $k > 0$. When $k = 0$ we have the OLS estimator $b_0 = b$, which is known to be maximum likelihood, best linear unbiased estimator (BLUE). A mixed regression interpretation of this estimator was indicated earlier in equation (52) of Chapter 3.

Among the key properties of b_k, note that it is biased:

$$\text{Bias}\ (b_k) = Eb_k - \beta = -k(X'X + kI)^{-1}\beta, \tag{2}$$

which depends on the unknown β. Many of the theoretical problems with ORE arise from this dependence on β. Also, note that as $k \to \infty$, $b_k \to 0$, the null vector.

The idea that adding a small constant to the diagonal of a matrix will improve the "conditioning" of a matrix has been long recognized by numerical analysts, because this would dramatically decrease its "condition number" (see equation 4 in Chapter 5).

Riley's [1955] method for avoiding numerical difficulties associated with inverting a square matrix Q, which is $X'X$ in the least squares context, is similar to ridge regression. Let z denote the vector $X'y$. Instead of the usual Q^{-1} Riley proposes the following estimator whose first term is the ridge estimator b_k.

$$\hat{b} = (Q + kI)^{-1}z + k(Q + kI)^{-2}z \tag{3}$$

$$= b_k + k(Q + kI)^{-2}z.$$

In solving nonlinear least squares problems, a Taylor series approximation is often used to linearize the problem, so that OLS can be used to estimate the linearized regression equation. When the underlying matrix is ill-conditioned, the initial solution is not improved because of "overshooting". Levenberg [1944] suggested a method of "damped least squares" to damp the absolute values of the increments when Taylor approximations are successively improved. This involved using arbitrary positive "weighting factors" which are added to the principal diagonal. Levenberg goes on to show that the directional derivative of the residual sum of squares has a minimum when these weighting factors are equal. This provides an interesting justification for using a common biasing parameter k in ORE. The idea of adding k to the principal diagonal was later used by Marquardt [1963] to propose his useful nonlinear least squares algorithm. A. E. Hoerl's [1959] ridge analysis for nonlinear response surfaces was adapted for regression problems by Hoerl [1962]. The name ridge regression was first used in Hoerl and Kennard [1970a,b]. A new explanation of the term "ridge" is that the ridge solutions are along a ridge closest to the origin in the likelihood surface. (See Figures 1.4a and 1.4b where the ridge is depicted.)

7.1.1 Bias in Shrinking Towards Zero

We have already noted that ridge estimator b_k is biased, and shrinks the estimates towards zero. This is similar to R. A. Fisher's "null hypothesis," where the assumption is as follows. In the absence of specific evidence to the contrary, the assumption is that a given regressor has no effect on the dependent variable. This is a scientifically conservative position, even though the very suggestion to use anything "biased" seems blasphemous to the uninitiated. The bias can in fact make the final conclusions more conservative, and can be traded off against large reductions in the variance.

Many researchers tend to forget that the term "unbiased" means that in a large number of repetitions the average of the deviations of the estimator from the true parameter is zero. In practice, we do not generally have a large number of replications, and mere averaging to zero of large positive and negative deviations may not be satisfactory. Theil [1971, p. 91] states "it is good not to attach too much importance" to the unbiasedness criterion.

7.1.2 Uncorrelated Components of b

For a discussion of additional properties of RR we consider the singular value decomposition discussed in Chapter 1. The $T \times p$ matrix X can be written as

$$X = H\Lambda^{1/2}G', \tag{4}$$

where H is a $T \times p$ matrix of the coordinates of the observations along the principal axes of X standardized in the sense that $H'H = I$. The matrix Λ is a diagonal matrix of eigenvalues $\lambda_1 \geq \lambda_2 \geq \cdots \geq \lambda_p$, and G is the $p \times p$ matrix of eigenvectors g_i satisfying $X'X = G\Lambda G'$, and $G'G = I$.

From (4) consider the canonical model

$$y = H\Lambda^{1/2}G'\beta + u = H\Lambda^{1/2}\gamma + u \tag{5}$$

which defines a parameter vector $\gamma = G'\beta$. The OLS estimate of γ is denoted by c.

From $c = G'b$ we premultiply both sides by G and write a decomposition of b into p components as:

$$b = Gc = G\Lambda^{-1}\Lambda^{1/2}H'y = G\Lambda^{-1/2}H'y = (X'X)^{-1}X'y \tag{5a}$$

$$b_i = \sum_{j=1}^{p} g_{ij}c_j$$

for $i=1, \cdots, p$. As shown in Chapter 1, the variance covariance matrix of b is

$$V(b) = \sigma^2(X'X)^{-1} = \sigma^2 G\Lambda^{-1}G'. \tag{6}$$

Since $b = Gc$, $V(b) = V(Gc) = GV(c)G'$ implies that

$$V(c) = \sigma^2\Lambda^{-1}. \tag{7}$$

The elements c_i are called "uncorrelated components" because $V(c) = \sigma^2\Lambda^{-1}$ is diagonal.

The ordinary ridge estimator for γ in (5) is

$$c_k = (\Lambda + kI)^{-1}\Lambda^{1/2}H'y.$$

Further

$$b_k = (X'X + kI)^{-1}X'y = (G\Lambda G' + kI)^{-1}G\Lambda^{1/2}H'y$$

$$= (G\Lambda G' + kGG')^{-1}G\Lambda^{1/2}H'y$$

$$= G(\Lambda + kI)^{-1}\Lambda^{1/2}H'y = Gc_k.$$

Alternatively, from the last line we have

$$b_k = G(\Lambda + kI)^{-1}\Lambda^{1/2}H'y$$
$$= G(\Lambda + kI)^{-1}\Lambda G'G\Lambda^{-1}\Lambda^{1/2}H'y \quad (7a)$$
$$= G\Delta G'b \ ,$$

where

$$\Delta = \text{diag}(\delta_i); \ \ \delta_i = \lambda_i(\lambda_i + k)^{-1}, \ i = 1, \cdots, p,$$

is a diagonal matrix of "shrinkage factors". For strictly positive k and strictly declining eigenvalues we have "declining deltas," with the smallest delta for the minor axis, [Vinod 1974 and 1976j, p838]. In the following section we will clarify why "declining deltas" are useful for multicollinear data. (See also Section 5.3 of Chapter 5.)

7.1.3 Definition of Generalized Ridge Estimator

The "generalized" ridge estimator (GRE) of γ in (5) is obtained by augmenting the ith diagonal element of Λ by a positive constant k_i, and is given by

$$c_K = (\Lambda + K)^{-1}\Lambda^{1/2}H'y,$$

where $K = \text{diag}(k_i)$ is a diagonal matrix of (distinct) biasing factors k_i.

The GRE of β in $y = X\beta + u$ can now be written as

$$b_K = Gc_K = G(\Lambda + K)^{-1}\Lambda^{1/2}H'y \quad (8)$$
$$= G(G'X'XG + K)^{-1}\Lambda^{1/2}H'y$$
$$= G(G'X'XG + G'GKG'G)^{-1}\Lambda^{1/2}H'y$$
$$= (X'X + GKG')^{-1}X'y \ .$$

We can also write the generalized ridge estimator

$$b_K = (X'X + GKG')^{-1}X'y = G\Delta G'b \ ,$$

where $\Delta = \text{diag}(\delta_i)$, the diagonal matrix of "shrinkage fractions" $\delta_i = \lambda_i(\lambda_i + k_i)^{-1}$ noted earlier. The ridge estimator b_k of (1) is a special case of GRE when all $k_i = k$.

We recall the property of ordinary ridge estimator $b_k = G\Delta G'b$ that the deltas, $\delta_i = \lambda_i(\lambda_i + k)^{-1}$, are declining (nonincreasing, to be precise) for any fixed value of $k > 0$. Since the variance of $c_p (= \sigma^2\lambda_p^{-1})$ is the largest (recall, $\lambda_1 \geq \cdots \geq \lambda_p$), it makes intuitive sense that the shrinkage fraction δ_p be the smallest. The δ_i's provide a unifying thread among various shrinkage estimators. For example, Stein-rule estimators (Chapter 6) have equal δ_i for all i.

The principal component estimator (PCE) discussed by Kendall [1957],

Marquardt [1970], among others deletes a certain number, say m, of the last terms from the summation in (5a). Thus PCE is a special case of the GRE, b_K where $\delta_1 = \delta_2 = \cdots = \delta_{p-m} = 1$ and $\delta_{p-m+1} = \delta_{p-m+2} \cdots = 0$ which satisfies "declining deltas" of the matrix X and allows m to be a noninteger. Marquardt [1970] refers to $p-m$ as the "assigned rank." The PCE solutions at m=1 ($\delta_1 = 1$, and $\delta_2 = 0$) is marked as PCR in Figures 1.4a and 1.4b given in Chapter 1. Similarly, PCE solutions at m=0.5 has $\delta_1 = 1$, as before, but $\delta_2 = 0.5$, and it is marked by PCR* in the same figures of Chapter 1.

7.2 PROPERTIES OF THE RIDGE ESTIMATOR AND CHOICE OF BIASING PARAMETERS k

Gauss in 1809 suggested MSE as the most relevant criterion for choice among estimators. The MSE matrix for an estimator b of β is defined by $MtxMSE(b) = E(b - \beta)(b - \beta)'$. We measure the "closeness" of b to β in terms of squared Euclidean distance by the trace of MtxMSE.

$$MSE(b) = E(b - \beta)'(b - \beta) \tag{9}$$

$$= TrV(b) + \text{Bias}(b)'\text{Bias}(b) \ ,$$

where Tr denotes the trace (see further details in Chapter 2).

In choosing MSE(b) as our basic criterion we are ignoring the off-diagonal elements of MtxMSE, and are weighting all $(b_i - \beta_i)^2$ equally. If we consider b as the OLS estimator, then, although b is BLUE it is not necessarily "closest" to β, because linearity and unbiasedness are irrelevant for closeness.

Assuming that k_i is nonstochastic, the expected value of the GRE is $Eb_K = G\Delta G'Eb = G\Delta G'\beta$.

Thus $b_K - Eb_K = G\Delta G'(b - \beta)$, and $E(b_K - Eb_K)(b_K - Eb_K)' = G\Delta G' \cdot E(b - \beta)(b - \beta)'G\Delta G' = \sigma^2 G\Delta G'(X'X)^{-1} G\Delta G' = \sigma^2 G\Delta G'(G\Delta G')^{-1}G\Delta G'$, where $E(b - \beta)(b - \beta)' = \sigma^2(X'X)^{-1}$ is the variance covariance matrix of the OLS. After simplification, we get

$$V(b_K) = \sigma^2 G \ \text{Diag}\ [\delta_i^2/\lambda_i]G' \ . \tag{10}$$

When we use $k_i > 0$ in (8), we use fractional $\delta_i = \lambda_i/(\lambda_i + k_i) < 1$ to obtain possibly large reductions in the variance in exchange for bias. The $V(b_k)$ of the ordinary ridge estimator is the same as (10), except that all k_i are equal to k.

The expression for the bias of GRE is

$$\text{Bias}(b_K) = G \ \text{Diag}(\delta_i - 1)G'\beta \tag{11}$$

$$= G \ \text{Diag}(\delta_i - 1)\gamma \ .$$

Now,

$$MSE(b_K) = \sigma^2 \sum_{i=1}^{p} \delta_i^2/\lambda_i + \gamma' \text{Diag}(\delta_i - 1)G'G \text{ Diag}(\delta_i - 1)\gamma \tag{12}$$

$$= \sigma^2 \sum_{i=1}^{p} \delta_i^2 \lambda_i^{-1} + \sum_{i=1}^{p} (\delta_i - 1)^2 \gamma_i^2 .$$

The MSE matrix of the generalized ridge estimator is:

$$MtxMSE(b_K) = E(b_K - \beta)(b_K - \beta)' \tag{12a}$$

$$= V(b_K) + \text{Bias}(b_K)\text{ Bias}(b_K)'$$

$$= \sigma^2 G \text{ Diag}(\delta_i^2/\lambda_i)G' + G \text{ Diag}(\delta_i - 1)G'\beta\beta'$$

$$\times G \text{ Diag}(\delta_i - 1)G'$$

$$= \sigma^2 G \text{ Diag}(\delta_i^2/\lambda_i)G' + G \text{ Diag}(\delta_i - 1)\gamma\gamma'$$

$$\times \text{Diag}(\delta_i - 1)G' .$$

The expressions for MSE and MtxMSE for the ORE follow by substituting $k_i = k$ for all i.

The GRE is a Bayesian estimator given in (87) of Chapter 3, for $\Omega = GK^{-1}G'$ and $\beta_0 = 0$. Thus from the result in (97) it is clear that the GRE estimator is to be preferred to the OLS estimator b, that is,

MtxMSE(b) − MtxMSE(b_K) = a positive definite matrix so long as

$$\beta'[(X'X)^{-1} + 2GK^{-1}G']^{-1}\beta \leq \sigma^2 , \tag{12b}$$

or $\beta'G[\Lambda^{-1} + 2K^{-1}]^{-1}G'\beta \leq \sigma^2$, where we use $X'X = G\Lambda G'$.

For $K = kI$, $MtxMSE(b) - MtxMSE(b_k)$ is a positive definite matrix if

$$\beta'G[\Lambda^{-1} + \frac{2}{k} I]^{-1}G'\beta \leq \sigma^2$$

or if $\frac{k}{2}\ \beta'G[I - \text{Diag}(\frac{k}{k + 2\lambda_i})]G'\beta \leq \sigma^2$. Since the left-hand side of the inequality is positive and $GG' = I$, the value of k for which the above inequality holds for any λ_i is given by $\frac{k}{2}\ \beta'GG'\beta = \frac{k}{2}\ \beta'\beta \leq \sigma^2$ or

$$k \leq \frac{2\sigma^2}{\beta'\beta} . \tag{12c}$$

Note that this condition is sufficient for superiority of b_k, but it is not necessary. In practice, this condition may be too conservative.

A necessary and sufficient condition for $MtxMSE(b)$ ">" $MtxMSE(b_k)$ given by Swindel and Chapman [1973] is

$$0 < k < 2/[-\min(0,\zeta)] \tag{12d}$$

where ζ is the minimum eigenvalue of $(X'X)^{-1} - (\beta\beta'/\sigma^2)$. If the minimum eigenvalue is positive (12d) suggests that any nonstochastic k in the open interval $(0, +\infty)$ will reduce the MSE of OLS. Note that $\beta\beta'$ is a $p \times p$ matrix of rank 1 with the only nonzero eigenvalue given by $\|\beta\|$, the Euclidian length. Hence $\zeta = [\lambda_1^{-1} - \|\beta\|\sigma^{-2}]$ will be large and negative for regression problems having large "signal to noise ratio." Since we divide by its absolute value in (12d) the upper bound will be small. Thus a small nonstochastic biasing parameter k is usually appropriate for problems having large signal to noise ratio.

7.2.1 Ridge Existence Theorems

The main theoretical justification for RR given by H-K [1970a,b] is their theorem that a strictly positive k exists for which the trace of the MSE matrix satisfies

$$MSE(b_k) < MSE(b). \tag{13}$$

The original proof of this theorem by Hoerl-Kennard [1970a, Theorem 4.3] involves differentiating MSE with respect to k and showing that the derivative is negative when evaluated near $k = 0$. Thus strictly positive k's exist which will reduce the MSE of OLS. Rather than reproducting that proof we shall define an "acceptable range"

$$0 < k < k_{max} \tag{14}$$

of the biasing parameter k values. Maquardt and Snee [1975] define an "admissible" range of k wherein $MSE(b_k) < MSE(b)$. The term "acceptable" is used here to avoid confusion with decision theoretic admissibility concepts. The "existence" theorem may be proved by showing that the acceptable range (14) is non-empty.

For $K = kI$, it follows from (12c) that as long as $0 < k < k_{max}$, where $k_{max} = 2\sigma^2/\beta'\beta$, $MtxMSE(b) - MtxMSE(b_k)$ is positive definite, (also see Theobald [1974]). That is, $E(b_k - \beta)'\, W(b_k - \beta)$ is smaller than $E(b - \beta)'\, W(b - \beta)$ for all W. Since k_{max} must be strictly positive for all σ^2 and β, this proves the existence theorem for the most general MtxMSE criterion, and without any reference to multicollinearity. The existence theorem is true even when multicollinearity is absent. However, in that case there will be very little scope for reducing MSE(b), and the positive k will be very close to zero.

The expression $k_{max} = 2\sigma^2/\beta'\beta$ involves true unknown parameters, hence estimating k_{max} seems to be difficult if not impossible. Without an operational estimate of k_{max} in (8) the existence theorem cannot fully justify ridge regression. Once stochastic values of k are used in ORE (or GRE) the expression (10) for its covariance matrix is no longer valid, so that the intuitively tempting estimate of k_{max} based on OLS estimates of σ^2 and β is questionable.

7.2.2 Estimating the Acceptable Range of k_i and Shrinkage Factor of δ_i

We have defined the "acceptable range" of k values in (14) which is relevant only for the "ordinary" ridge estimator b_k. For the "generalized" ridge there is, in general, a separate range for each k_i given by

$$0 < k_i < k_{i,max} . \tag{15}$$

Now the shrinkage fractions $\delta_i = \lambda_i(\lambda_i + k_i)^{-1}$ has the following range associated with each k_i

$$\delta_{i,min} < \delta_i < 1 . \tag{16}$$

First, note that working with the canonical model and uncorrelated components does not change the MSE of OLS.

$$MSE(b) = E(Gc - G\gamma)'(Gc - G\gamma) = MSE(c) . \tag{16a}$$

Now, in terms of individual components note that $\text{MSE}(c) = \sum_{i=1}^{p} \text{MSE}[c_i]$.

For non-stochastic δ_i we can directly write

$$MSE[\delta_i\ c_i] = \sigma^2\delta_i^2/\lambda_i + (1 - \delta_i)^2\gamma_i^2 , \tag{17}$$

where the first term is the variance and the second the squared bias. It is easy to verify from (12) that

$$MSE(b) = \sum_{i=1}^{p} MSE[\delta_i\ c_i] \tag{18}$$

$$MSE[c_i] - MSE[\delta_i\ c_i] = (\sigma^2/\lambda_i)\ [1 - \delta_i^2] - (1 - \delta_i)^2\gamma_i^2 \tag{19}$$

$$= (1 - \delta_i)\ [(\sigma^2\lambda_i^{-1})\ (1 + \delta_i) - \gamma_i^2(1 - \delta_i)] .$$

Thus, $\delta_i\ c_i$ will reduce the MSE of c_i if and only if (19) is positive, hence the "acceptable" range of δ_i wherein $\text{MSE}[\delta_i\ c_i] < \text{MSE}[c_i]$ requires that both terms on the right-hand side of (19) are positive. That is,

$$1 > \delta_i > \delta_{i,min} = max[0,\ 1 - 2\sigma^2(\sigma^2 + \lambda_i\ \gamma_i^2)^{-1}] , \tag{20}$$

which defines $\delta_{i,min}$ as the larger of the two numbers in brackets. This result was given in Vinod [1978r].

The implication of the component-wise bounds on k_i and $\delta_{i,min}$ for the "ordinary" ridge may be stated by defining $k_{max} = min(k_{i,max})$ and $\delta_{min} = max(\delta_{i,min})$.

From the definition of $\delta_{i,min}$ in (20), Vinod's $k_{i,max}$ is as follows.

$$k_{i,max} = \begin{cases} \infty, \text{ when } \sigma^2\gamma_i^{-1} \geq \gamma_i^2 \\ 2\sigma^2(\gamma_i^2 - \sigma^2\lambda_i^{-1})^{-1}, \text{ otherwise.} \end{cases} \tag{21}$$

Note that this is always strictly positive, as it should be. Theobald's [1974] result would imply $k_{i,max} = 2\sigma^2\gamma_i^{-2}$ for the ith component, which is always smaller than the above expression. Smaller k_i means less shrinkage, and solutions closer to OLS. Now a larger upper bound of (21) is more useful than a smaller one because it allows more leeway to the researcher.

Two important advantages of Theobald's $k_{max} = 2\sigma^2/\gamma'\gamma$ are that he requires the knowledge of only the squared length $\beta'\beta = \gamma'\gamma$ of true unknown β_i (or γ_i for i=1, $\cdots$, p), and that his MtxMSE criterion is more general. See Farebrother [1976] also.

Vinod [1976 p] derives an expression for k_{max} which also depends on $\beta'\beta$, and which is *efficient* in the sense that the bound on the ORE bias may be reached when β is parallel to g_p the eigenvector of $X'X$ associated with λ_p. This is a pessimistic configuration for ORE. Vinod's k_{max} is a positive solution of a nonlinear (fixed point type) equation

$$\sum_{i=1}^{p} \lambda_i^{-1} - \sum_{i=1}^{p} \lambda_i(\lambda_i + k_{max})^{-2} = (\beta'\beta/\sigma^2)[k_{max}^2(\lambda_p + k_{max})^{-2}] . \tag{22}$$

In particular, when $k_{\max} = 0$ the right hand side is zero. Therefore for OLS, we assume that $\beta'\beta/\sigma^2$ can be very large without bound. For small $k_{\max}$ we must not let the right hand side (squared bias) become larger than the left (variance reduction).

Two major difficulties with the above theoretical expressions for δ_{min} or k_{max} are: (a) they depend on (pessimistic configurations of) unknown parameters; and (b) the expressions are derived under the assumption of nonstochastic biasing parameters k_i (or δ_i), hence these theoretical estimates of acceptable ranges of k or δ values are mainly heuristic. They are useful in determining whether a given application of ridge regression involves excessive shrinkage.

Vinod [1976p] attempts to solve the two difficulties mentioned above by considering a sampling distribution of $\mu^0 = b'b/s^2$ as an estimate of $\mu = \beta'\beta/\sigma^2$. He proves that μ^0 is distributed as:

$$\mu^0 \sim \sum_{i=1}^{p} \lambda_i^{-1} F_{nc}(1, T-p-1, \gamma_i^2\lambda_i\sigma^{-2}), \tag{22a}$$

a weighted sum of noncentral F variables with the degrees of freedom and noncentrality indicated in the parentheses. He also notes that using a large value of $\beta'\beta/\sigma^2$ (=μ) in (22) is relatively safe because it leads to a conservative (small) estimate of $k_{\max}$.

Assume the researcher is willing to specify the probability of the unfortunate occurrence, when a proposed value μ^* for μ is not large enough for the true unknown μ:

$$Pr^* = Pr(\mu^* < \mu) = \alpha, \tag{22b}$$

where $\alpha = 0.05$, say. Now some approximations to (22a) similar to Watson [1955, p. 340] are used by Vinod to find μ^* and hence k_{max} via (22).

Vinod writes Pr^* as

$$Pr^* = Pr[F^* p^{-1} < \mu p^{-1}],$$

where $F^* \sim F_{nc}(p, T-p-1, \mu)$. This is further approximated by a central F variable (F_c) as:

$$F^* = \frac{p+\mu}{p} F_c(\frac{p+\mu}{p+2\mu}, T-p-1), \tag{22c}$$

with the indicated degrees of freedom.

Vinod's [1976p] plots of Pr^* against $\beta'\beta/p\sigma^2$ $(=\mu p^{-1})$ indicate that Pr^* is almost always less than 0.5, and much smaller for a smaller $\beta'\beta/p\sigma^2$. Recently Oman [1981] has resurrected expressions similar to (22), (22a) and a refinement to (22c).

The main advantage of these methods based on (22) to (22b) is that they permit a larger k for highly multicollinear data. (See Exercise 7.7). The following chapter will discuss the implications of stochastic k_i. Before we consider these questions let us discuss the RIDGE TRACE graphical methods for finding a specific solution (k) within the ridge family, and hopefully satisfying $0 < k < k_{max}$.

7.2.3 The RIDGE TRACE and Stability of ORE

A RIDGE TRACE is a plot of b_{ki} (the ith component of b_k) against k suggested by Hoerl and Kennard [1970] to choose k in a ridge regression. Theil [1963] seems to have independently suggested a similar plot. Even if actual use of ridge solutions is not intended, these inexpensive plots should be routinely produced as "data analytic" tools, because unstable plots may suggest seriously deficient data or model specification.

For example, a RIDGE TRACE having two lines for two regressors b_{ki} and b_{kj} (see Figure 7.1) which move in a roughly opposite direction, should suggest that b_{ki} and b_{kj} are negatively correlated with a large correlation coefficient. Thus the instability of the RIDGE TRACE indicates intercorrelations among regressors arising from multicollinearity. Clearly, the ridge solutions in the unstable region of RIDGE TRACE are more seriously affected by multicollinearity. In Figure 7.1 the curve marked j changes sign from negative to positive. Various sign changes can be quickly seen on a RIDGE TRACE.

Hoerl and Kennard [1970b] recommended standardizing the data before plotting on a RIDGE TRACE, so as to retain numerical comparability of regression coefficients. This is discussed in the following section.

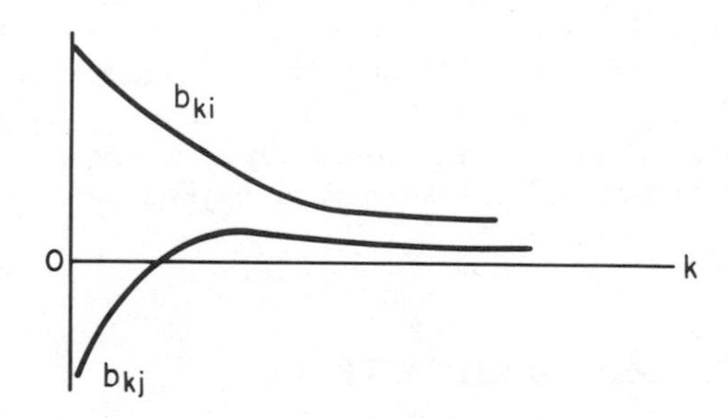

Figure 7.1 RIDGE TRACE for Negatively Correlated Regression Coefficients

7.2.4 Data Standardization in Ridge Regression

One source of controversy in ridge methods is the advisability of re-centering and re-scaling the original data to make $X'X$ a correlation matrix in (1). Standardization is unnecessary for most theoretical results, and not advisable for those (rare) cases when the investigator is committed to the centering, scaling and MSE computations in original units. Therefore, it may be confusing when Hoerl and Kennard [1970b] and Marquardt and Snee [1975] recommend standardization without ruling out these (rare) cases.

Often there is nothing fundamentally different in a model where (i) a temperature variablc is measured in centigrade or Fahrenheit degrees; (ii) a time variable starts at 1 or 1951; (iii) a money variable is measure in U.S. dollars, British sterling, or a linear combination of the prices of various currencies reflected by, say, the Special Drawing Rights; and (iv) the base year of an index number deflator variable is 1961 or 1971, etc. For these "essentially similar" models the investigators want the regression coefficients to be "essentially similar," i.e., "equivariant" (Berk, 1967).

For example, two econometricians using slightly different centering and scaling of the regressors for the same basic model specification will be surprised to find different numerical estimates of underlying economic entities (e.g., elasticities) if they apply RR directly to (1). For some models, it may be possible to estimate arbitrarily different elasticities by measuring certain regressors in billions (say) instead of millions or thousands.

Thus the empirical practitioner may be better off using standardization conventions rather than be faced with such arbitrariness in his ridge estimates.

We are usually not concerned about equivariance because it always holds for OLS. If we want any of the ridge estimators to be equivariant the $X'X$ matrix should not change after any re-centering or re-scaling of regressors.

An additional advantage of the standardization is that it makes the numerical magnitudes of β_i comparable with each other. If a RIDGE TRACE does not plot standardized regression coefficients, its appearance can be dramatically altered by a simple change of origin and scaling of the variables. There is then the danger of naively misinterpreting the meaning of the plot. From a Bayesian viewpoint Lindley and Smith [1972] show that ORE implicitly assumes that β_i are "exchangeable *a priori*". It can be argued (see Vinod [1978c] for details) that exchangeability is more readily satisfied by a standardized model. To discuss this point (see Section 7.3) we need a Bayesian interpretation of ridge methods.

7.2.5 Horizontal m Scale for RIDGE TRACE

Tukey [1974], Thisted [1976] and Smith and Campbell [1980] have suggested logarithmic transformations of the horizontal scale of a RIDGE TRACE to obtain meaningful plots. Obenchain and Vinod [1974], Vinod [1974], Vinod [1976] and Obenchain [1976], discuss the advantages of choosing a new scaling on the horizontal axis of a RIDGE TRACE called the "multicollinearity allowance," m. It is defined by

$$m = p - \sum_{i=1}^{p} \lambda_i/(\lambda_i + k_i) = p - \sum_{i=1}^{p} \delta_i \,. \tag{23}$$

Note that when $k = k_1 = \cdots = k_p = 0$, $m = 0$ and when $k = \infty$, $m = p$. The transformation from k to m defined by (23) can be readily tabulated for any given set of eigenvalues λ_i. The tabulation for "ordinary" ridge is simple because all k_i equal k. We can simply increase k from 0 in appropriate intervals, and note the value of the corresponding m in a table. There is a unique value of m for each k and vice versa. It is more interesting to obtain k from a specified value of m in the range 0 to p by reading the above mentioned table in a reverse fashion (with possible interpolation). An Explanatory Note at the end of this chapter gives a computer program which uniquely determines a k from a given m. The computer program uses Newton's method to search for a unique k from m. A computational shortcut to get m for ORE without knowing eigenvalues is discussed in Section 7.4 below.

Now we discuss some of the advantages of the m scale.

(i) Finite Range

In general, the k on the horizontal axis of a conventional RIDGE TRACE can have an infinite range $0 \leq k \leq \infty$. Any truncation of the range for plotting convenience such as the range $0 \leq k \leq 1$ recommended by Hoerl and

Kennard can be misleading for some examples, because some of the "action" may be outside the truncated range. For the m-scale the range is $0 \leq m \leq p$, which is finite, and obviously need not be truncated at all.

(ii) Generality

The k-scale RIDGE TRACE cannot be plotted for "generalized" ridge regressions when the k_i's are distinct. The m-scale TRACE has been plotted for principal components regression in Vinod [1974]. Once the rule for defining k_i values in a "generalized" ridge is specified, it is quite simple to plot an m-scale RIDGE TRACE for GRE as in Obenchain and Vinod [1974]. Furthermore, these plots can be compared to each other.

(iii) Rank Deficiency Interpretation

For the example in Vinod [1976j] where there are 3 regressors, the eigenvalues are $\lambda_1 = 2.66$, $\lambda_2 = 0.33$ and $\lambda_3 = 0.01$, respectively. From a geometrical viewpoint the available data may be thought to be "hammered-flat cigar-shaped" having most of the spread in one principal direction (cigar length) and the least spread in the direction of the flat part of the cigar after it is hammered flat. Thus, from the relative smallness of λ_3 we conclude that it would be appropriate to treat the data as approximately two dimensional rather than three. The choice $m = 1$ in some sense recognizes that the available data are two dimensional, as if "hammered-flat cigar-shaped." Thus we give the multicollinearity allowance m a clear interpretation as the assigned deficiency in the rank of $X'X$.

For the principal components regression (PCR) having assigned rank r, one specifies $\delta_1 = \delta_2 = \cdots = \delta_r = 1$, and $\delta_{r+1} = \delta_{r+2} = \cdots = \delta_p = 0$. Upon substituting in (23) verify that $m = p - \sum_{i=1}^{p} \delta_i = p - r$, which is the rank deficiency.

(iv) More Reliable Stable Region

Hoerl and Kennard [1970] suggest choosing k graphically in the "stable region" of the RIDGE TRACE having k as the horizontal scale. However, from (7a) we have

$$db_k/dk = -G \text{ Diag}[\lambda_i/(\lambda_i + k)^2]G'b \ . \tag{24}$$

Since $\lambda_i(\lambda_i + k)^{-2} = \delta_i^2\lambda_i^{-1}$, write $|db_k/dk| = \sigma^{-2}(Vb_k)b$ from (10). For orthogonal data $X'X = I$, all $\lambda_i = 1$ and hence $V(b_k)$ simplifies to $\sigma^2(1+k)^{-2}I$, because $G'G = I$. Hence, the orthogonal data $|db_k/dk| = b/(1+k)^2$, and the absolute value of the change in b_k for a given change in k is smaller for larger k. Thus, the k scale has the unfortunate property that the ridge trace may appear to be more stable for larger k even for completely orthogonal data. Now we shall show that the m scale does not have this property.

From (23), $dm/dk = \sum_{i=1}^{p} \lambda_i/(\lambda_i + k)^2 = \bar{S}$ which defines the notation $\bar{S}$

for the scalar involved. Now from (24)

$$db_k/dm = (db_k/dk)(dk/dm) = -(b/\sigma^2)\ V(b_k)/\bar{S}\ . \tag{25}$$

Again, for completely orthogonal data $\bar{S}$ becomes $\sum_{i=1}^{p} 1/(1+k)^2$ $= p/(1+k)^2$, $(p/\bar{S})V(b_k) = \sigma^2 I$ and $db_k/dm = -\,b/p$ which does not change with m. This shows that having m on the horizontal axis of the RIDGE TRACE will not give an appearance of greater stability at larger m. The RIDGE TRACE for orthogonal data with m on the horizontal axis consists of p straight lines from b to zero. A graphical indication of whether the stable region has been achieved is that the lines become straight converging at zero when $m = p$.

It is also possible to give an analytical expression for closeness to orthogonality implied by the "stable region" following Vinod [1976j]. We may simply compare the $db_k/dm = -b/p$ for orthogonal data with $-\sigma^{-2}\ V(b_k)b/\bar{S}$ from (25). The sum of squares of differences between these two expressions for db_k/dm can be written as:

$$\left[\frac{b}{p} - \sigma^{-2}\frac{V(b_k)}{\bar{S}}\,b\right]'\left[\frac{b}{p} - \sigma^{-2}\frac{V(b_k)}{\bar{S}}\,b\right]$$

$$= b'\left[\frac{GG'}{p} - \frac{G\ Diag(\delta_i^2\lambda_i^{-1})G'}{\bar{S}}\right]'\left[\frac{GG'}{p} - \frac{G\ Diag(\delta_i^2\lambda_i^{-1})G'}{\bar{S}}\right]b$$

$$= b'G\left[\frac{1}{p} - \frac{Diag(\delta_i^2\lambda_i^{-1})}{\bar{S}}\right]'\left[\frac{1}{p} - \frac{Diag(\delta_i^2\lambda_i^{-1})}{\bar{S}}\right]G'b$$

$$= \sum_{i=1}^{p}\left[\frac{1}{p} - \frac{\delta_i^2\lambda_i^{-1}}{\bar{S}}\right]^2 c_i^2 = \sum_{i=1}^{p}\left[1 - \frac{p\,\delta_i^2\lambda_i^{-1}}{\bar{S}}\right]^2\frac{c_i^2}{p}\,,$$

where $c = G'b$, and $G'G = I$ as before. Further, $\bar{S} = \sum_{j=1}^{p}\delta_j^2\lambda_j^{-1}$. This expression can be simplified by cancelling c_i^2/p to yield Vinod's non-stochastic index of stability of relative magnitudes

$$ISRM = \sum_{i=1}^{p}\left[\frac{p\,\delta_i^2\lambda_i^{-1}}{\sum_{j=1}^{p}\dfrac{\delta_j^2}{\lambda_j}} - 1\right]^2. \tag{26}$$

In practice, one chooses certain k values: 0.0001, 0.001, 0.01, 0.02, ... etc. Each of these imply p values of $\delta_i = \lambda_i\ (\lambda_i + k)^{-1}$. Upon substituting these in (26) one obtains one ISRM (k) for each choice of k. If the regression coefficients are becoming more "stable" as k increases, we expect ISRM (k) to decrease. If OLS is unstable, it will make $k = 0$ (or $m = 0$) solution on a RIDGE TRACE noticeably unstable and ISRM ($k = 0$) large. It will be shown in Section 7.3.4 that any rule for selection of k based on a non-stochastic criterion (such as ISRM) will give "admissible" (Bayesian) estimators. ISRM criterion is such that the corresponding Bayesian prior attaches considerable importance to stability. However, in the classical framework there is no guarantee that MSE will be reduced. The admissibility property disappears if the stable region is selected graphically from a RIDGE TRACE which is obviously affected by y. A simulation in Wichern and Churchill [1978] suggests that the choice of k based on ISRM may give values of k that are too large. Vinod's [1979] letter on the subject discusses a deficiency in the simulation.

In any case, ISRM does provide a quantification of the concept of "stable region," especially for the m-scale. If we plot ISRM (k) as a function of k, it can have multiple local minimums. The global minimum of ISRM (k) tends to emphasize "stability" without regard to the bias. Since bias is important, we suggest a solution where the bulk of the potential reduction in ISRM is achieved, say at the first local minimum, or at a pre-specified percentage (say 50%) of the potential reduction.

7.3 BAYESIAN AND NON-BAYESIAN INTERPRETATIONS OF RIDGE METHODS

Under appropriate normality assumptions (and for certain non-normal distributions) the Bayesian methods imply (see Section 7.3.4 below) that the posterior mean is the optimal estimator when using MSE as expected loss. Giving a Bayesian interpretation to the GRE estimator b_K amounts to deriving the prior distribution for which b_K is the posterior mean. This exercise is illuminating even though one may not accept the philosophical implication of treating the parameter vector β as a random variable, or the subjective aspects of the prior distribution. Some Bayesians feel that there is nothing new with the GRE because these estimators can be given a Bayesian interpretation. However, there are non-Bayesian motivations also for ridge methods. We derive the ridge estimator as the solution to a constrained optimization problem in the following subsection.

7.3.1 Constrained Optimization Interpretation

Let β^* be any known value of β. Then the sum of squared errors due to β^* is

$$
\begin{aligned}
SSE(\beta^*) &= (y - X\beta^*)'(y - X\beta^*) \\
&= (y - Xb + Xb - X\beta^*)'(y - Xb + Xb - X\beta^*) \\
&= (y - Xb)'(y - Xb) + (\beta^* - b)'X'X(\beta^* - b) \\
&= SSE(b) + F ,
\end{aligned}
$$

where $F = (\beta^* - b)' X'X(\beta^* - b)$ is the square of the bias due to using β^* instead of b, the OLS estimator. The SSE is at its minimum when $\beta^* = b$; therefore, SSE(β^*) can be viewed as the minimum value, $(y - X b)'(y - X b)$, plus the quadratic form in $(\beta^* - b)$. We have seen that the distance between b and β increases with the ill-conditioning of $X'X$; therefore, we would like to minimize the squared distance of an estimator $\beta^{*\prime}\beta^*$. But this minimization cannot be done arbitrarily. We try to minimize the squared distance for a given level of the residual sum of squares (i.e., we can get many estimators that give the same SSE, but we would like to choose the one that has the smallest squared distance). Thus Hoerl and Kennard [1970] derived the ORE as the solution to the following problem:

Minimize $\beta^{*\prime}\beta^*$

Subject to

$$(\beta^* - b)'X'X(\beta^* - b) = F , \tag{27}$$

where F is a constant. That is we are trying to find an estimator whose inner product is a minimum for a given level of the SSE. An equivalent statement to (27) is:

Minimize $(\beta^* - b)'X'X(\beta^* - b)$

Subject to

$$\beta^{*\prime}\beta^* = r^2 , \tag{28}$$

where r^2 is a constant.

The solution to (27) and (28) is

$$b_k = (X'X + kI)^{-1}X'y .$$

where $\frac{1}{k}$ is the Lagrangian multiplier of (28).

7.3.2 Inequality Constraints Interpretation

Suppose the two sided inequality constraints on the elements of β are

$$\underline{\beta} \leq \beta \leq \overline{\beta} , \tag{29}$$

where $\underline{\beta}$ and $\overline{\beta}$ are the lower and upper bounds of the elements of β. We can formulate this inequality constraint as follows. It may be observed that the region bounded by the inequality constraints (29) is contained in the ellipsoid

$$(\beta - \beta_0)'B^{-1}(\beta - \beta_0) = 1 ,$$

where B is a nonsingular matrix and β_0 is a fixed vector. Further, this ellipsoid has minimum volume when $B = pD$, where D is a diagonal matrix. Thus the constraints (29) can be formulated as

$$(\beta - \beta_0)'D^{-1}(\beta - \beta_0) = p , \qquad (30)$$

Now the estimation of β subject to (30) can be carried out by using the minimax principle [Rao, 1973, pp. 340-43], Bibby and Toutenburg [1977, pp. 150-53]. This involves the estimation of a linear function $\lambda'\beta$ by a vector

$$a = \lambda'[\beta_0 + Q(y - X\beta)] , \qquad (31)$$

where Q is an arbitrary matrix to be chosen, and where β_o and λ are specified vectors.

Note that $E(a) = \lambda'\beta_o + \lambda'(QX - I)\beta$. Hence it is a biased estimator of $\lambda'\beta$, and the MSE(a) is

$$MSE(a) = \sigma^2\lambda'QQ'\lambda + [(\beta - \beta_0)'(QX - I)'\lambda]^2 . \qquad (32)$$

The minimax estimator a_0 of $\lambda'\beta$ is characterized by

$$\underset{Q}{Min}\left[\underset{\beta}{Max}\ MSE(a)\right] \qquad (33)$$

when $(\beta - \beta_0)'D^{-1}(\beta - \beta_0) = p$. If we maximize the quantity $[(\beta - \beta_0)'(QX - I)'\lambda]^2$ with respect to β satisfying (30), then its maximum value is

$$p\lambda'(QX - I)'D(QX \quad I)\lambda ,$$

whence

$$\underset{\beta}{Max}\ MSE(a) = \sigma^2\lambda'QQ'\lambda + p\lambda'(QX - I)'D(Q - I)\lambda .$$

subject to

$$(\beta - \beta_0)'D^{-1}(\beta - \beta_0) = p.$$

Minimizing it with respect to Q yields

$$Q = p[\sigma^2 I + pXDX']^{-1}XD = (X'X + \frac{\sigma^2}{p}D^{-1})^{-1}X' ,$$

so that the minimax estimator of β is given by

$$\hat{b} = \beta_0 + [X'X + \frac{\sigma^2}{p}D^{-1}]^{-1}X'(y - X\beta_0) \qquad (34)$$

$$= (X'X + \frac{\sigma^2}{p}D^{-1})^{-1}(X'y + \frac{\sigma^2}{p}D^{-1}\beta_0) .$$

For $\beta_0 = 0$ and $\frac{\sigma^2}{p}D^{-1} = kI$, the estimator $\hat{b} = b_k$, the ORE. If $\beta_0 = 0$ and $\frac{\sigma^2}{p}D^{-1} = GKG'$ then $\hat{b} = b_K$, the GRE.

Lin and Kmenta [1980] show that using bounds $\pm 2\sigma\ k^{-1/2}$ in (29) and combining this prior knowledge in the form of a "mixed regression" estimator of Theil and Goldberger leads to ridge regression.

7.3.3 Restricted Least Squares Interpretation

Suppose the prior information on the true value of β lie in or on an ellipsoid given by

$$(\beta - \beta_0)'D^{-1}(\beta - \beta_0) \leq r^2 , \tag{35}$$

where D^{-1} is a symmetric matrix as before, r^2 is a known scalar constant, and β_0 is a fixed vector as before. Assume that the OLS estimator b does not satisfy the constraint in (35). This is a plausible assumption when there is a high degree of collinearity among the regressors.

A restricted least squares estimator can now be used to estimate $y = X\beta + u$. It chooses an estimator for β which minimize $(y - X\beta)'(y - X\beta)/\sigma^2$ subject to the inequality constraint in (35). Since D^{-1} is symmetric and positive definite, we can write $P'D^{-1}P = I$ or $D = PP'$. Let $\delta = P^{-1}(\beta - \beta_0)$. Then to implement the above restricted least squares procedure we choose the estimate of δ to minimize

$$\frac{1}{\sigma^2}[(y - X\beta_o) - XP\delta]'[(y - X\beta_0) - XP\delta] ,$$

subject to the constraint $\delta'\delta \leq r^2$. It follows from the result in Meeter [1966] that this estimate (d) is the solution of the equations

$$(\frac{1}{\sigma^2}P'X'XP + \mu I)d = \frac{1}{\sigma^2}P'X'(y - X\beta_0) , \tag{36}$$

where μ is chosen so that $d'd = r^2$. The notation δ used in this sub-section should not be confused with our usual notation for shrinkage factors.

The estimator of β implied by d is $b_\mu = Pd + \beta_0$, or

$$b_\mu = (\frac{X'X}{\sigma^2} + \mu D^{-1})^{-1}(\frac{X'y}{\sigma^2} + \mu D^{-1}\beta_0) \tag{37}$$

$$= (X'X + \sigma^2\mu D^{-1})^{-1}(X'y + \sigma^2\mu D^{-1}\beta_0) .$$

For $\beta_0 = 0$, this estimator is in the form of a ridge estimator.

7.3.4 Bayesian Interpretation

Consider the prior distribution of β as a multivariate normal with mean vector zero and variance covariance matrix $\sigma^2GK^{-1}G'$, that is $\beta \sim N(\beta_0,\ \sigma^2GK^{-1}G')$. Further, $y \sim N(X\beta,\ \sigma^2 I)$. Then, it follows from the results in Chapter 3 that the posterior density of β is also a multivariate normal with the posterior mean (equation (87), Chapter 3):

$$b^* = (X'X + GKG')^{-1}(X'y + GKG'\beta_0) \qquad (38)$$

$$= (X'X + GKG')^{-1}(X'Xb + GKG'\beta_0) ,$$

where b is the OLS estimator.

If the prior of β is $\sim N(0, \sigma^2 GK^{-1}G')$, then $b^* = (X'X + GKG')^{-1}X'y$, which is identical with the GRE given earlier. The ORE implies that the prior of β is $\sim N(0, \sigma^2 k^{-1} I)$. Some Bayesians feel that this prior is unrealistic, and a prior mean other than the null vector should be used. In the absence of specific prior knowledge it is often scientifically conservative to shrink toward the zero vector. We have stated in an earlier section that the usual null hypothesis also implies shrinking toward zero. In (rare) cases when such prior knowledge about β_0 is available, the modification of ORE or GRE is that one shrinks toward this known prior, as in (38). There is no difficulty in subjectively specifying a non-zero β_0. The problem is that the others may prefer a different β_0, and the solution can be so sensitive to this choice, that almost any solution can be obtained. The existence of minimax ridge estimators discussed in Chapter 8 shows that ORE can reduce the MSE of OLS everywhere. This avoids unscientific discussions of the "realism" of priors.

7.3.4.1 Admissibility of ORE for Non-stochastic k

When k is based on a non-stochastic criterion, the above mentioned prior is consistent with the so called "proper" Bayesian framework. It then means that ORE is "admissible" (unbeaten somewhere, Section 2.7.2) simply because it belongs to the "proper" Bayesian framework. No further proof of "admissibility" is needed.

Lindley and Smith [1972] have given a hyperparameter interpretation of the ridge estimators. According to them if the prior of β is $\sim N(\beta_0, \sigma^2 GK^{-1}G')$ and further, the prior on the hyperparameter of β_0 is a diffuse prior, then the resulting mean of the posterior density of β is the GRE. For proofs and further details see Chapter 3.

We should note here that for the Stein-rule estimator discussed in Chapter 6, the prior of β is $\sim N(0, \sigma^2(X'X)^{-1})$. Thus, the GRE differs from the Stein-rule estimator with respect to the specification of the variance covariance matrix of the prior distribution of β.

7.3.5 Prior Exchangeability and Standardization

Lindley and Smith [1972] discuss $\beta \sim N(0, \sigma^2 Gk^{-1}IG')$ and note that this implicit prior distribution for the ridge estimators implies "exchangeability" among regression coefficients, i.e., the regression coefficients are unaltered by a permutation of the suffixes (i =1, ..,p).

For the standardized model (one in which $X'X$ is a correlation matrix) the numerical magnitudes of β_i are comparable to each other. In fact, Marquardt and Snee [1975, p. 6] state that "In this scaling it is exceedingly rare for the population value of any regression coefficient to be larger than three in a real problem." When there is only one regressor ($p = 1$), the standardized regression coefficient β_1 equals the correlation coefficient, and varies over the range -1 to $+1$, whereas the unstandardized coefficient can be as large as a million. In general, the vector of unstandardized regression coefficients may vary over a much wider range of values than β. Since the range of numerical values for standardized regression coefficients β_i is usually much smaller, it is safe to assume that "exchangeability" is more readily satisfied by β_i. Of course, the centering and scaling of the regressors may happen to be such that the unstandardized coefficients are already "exchangeable." Whether model $y = X\beta + u$ satisfies "exchangeability" or not depends too much on the units of measurement of all regressors. This is first noted in Vinod [1978c]. Recently, Draper and Van Nostrand [1979, p. 458] have reviewed the standardization controversy and concluded that standardization is generally desirable in practical situations. The discussion following Smith and Campbell [1980] clarifies the issues involved.

7.3.6 Bayesian Role of Prior Measured by m

In this section we consider the Bayesian viewpoint of the m-scale method of selecting k in the ORE. We show here that m/p measures the share of the prior information in the posterior precision. Theil [1963, pp. 409-12] shows that there is a unique measure $f(A, B)$ of the relative contribution of the sample (A) and a priori (B) information to the posterior precision (inverse variance) of an estimator. Theil also shows that $f(A, B)$ satisfies four intuitive criteria:

(i) $f(A, B) + f(B, A) = 1$

(ii) $f(O, B) = 0$, and $f(A, O) = 1$

(iii) f is invariant with respect to linear transformations, and

(iv) f is linear.

Theil's measure is

$$f(A, B) = Tr[A(A + B)^{-1}]/p \, , \tag{39}$$

where A^{-1} is the covariance matrix for the sample given by $\sigma^2 G \Lambda^{-1} G'$, and B^{-1} is the covariance matrix for the ridge estimator prior given by $\sigma^2 G K^{-1} G'$.

Now $A+B=G(\Lambda+K)G'$, $A(A+B)^{-1}=G\ \mathrm{Diag}(\delta_i)G'$, and $Tr\ \mathrm{Diag}(\delta_i) = \sum_{i=1}^{p}\delta_i = p-m$ implies that Theil's measure can be written in terms of our multicollinearity allowance as

$$f(A,B)=1-(m/p)\ . \tag{40}$$

When $m=0$ we have $f(A,B)=1$, and the share of the prior information in the posterior precision is zero. At $m=p$, $f(A,B)=0$, and the prior contributes 100% of the share. At any point on the m-scale RIDGE TRACE, the percentage contribution of the prior is readily computed from $100(p-m)/p$.

Among the disadvantages of the m scale we should mention is the additional computational burden. We recommend all users of regression to routinely compute the eigenvalues λ_i of $X'X$. Once the eigenvalues are known, computation of $m=p-\sum_{i=1}^{p}\lambda_i(\lambda_i+k)^{-1}$ is trivial. If the software for eigenvalues is unavailable or expensive, the practitioner can compute m by the following formula for the "ordinary" ridge estimator

$$m=p-Tr(X'X)(X'X+kI)^{-1}\ , \tag{41}$$

which involves a matrix multiplication of two p-dimensional known matrices, whose trace is computed simply by adding all diagonal elements.

7.4 COMPUTATIONAL SHORTCUTS FOR ORDINARY RIDGE

Sometimes a practitioner may wonder whether using ridge methods would "make a difference" to his regression problem. For this purpose a "quick and dirty" look at the ridge family may be needed as a first step before more careful study is undertaken. This is indeed possible by merely adding p fictitious observations to the data as follows.

Let X_a denote a $(T+p)\times p$ augmented X matrix having p additional rows (observations) represented by $k^{1/2}I_p$, where I_p is a $p\times p$ identity matrix. Also, let y_a denote the $(T+p)\times 1$ vector assembled from the vector y by simply adding a null vector 0 of dimension $p\times 1$. Now write a $p\times p$ symmetric square matrix

$$X_a'X_a=[X' \vdots \sqrt{k}\ I_p]\begin{bmatrix} X \\ \cdots \\ \sqrt{k}\ I_p\end{bmatrix}=(X'X+kI) \tag{42}$$

and a $p\times 1$ vector

$$X_a'y=[X' \vdots \sqrt{k}\ I_p]\begin{bmatrix} y \\ 0\end{bmatrix}=X'y\ .$$

Vinod [1980] suggests that the biasing parameter k may be thought to be the square of the measurement error typical to the set of standardized regressors. The p additional observations represent our extraneous knowledge that certain "small" changes in the regressors and zero changes are indistinguishable. If this interpretation is accepted, the whole difficulty of choosing k, MSE, etc. disappears. Then, we end up with the usual OLS estimation of all $T+p$ observations, and there is no claim that the MSE of OLS based on T observations is somehow reduced.

Thus note that the ordinary ridge estimator is

$$b_k = (X'X + kI)^{-1}X'y = (X_a'X_a)^{-1}X_a'y_a \, . \tag{43}$$

This shows that usual computer programs for multiple regression can be used to obtain ridge solutions accurately. Of course, the standard errors computed by the OLS computer programs applied to y_a and X_a will be from $(X'X + kI)^{-1}$ rather than the correct expression from $V(b_k)$ of (10). In the absence of a ready access to eigenvalue computational algorithms, $V(b_k)$ can be computed from the equivalent expression

$$V(b_k) = \sigma^2(X'X + kI)^{-1}(X'X)(X'X + kI)^{-1} \, . \tag{44}$$

It is interesting to note that the multicollinearity allowance $m = p - Tr(X'X)(X'X + kI)^{-1}$ can be readily computed in the process of computing (44) by merely adding diagonals before final matrix multiplication by $\sigma^2(X'X + kI)^{-1}$. This idea is from a private communication of D. B. Preston of Bell Laboratories.

In light of these computational shortcuts, we recommend a routine plotting of RIDGE TRACEs, even if the investigator does not wish to use ridge methods. Large negative correlation between b_i and b_j leads to a trade-off relation between the magnitudes of b_i and b_j. On the RIDGE TRACEs such a trade-off means that the lines for b_{ki} and b_{kj} generally move closer together as k increases. For a large positive correlation between b_i and b_j the corresponding lines move (generally) in the same direction. The reader may wonder: why should I care about the trade-off? From equations (7) and (8) of Chapter 5 recall that the sum of regression coefficients is relatively more reliably estimated (estimable), whereas their difference is less reliable. By trading off the magnitudes of such coefficients we reduce their difference without much effect on their sum. The overall effect is to find more reliable coefficients, their sums and differences in various pairs.

We may need a tedious study of 2^p regressions to obtain the information supplied visually by inexpensive RIDGE TRACEs. When Leamer and Chamberlain [1976, p. 86] show that the "ridge estimator is less informative" (in the sense of a search estimator) than the 2^p regressions, they ignore the costs of 2^p regressions (especially for large p) and the visual appeal of RIDGE TRACEs. By looking at these plots one knows the stability and interrelationships among relevant regression coefficients. One should not use least squares results (k=0) when unstable RIDGE TRACEs are observed for slightly larger k values near zero.

From a RIDGE TRACE it is clear that ridge regression methods do not yield a single automatic solution to the estimation problem, but rather, a family of solutions. Various rules for choosing k (or m) have been suggested and simulated in the literature, and we shall discuss them with the help of an example in the following section, and also in the following Chapter 8.

7.5 AN ILLUSTRATIVE EXAMPLE FOR RIDGE SOLUTION SELECTION

There are numerous heuristic and theoretically rigorous techniques available for choosing the biasing parameters in ridge regression. Some are discussed in the surveys by Vinod [1978] and Draper and Van-Nostrand [1979]. The theoretical MSE properties of some of these can be studied by thinking about them as special cases of "double f-class" or "double h-class" ridge estimators defined in equations (20) and (33) respectively of the following chapter. For some of the others only certain simulation results are available. Since it is convenient to discuss ridge solution selection in the context of an example, we will now turn to an example.

Consider the illustrative example of Chapter 1 having nine observations (T=9) plotted in Figure 1.1 as a three dimensional pillar chart. The OLS estimation yields the following fitted equation:

$$y = 12.40 - 0.17\,x_1 + 1.12\,x_2\,. \tag{45}$$

The R^2 for this fit is 0.975823. The sum of squares (SS) of residuals is 6.0338, whereas SS due to regression is 243.53. In terms of the standardized data, we have SS $Reg = R^2 = 0.975823$ and $SS\ Res = 1 - R^2 = 0.024177$ for the regression and residual SS, respectively. The estimate of σ^2 for the standardized model is $SS\ Res/(T-p-1) = 0.004028$. The standardized regression coefficients are: $std.b_1$=−0.08665 and $std.b_2$=1.06913. The SS of these $std.b_i$ values (i=1,2) is 1.1506

For the standardized model, $X'X$ becomes the correlation matrix which is 2×2 with ones along the diagonal and the off-diagonal elements (both) equal to 0.943031. The $X'y$ vector gives the correlation coefficients between y and x_1 and between y and x_2 respectively, which are 0.921573 and 0.987417 respectively. The eigenvalues of $X'X$ are: $\lambda_1 = 1.943031$ and $\lambda_2 = 0.056969$ whose reciprocals add up to 18.068066. Consider the matrix H of the coordinates of the observations along the principal axes of standardized X from our singular value decomposition. Let h_i denote the columns of H matrix, and let R_{yi} denote the usual correlation coefficient between the T pairs of values for y and h_i. For this example, we have $R_{y1} = -0.968387$ and $R_{y2} = 0.195066$ which satisfy the following interesting relation with the usual squared multiple correlation coefficient: (See Section 1.3.3 of Chapter 1)

$$R^2 = R_{y1}^2 + R_{y2}^2\,. \tag{46}$$

Let g_i denote the columns of the matrix of eigenvectors G. For this example, the two entries in g_1 are both 0.707107, whereas the entries in g_2 are -0.707107 and 0.707107 respectively.

The uncorrelated components of the vector of standardized regression coefficients are $std.c_1 = 0.694719$ and $std.c_2 = 0.817264$ respectively. The sum of squares of these $std.c_i$ values (i=1,2) is 1.1506 which is identical to the SS of standardized b_1 and b_2 values noted above. The reason for this is the fact that c is defined by the relation $c = G'b$, and $G'G = I$ because G is orthogonal. Note that c_1 and c_2 are uncorrelated with each other in the sense that their variance covariance matrix is proportional to a diagonal matrix, eq. (7), with elements 0.514660 and 17.5534 respectively. Thus the variance of c_2 is more than 34 times larger than the variance of c_1. In other words, the estimated c_2 lacks "precision." In order to deemphasize c_2, we will see that the shrinkage fraction δ_2 associated with c_2 will usually be a smaller number. Now $\delta_1 > \delta_2$ implies the "declining deltas" property due to Vinod [1974] mentioned at the end of Section 7.1.2 above. Ordinary Ridge estimators have "declining deltas," and it is up to the practitioner to decide whether this is appropriate for his data. For this purpose, Vinod [1976j, Section 2-4] suggests checking whether $|R_{y1}| \leq |R_{y2}|$. We suggest preparing an "Advisability of Declining Deltas" table along following lines.

TABLE 7.1

Advisability of Declining Deltas

Principal Axis i	Uncorrelated Components c_i	Eigenvalue λ_i	Variance of c_i s^2/λ_i	Students t value $\|t_i\|$	Relative Precision $\sqrt{\lambda_i}$	Correlation With y $\|R_{yi}\|$
1	-0.6947	1.943031	0.002073	15.25	1.3939	0.968
2	+0.8173	0.056969	0.070705	3.07	0.2387	0.195

Notes: High correlation between last two columns favors ridge. Declining $|t|$ values favor ridge type declining deltas.

Declining deltas implicit in ridge methods will relatively de-emphasize minor axis $i = p,\ p-2,\ ...$ and emphasize the major axes $i = 1,2...$. For $p = 2$, the minor axis at $i = 2$ in demphasized. This seems to be appropriate because the relative precision of the minor axis is smaller. For examples having large p Obenchain and Vinod [1974] define a correlation coefficient between $\lambda_i^{1/2}$ and $|R_{yi}|$ called "Positive Correlation Spread Association (PCSA)." Largeness of PCSA (for our example PCSA = 1) suggest that "declining deltas" are appropriate. If it turns out that declining deltas are untenable, Obenchain [1978] suggests generalized ridge regression based on an additional parameter to be noted later in (53).

An alternative method of Section 5.1.5 of Chapter 5 above involves computing eigenvalues λ_i^A and eigenvectors g_i^A with elements g_{ij}^A $(i,j=0,1,...p)$ of the augmented correlation matrix for [y, X]. If g_{0i}^A and λ_i^A are both declining with i we note that declining deltas are obviously appropriate because they de-emphasize only the "non-predictive multicollinearities."

The covariance matrix of OLS regression coefficients given by $s^2(X'X)^{-1}$ can be re-scaled so that it becomes a correlation matrix among regression coefficients having ones along the diagonal. Recall, Chapter 5, Eq. (11), that the ith diagonal element of $(X'X)^{-1}$ is the "variance inflation factor," VIF(i). The rescaling involves dividing i, jth element by the square root of the product $VIF(i) \times VIF(j)$. Some computer programs report these correlation matrices routinely. For our example, the off-diagonal elements (both) are -0.943931. The negative sign suggests a trade-off relationship between their values. Thus we can expect the RIDGE TRACE for this example to have lines going in the opposite direction, similar to Figure 7.1 above. When $k > 0$ we need to rescale the matrix (10) to be a correlation matrix. We denote this matrix of correlation coefficients among (ridge) regression coefficients by $R^b(k)$, defined at each k. Note that its "size" can be measured by sum of squares of $p(p-1)/2$ correlations used in Obenchain and Vinod [1974] and Obenchain [1975] as a measure of the instability of a solution along a RIDGE TRACE called sum of squares of correlations among beta coefficients (SSCBC). For our example, at the OLS value, SSCBC is $(-.943931)^2$ which decreases as k increases. One disadvantage of SSCBC is that $R^b(k)$ matrix needs to be actually computed for each k. Another measure of the "size" of R^b consistent with matrix algebra is its Euclidian norm, which is found as the square root of the sum of squares of its eigenvalues. For $k > 0$, this Euclidian Norm is $s[\sum_{i=1}^{p} \delta_i^4/\lambda_i^2]^{1/2}$. Note that this norm can be computed without actually computing $R^b(k)$ for each k.

Monitoring of ridge solutions for various values of k considered so far includes the multiple correlation coefficient, R^2, $\sum_{i=1}^{p} s^2 \, VIF(i)$, where $VIF(i) = i^{th}$ diagonal element of $G \, Diag(\delta_i^2/\lambda_i) G'$, SSCBC, ISRM, etc. We recommend that the user compute them for each choice of k (and/or m). Generally speaking, it is advisable to stay close to the OLS solution $(k = 0)$, where the bias is zero. We suggest rejection of large biasing parameters k which shrink the uncorrelated component c_i so much that the value $\delta_i c_i$ is outside the 95% confidence interval around the OLS value c_i. For our example, the degrees of freedom are 6 and tabulated t value is 2.447. The usual 95% confidence intervals based on adding and subtracting 2.447 times the standard errors (square root of variance of c_i from Table 7.1) are given in Table 7.2. The shrinkage in the Table is computed by dividing c_i by the respective lower and upper bounds. Thus our $\delta^{5\%}$ in the Table is simply the smaller of the two shrinkage factors. The corresponding excessive shrinkage in the Table is found by the definitional relation $k_i^{5\%} = (\lambda_i/\delta_i^{5\%}) - \lambda_i$.

TABLE 7.2
Avoiding Excessive Shrinkage

Principal	95% Confidence Internal			Shrinkage δ		Excessive	
Axis	Lower	c_i	Upper	Low	Up	$\delta_i^{5\%}$	$k_i^{5\%}$
1	-.8061	-0.6947	-.5833	1.1604	.8396	0.8396	0.3712
2	0.1666	0.8173	1.4680	0.2039	1.7961	0.2039	0.2224

Table 7.2 suggests that $k^{5\%} = \min_i (k_i^{5\%}) = 0.2224$ for the k in ORE will lead to a shrinkage of c_i on its 95% confidence boundary. Now we may regard 5% to be the acceptance level associated with $k = 0.2224$. For this example it is advisable not to use any k larger than 0.2224. Instead of fixing the 5% level we can evaluate it for each k of ORE and ridge solutions can be compared with each other with reference to the acceptance level. Similar ideas are found in Vinod [1977], Obenchain [1977] and McCabe [1978].

7.5.1 Further Techniques for Solution Selection Illustrated

In ordinary ridge regression, we shrink the OLS estimates c_i by the shrinkage fractions defined by $\delta_i = \lambda_i (\lambda_i + k)^{-1}$, where k is the biasing parameter of ridge regression used in equation (1). Figure 7.2 gives a Hoerl-Kennard style RIDGE TRACE for this example, with k on the horizontal axis. Figure 7.3 gives the RIDGE TRACE with respect to the m-scale discussed in Section 7.2.4 above. In both cases we have plotted the unstandardized ridge coefficients excluding the intercept.

For this example, one eigenvalue $\lambda_2 = 0.056069$ may be judged to be too small, and we may decide that there is m=1 deficiency in the rank of the $X'X$ matrix. This would suggest a choice of m=1 as the solution point. The corresponding value of k may be obtained by using a computer program given at the end of this chapter, as Explanatory Note 7.1. The computer program uses Newton's method to find k from given m and the eigenvalues. For m=1 we have k=0.3327 and we can directly verify whether the computer program works correctly by using the defining relation $m = p - \delta_1 - \delta_2 - \ldots - \delta_p$, where $\delta_i = \lambda_i(\lambda_i + k)^{-1}$. When $k = 0.3327$ we have $\delta_1 = 0.853803$ and $\delta_2 = 0.146197$ as the shrinkage fractions. The standardized regression coefficients are $std.b_1 = 0.33494$ and $std.b_2 = 0.50391$. The unstandardized $b_1 = 0.6713$ and $b_2 = 0.5264$, with the correlation coefficient between them of only 0.00000030. The Euclidian norm of this correlation matrix is 1.4142, whereas SSCBC = 2. For this choice of $k = 0.3327$ (m=1) the multiple correlation $R^2 = 0.9260$, the ISRM defined by Eq. (26) is almost zero.

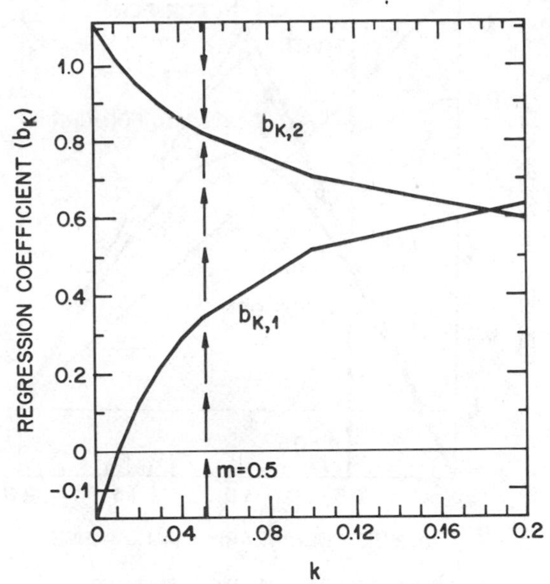

Figure 7.2 Hoerl-Kennard Ridge Trace for Ordinary Ridge Regression

Hoerl, Kennard and Baldwin [1975] suggested the following choice of k

$$k_{HKB} = p\, s^2 / \sum_{i=1}^{p} b_i^2 \,, \tag{47}$$

where $s^2 = (y-Xb)'(y-Xb)(T-p-1)^{-1}$, and where b_i are the unstandardized regression coefficients. It is sometimes convenient to count the intercept as one of the regressors. When this is done, the denominator of s^2 is $(T-p)$ instead of $(T-p-1)$. We expect that the reader will infer our treatment of the intercept depending on the context. For our examplc, $s^2 = 1.0056$, and the sum of squares of b_i is 1.277634. Hence, $k_{HKB} = 1.5742$. An iterative version of (47) is suggested by Hoerl and Kennard [1976] whereby a new denominator is used for each iteration.

Lawless and Wang [1976] suggested a Bayesian modification of k_{HKB} as

$$k_{LW} = ps^2 / \sum_{i=1}^{p} \lambda_i c_i^2 \,, \tag{48}$$

where $c = G'b$ gives the elements of unstandardized uncorrelated components of b as $c_1 = 0.9126$ and $c_2 = 0.66695$, and λ_i are the eigenvalues of $X'X$. For this example, we have $k_{LW} = 1.6435$.

A generalization of k_{LW} is suggested by Thisted [1976] which is guaranteed to reduce the MSE of OLS if σ^2 is known, and $p \geq 3$

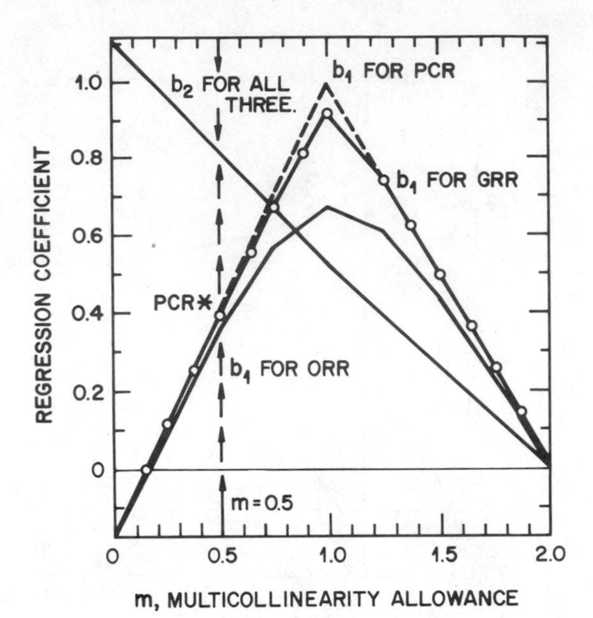

Figure 7.3 Three RIDGE TRACES

$$k_{THD} = \sigma^2 / \sum_{i=1}^{p} c_i^2 \lambda_i^{-1} \lambda_p^{-1} W_i^{-1} \ , \tag{48a}$$

where $W_i = \max\,[0, \mathit{Trace}\ \Lambda^{-2} - 2\lambda_i^{-1}\lambda_p^{-1}]$. For our example, $W_2 = 0$ implies $k_{THD} = 0$. This means that the OLS solution is recommended. If $\lambda_i = 1$ for all i we have $k_{THD} = (p-2)s^2/\sum c_i^2$, which can be seen to be a conservative generalization of k_{HKB} given in (47).

Mallows [1973] discusses his C_p plotting technique designed for selecting a subset of regressors from a larger set of candidate regressors. A good discussion of this is available in Daniel and Wood [1971], with computer programs and examples where the choice of the best subset is indicated by low values of C_p. For ridge regression, Mallows [1973] suggests a modification of C_p defined as:

$$C_L = (y - Xb_k)'(y - Xb_k)/s^2 - T + 2 + 2(\sum_{i=1}^{p} \delta_i) \ . \tag{49}$$

For our example, when m=1, C_L = 12.8582 which is considerably higher than the corresponding value at m=0 (i.e., OLS) where $C_L = p + 1 = 3$, which means that m=1 is not good solution by this criterion. Even for m = 0.25, k = 0.018 we find that C_L = 3.0673 is larger than its value at m=0. It is a common occurrence that the value of k suggested by Mallows' C_L criterion is smaller than k_{HKB} or k_{LW}. An attractive property of C_L criterion is that it tends to minimize an almost unbiased estimate of the "predictive" mean squared error defined by:

$$PMSE = E(y - Xb_k)'(y - Xb_k) .$$

Obenchain [1975] shows that the slight bias can be removed by redefining $s^2 = (y-Xb)'(y-Xb)\,(T-p-3)^{-1}$ before substituting it in (49) and adding 2.

Obenchain [1975] also shows that minimizing estimated PMSE yields the following shrinkage factor denoted by a superscript MAL for Mallows:

$$\delta_i^{MAL} = \max(0,\ 1 - \frac{1-R^2}{(T-p-1)R_{yi}^2}) \tag{50}$$

which will minimize Mallows' C_L statistic. Recall from (46) that our $R_{y1}^2 = 0.9378$, $R_{y2}^2 = 0.0380$, and $R^2 = 0.9758$. Therefore, $\delta_1^{MAL} = 0.9957$ and $\delta_2^{MAL} = 0.8939$. It is interesting to note that $\delta_i^{MAL} = \max(0,\ 1 - F_i^{-1})$ where F_i is Snedecor's F statistic for testing the significance of c_i. In Table 7.1 these F_i are found by squaring the $|t_i|$ values.

Dempster, Schatzoff and Wermuth [1977] report a major simulation of various ordinary ridge estimators, according to which the following choice called RIDGM was a winner. Their choice of k denoted by k_{DSW}, where the subscript refers to the first letter in their names, solves the following nonlinear equation for a positive k_{DSW}:

$$p = \sum_{i=1}^{p} c_i^2/(s^2\, k_{DSW}^{-1} + s^2\, \lambda_i^{-1}) . \tag{51}$$

The derivation of this rule is motivated by Bayesian methods. The prior distribution of γ_i is $N(0,\omega^2)$, the observable c_i is independently and identically distributed as $N(0,\omega^2 + \sigma^2/\lambda_i)$ *a priori*. It can be shown that the prior expectation of $\sum_{i=1}^{p} c_i^2/(\omega^2 + \sigma^2/\lambda_i)$ is simply p. It is an interesting exercise to the reader to compute k_{DSW} for our example.

Efron and Morris' comment on Page 91 following the Dempster et al. [1977] includes an "Empirical Bayes" estimator which focuses attention on a so-called "marginal distribution." The word "empirical" suggests that the prior distribution is obtained from the available data rather than being truly "prior". In "Empirical Bayes" approach the prior is integrated out. In the ridge regression context, the marginal distribution of b (OLS coefficients) is an average of the "conditional" distribution of $b \sim N(\beta, \sigma^2(X'X)^{-1})$, and the ridge prior $\beta \sim N(0,\ \sigma^2 G K^{-1} G')$. Writing $X'X = G \Lambda G'$, the marginal is $b \sim N(0,\ \sigma^2 G[K^{-1} + \Lambda^{-1}]G')$. Obenchain [1981] explains that the optimal shrinkage factor from this rule reduces to Mallows' rule $\delta_i^{MAL} = \max(0,\ 1 - F_i^{-1})$, where $F_i = c_i^2 \lambda_i / s^2$ is the F statistic for testing the significance of c_i mentioned above.

McDonald and Galarneau [1975] use the following result to suggest k_{MG} which equates the squared length of the regression vector to its expectation. The basic result is

$$Eb'b = \beta'\beta + \sigma^2\ \mathrm{Trace}(X'X)^{-1} . \tag{52}$$

As k increases we know that $b_k' b_k$, the squared length of the ridge estimator, decreases. The idea is to reduce the length by $s^2 \sum_{i=1}^{p} \lambda_i^{-1}$. The determination of k_{MG} for our example is left as an exercise. This procedure defaults to the OLS estimator if $b'b < s^2 \sum_{i=1}^{p} \lambda_i^{-1}$.

The solution which corresponds to the choice m=0.5 (i.e., k = 0.05139) is marked for special attention in Figures 7.2 and 7.3. For this choice the unstandardized regression coefficients are $b_1 = 0.3503$ and $b_2 = 0.8174$, with shrinkage fractions $\delta_1 = 0.974235$ and $\delta_2 = 0.525765$. The correlation coefficient between b_1 and b_2 is -0.817074, which is less than it is for OLS. The Euclidian norm of this correlation matrix is 1.8263, SSCBC is 3.3352, C_L is 4.2783, R^2 is 0.9666 ISRM from Eq. (26) is 1.3352 which is smaller than a comparable OLS value of 1.93607 but larger than the nearly zero value at m=1 reported earlier. From the simulation in Wichern and Churchill [1978] and Vinod's letter [1979] one may conclude that minimizing ISRM may give an overestimate of m (or k) and one should try to use a lower k than the one indicated by ISRM, to be on the safe side. In light of these considerations, we give special attention to the solution at m=0.5. In the three dimensional plots of the likelihood function given in Chapter 1 (Figures 1.4a and 1.4b) the place marked by the symbol ORR represents the ordinary ridge regression solution for m=0.5 (k = 0.05139). See also the RIDGE TRACE of Figure 7.3.

The generalized ridge estimator b_K of Eq. (8) above is distinguished from the ordinary ridge estimator b_k by the fact that the biasing parameters can be distinct. Consider a solution where $k_1 = 0.010794$ and $k_2 = 0.055724$, which imply that $\delta_1 = 0.994475$, $\delta_2 = 0.505526$, and $m = p - \delta_1 - \delta_2 = 0.5$. The corresponding unstandardized regression coefficients are $b_1 = 0.393642$ and $b_2 = 0.815567$, with the multiple $R^2 = 0.966491$, Mallows' $C_L = 4.31602$. The symbol GRR marks this in Figs. 1.4a, 1.4b and 7.3.

One motivation for suggesting generalized ridge estimator is Hoerl and Kennard's [1970a] result that the "MSE optimal" values of k_i (denoted by k_i^{MSE} in this section) are distinct for different i. Assuming that $\gamma = G'\beta$ and σ^2 are known, and minimizing the theoretical expression for MSE gives $k_i^{\text{MSE}} = \sigma^2/\gamma_i^2$. The corresponding MSE-optimal shrinkage factors for our example are 0.995722 and 0.648008 respectively, if we replace the true unknowns by their OLS (maximum likelihood) estimates. Vinod [1976u] argues that if σ^2 and γ_i are known, the correct globally optimal estimator of δ_i is γ_i/c_i rather than $k_i^{\text{MSE}} = \sigma^2/\gamma_i^2$. Although the globally optimal choice is impractical, Kennard [1976], its existence suggests that it is incorrect to call k_i^{MSE} as "the optimal" choice. In the process of using the stochastic OLS estimates of σ^2 and γ_i^2 one loses the remaining optimality properties, and ends up with a heuristic estimator. Dwivedi, Srivastava and Hall [1980] have developed explicit expressions for the MSE of this heuristic estimator and tabulated them for various specific values of the noncentrality parameter $\lambda_i \gamma_i^2/2\sigma^2$. They find that ridge regression is more efficient (reduces the MSE of OLS) as long as

$\lambda_i \gamma_i^2 / 2\sigma^2 < 1$. They recommend a preliminary test for this. Further theoretically optimal ("maximum likelihood ridge") results are extensively studied by Obenchain [1975, 1978, 1981].

Obenchain discusses a two parameter estimator mentioned by Goldstein and Smith [1974] and others, and defined from $k_i = k\lambda_i^q$ as:

$$b^*(k,q) = \left[(X'X)^{1-q} + kI\right]^{-1}(X'X)^{-q}X'y \ . \tag{53}$$

When $q = 0$ we have ORE. For problems lacking "declining deltas" property, one can reject ridge regression. A more bold strategy is to use stochastic estimates of q by monitoring the cosine of the angle between the vector $|R_{y1}|,...,|R_{yp}|$ and the vector: $\lambda_1^{(1-q)/2},...,\lambda_p^{(1-q)/2}$

$$\cos(q) = \frac{\sum_{j=1}^{p} |R_{yi}| \lambda_j^{(1-q)/2}}{\left[\sum_{j=1}^{p} R_{yi}^2 \sum_{j=1}^{p} \lambda_j^{(1-q)}\right]^{1/2}} , \tag{54}$$

and choose the q which will maximize $\cos(q)$. Obenchain [1981] then recommends the biasing parameter after averaging of corresponding k_i values.

$$k^{OBN} = \frac{\left(\sum_{j=1}^{p} \lambda_j^q\right)\left(\sum_{j=1}^{p} \lambda_j^{(1-q)}\right)[1-R^2\cos^2(q)]}{pnR^2\cos^2(q)} . \tag{55}$$

The MSE properties of the resulting estimators are unknown and appear to be intractable. Note that large and positive q implies increasing deltas, q=1 corresponds to Stein-Rule type equal deltas, q=0 implies declining deltas as in ORE, and q<1 corresponds to rapidly declining deltas. When q is large and negative the deltas are similar to the principal components regression (PCR) and decline rapidly from one to zero. Although somewhat encouraging simulation results for (53) are reported by Gibbons [1981], it is not clear whether one and two parameter estimators can be directly compared; and whether the simulation results depend on assumed intra-class correlation coefficient structure of the regressors. More research is needed before (54) and (55) can be recommended. For our example, $q = 0.092$ appears to maximize $\cos(q)$ of (54) which is close to the ORE value of q=0. This confirms our earlier conclusion that "declining deltas" are appropriate for this example.

The principal components regression (PCR) noted above is a special case of generalized inverse estimator in Marquardt's [1970] terminology. For example, let the two shrinkage factors for PCR be 1 and 0.5 respectively. Note that this means m=0.5, and the corresponding solution can be plotted on a RIDGE TRACE provided we have the m-scale on the horizontal axis, as in Figure 7.3. This solution is marked PCR* to be distinguished from the other solution where one principal component is eliminated, m=1 with $\delta_1 = 1$ and $\delta_2 = 0$. When integer number of principal components are eliminated, our m values are also integers, and we have a special case called "generalized inverse" estimator, Marquardt [1970], Rao [1973] and others. This is plotted by the symbol PCR in Figure 7.3. The multiple R^2 for PCR* is 0.966311 and corresponding Mallows' C_L is 4.36081 (m=0.5). The Euclidian norm of the correlation matrix among regression coefficients is 1.8023, with the correlation coefficient between b_1 and b_2 of -0.790064, SSCBC = 3.2484 and ISRM = 1.2484. When m is increased from 0.5 to 1 at the solution marked PCR, the $R^2 = 0.937773$, $C_L = 1$, the Euclidian norm becomes 1.5069, the correlation coefficient between b_1 and b_2 is -0.367977, which is almost halved in magnitude, SSCBC = 2.2708 and ISRM = 2.

In Section 5.4 of Chapter 5 we have discussed the effects of perturbation in the available data *beyond the published digits*. It is clear from Table 5.3 that ridge solutions are considerably more stable than the OLS solutions. In many practical situations such stability is desirable. For these problems the choice of k for ridge regression may be based on a stability criterion. Allen [1974] suggests a two-step stability criterion. The first step is to omit the $t-th$ observation and find a "prediction" $(Xb_k)_t$ of the omitted y_t value for a certain range of k to minimize the "predictive" error from omitting one observation. This approach requires a large number of iterative solutions of the regression problem. It is not clear why one should omit only one observation at a time, rather than two or three (say). The solutions resulting from omitting two observations need not be similar to the solutions for k resulting from omitting only one observation. This technique is sometimes called cross-validation. It has been refined recently by Wahba [1976] and Golub, Heath and Wahba [1979] by introducing invariance with respect to rotation of axes. Obenchain [1981] shows that the resulting choice of k is generally close to Mallows' from (50) or Empirical Bayes methods.

From the discussion following Smith and Campbell [1980] it is clear that ridge regression should not be applied blindly to any and all regression problems. There are problems where ridge type shrinkage may yield a poorer result than OLS. Using Jensen's inequality Brook and Moore [1980] have shown that some kind of shrinkage of OLS is usually advisable. If the user is willing to check whether "declining deltas" are appropriate for his problem before applying ridge regression, he can be successful. In fact, we recommend constructing both Tables 7.1 and 7.2 to make sure that the shrinkage is not excessive.

Since our example has only nine observations and two regressors, it is obviously not intended to be a serious application of ridge methods. It should, however, help clarify the various formulas and concepts.

Explanatory Note 7.1

A FORTRAN program to Obtain k form m

The purpose of this note is to give a self-explanatory FORTRAN program for obtaining from the multicollinearity allowance m defined in equation (23) the appropriate k for "ordinary" ridge. The program is illustrated with the 3 regressors example in Vinod [1976j]. The notation in the computer program is:

S(I)	=	eigenvalues, λ_i
EM	=	m, multicollinearity allowance
EK	=	k, ridge biasing parameter
NREG	=	p, number of regressors
ACC	=	accuracy level desired, say 0.001
IPRINT	=	extent of printing desired.

```
//SRCHK JOB
//          EXEC FORTGCLG,RGNG=1
//FORT.SYSIN DD *
C MAIN CALLS SRCHK
            IMPLICIT REAL*8(A-H,P-Z)
            DIMENSION S(10)
            S(1)=2.66028
            S(2)=.327707
            S(3)=.012010
            EM=0.0
            DO 10 IE=1,13
            IF(IE.EQ.1) GO TO 10
            EM=DFLOAT(IE-1)/4.DO
            CALL SRCHK(S,EK,3,EM,0.001D0,2)
            WRITE(6,111) EM, EK
10          CONTINUE
111         FORMAT(' DESIRED EM= ',G12.6,' IMPLIES EK= ',G12.6)
            STOP
            END
            SUBROUTINE SRCHK(S,EK,NREG,EM,ACC,IPRINT)
            IMPLICIT REAL*8(A-H,P-Z)
C INPUT:    S IS A VECTOR OF EIGENVALUES OF X TRANSPOSE X MATRIX
C INPUT:    NREG= NUMBER OF REGRESSORS, I.E. DIMENSION OF
C           X TRANSPOSE X.
C INPUT:    ACC= LEVEL OF ACCURACY
C           IF ACC IS OUTSIDE THE RANGE .000001,20 ; ERROR IS ASSUMED.
C INPUT:    IPRINT IS LEVEL OF DETAIL IN PRINTING THE RESULTS
C           E.G. IPRINT< OR =1 WILL PRINT DETAILS, IPRINT > 1 WILL NOT.
C           OUTPUT: EK= THE BIASING FACTOR,RIDGE REGRESSION,
C           HOERL-KENNARD
C           TECHNOMETRICS 1970 FEBRUARY.
C THIS SUBROUTINE USES NEWTON'S METHOD TO OBTAIN EK FROM EM AS IN
C VINOD, H. D., JOUR. AMER. STATIST. ASSOC., DECEMBER 1976, PAGE 837.
            DATA ZERO/10.D-14/, FINITY/10D9 /
            DATA IBEGIN/0/
            DIMENSION S(1)
C DUMMY DIMENSION FOR VECTOR S IS 1. CALLING PROGRAM MUST
C HAVE DIMENSION=NO. OF REGRESSORS
            IBEGIN=IBEGIN+1
```

```
              P=NREG
C NREG MEANS NO OF REGRESSORS
              IF(IPRINT.LE.1) WRITE(6,100) (S(I),I=1,NREG)
100           FORMAT(' FIND K FROM EM SO THAT SUM S(I)/(SI+K)=P-EM'/(1X,
     X(G12.6))
              IF(ACC.LE.0.000001D0.OR.ACC.GT.20.D0)WRITE(6,107)ACC
              IF(ACC.LE.0.000001D0.OR.ACC.GT.20D0)RETURN
107           FORMAT(' ACC LE .000001 OR ACC GT 20 ',G16.6)
               EK=0.0
              IF(IBEGIN.GT.8) GO TO 200
114           FORMAT(' EM, EK ',2G12.6)
              IF(EM.EQ.0D0) WRITE(6,114) EM,EK
200           CONTINUE
              IF(EM.EQ.0D0) RETURN
              IF(EM.GT.P) WRITE(6,112) EM, P
112           FORMAT(' DESIRED EM IS .GE. NO OF REGRESSORS ',2G12.6)
              IF(EM.GE.P)EK=FINITY
              IF(EM.GT.P) RETURN
              ITER=0
               JERR=0
9              CONTINUE
               ITER=ITER+1
               SUM=0.0
               SUM2=0.0
               DO 10 I=1,NREG
               IF(EK.LT.0D0 ) EK=ZERO
               IF(EK.LT.0D0 ) JERR=JERR+1
               IF(JERR.GT.10) RETURN
              DELTAI= (S(I)/ (S(I)+EK ))
               SUM=SUM+DELTAI
               SUM2=SUM2+(S(I)/ (S(I)+EK)**2)
10            CONTINUE
              IF(SUM2.LE.ZERO) WRITE(6,113) SUM2
113           FORMAT(' SUM2 IS TOO SMALL ',G12.6,' EK=INFINITY ')
              IF(SUM2.LE.ZERO)EK=FINITY
              IF(SUM2.LE.ZERO) RETURN
              ACHIV=SUM-P+EM
C WE SHALL ATTEMPT TO MAKE ACHIV =0.0 USING NEWTON'S METHOD
              EK3 = ( ( SUM- P+EM)/ SUM2)
              IF(EK3 .LT.ACC) WRITE(6,108) EK,EK3 , SUM,P,EM,SUM2
108           FORMAT(' TAYLOR ADJUSTMENT HAS WRONG SIGN ',6G12.6)
               EK2=EK + EK3
              IF(IPRINT.LE.1) WRITE(6,1020) EK3,SUM,SUM2
1020           FORMAT(' ADJUSTMENT TO EK, SUM DELTAI,SUM ITS
     XPARTIAL DERIVATIVE ',3G12.6)
              IF(IPRINT.LE.1)
              WRITE(6,102) ITER, EM, EK, EK2
102           FORMAT(' ITERATION NO.,DESIRED EM,OLD EK,NEW EK '
     X,I4,2X3G12.6)
              IF(ACC.GT.20.D0) STOP
              EK=EK2
              IF( ITER.GT.25) RETURN
C IF YOU WNAT THE SUBR. TO TRY OTHER THAN 25 TIMES CHANGE THIS 25
              IF(DABS(ACHIV).GT.ACC) TO TO 9
```

```
          IF(ITER.LE.25) GO TO 9
          RETURN
          END
//*
//GO.SYSLIB DD DSN=SYS1.FORTLIB,DISP=SHR
//*
```

The corresponding k-scale and m-scale RIDGE TRACES are also included as Figures. 7.4 and 7.5 respectively. Although the example of Vinod [1976j] was originally intended to be illustrative, it assumed added importance when it was submitted as Exhibit No. 2073 in a Federal District Court in Chicago, Illinois, on May 19, 1980, by the American Telephone and Telegraph Co. in an anti-trust lawsuit by MCI Communications Corporation involving $2.7 billion.

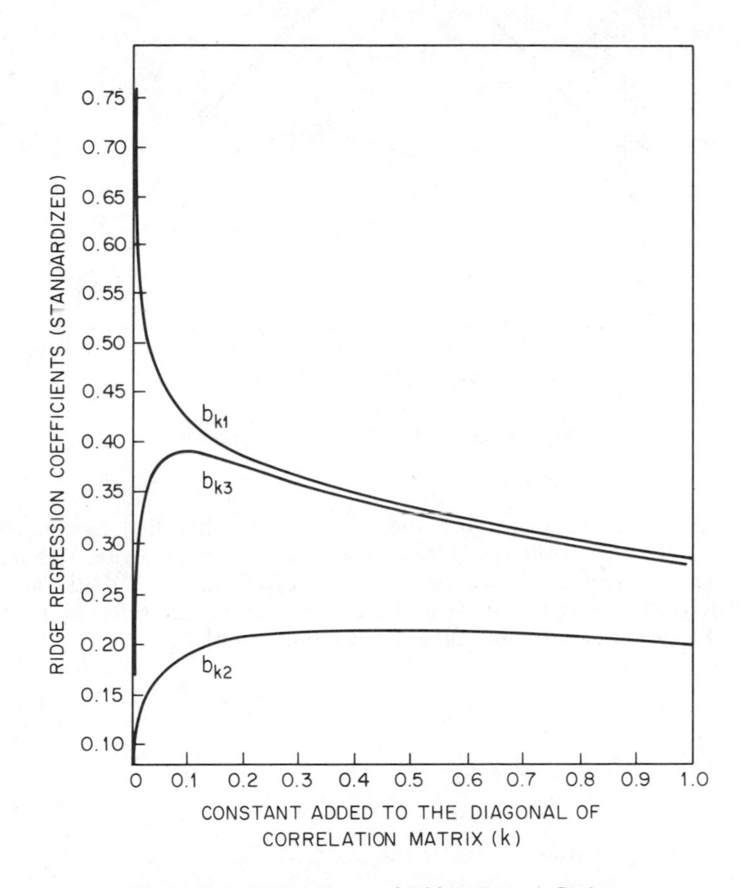

Figure 7.4 Hoerl-Kennard Ridge Trace k-Scale

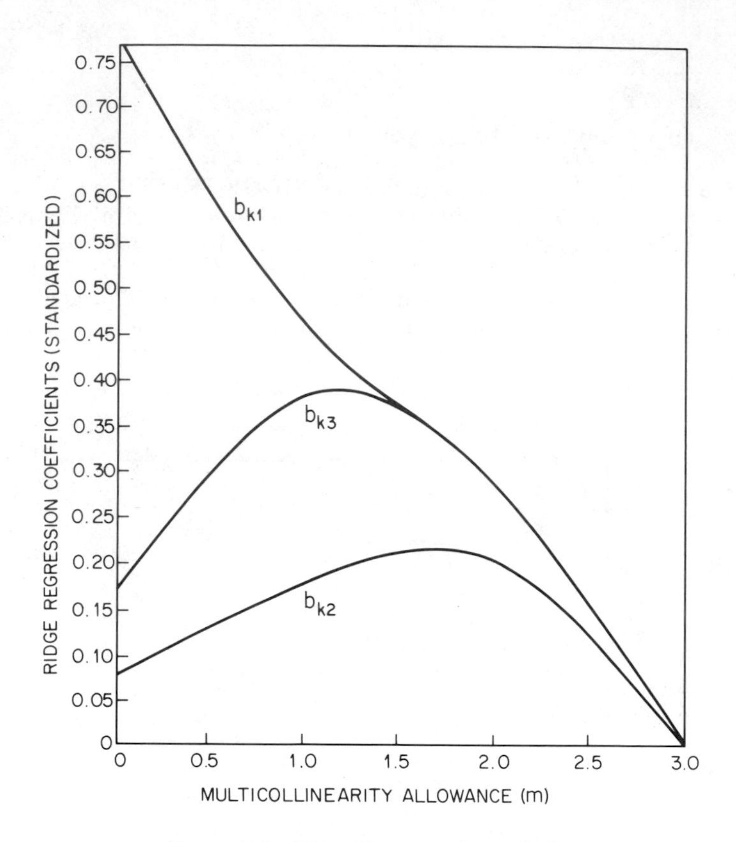

Figure 7.5 Ridge Trace on the m-Scale

One issue before the court and the jury was whether Bell System enjoys economies of scale. Although this issue was not the major issue, this may be one of the most serious applications of ridge regression. For additional examples of RIDGE TRACES on the m-scale, the reader is referred to Vinod [1974], Obenchain [1975] Obenchain [1976a and 1976b].

EXERCISES

7.1 Use the Bell System production function data of Exercise 1.6 and make a complete run of all ridge regression methods (including graphical, heuristic and minimax) discussed in this chapter. Show that declining deltas are appropriate.

7.2 Using the computer program for ordinary regression available to you follow the augmentation of Section 7.4 of the y and X data, and verify that regression coefficients for ORE are obtained correctly, but standard errors are incorrect, residual variance is also incorrect. Why? What are the correct expressions?

7.3 List the various non-stochastic methods of obtaining k. What is the common property shared by these methods?

7.4 Show that many of the ridge estimators are linear functions of the corresponding maximum likelihood estimators. Hence show that ridge estimators are "sufficient" statistics. Show that the constant estimator has zero variance, but does not belong to the class of sufficient estimators.

7.5 For the case of ordinary regression, show that Mallows' C_L statistic of (49) can be used for selection of subsets of regressors. Show that it involves an adjustment to R^2. Derive (50).

7.6 Develop a "preliminary" test for size of the non-centrality required by Dwivedi et. al. [1980]. Can you make a probabilistic statement that a ridge estimator is superior to OLS? In what sense are the minimax ridge rules pessimistic?

7.7 From the sampling distribution of $b'b/s^2$ insert a non-stochastic tabulated value in equation (22) and solve for a non-negative k_{max}.

7.8 From Bhattacharya's [1966] version of the Stein-Rule estimator for the regression problem construct a set of stochastic k_i which are guaranteed to reduce the MSE of OLS everywhere in the parameter space (Hint: See Vinod [1978i]). Show that this amounts to minimax estimation and that it does not necessarily yield declining deltas.

8

Further Ridge Theory and Solutions

8.1 INTRODUCTION

The ridge existence theorems and expressions for mean squared error (MSE) usually based on the assumption of nonstochastic biasing constants k_i were discussed in the previous chapter. In this chapter we discuss certain recent extensions of the theory for stochastic k_i and resulting implications for the selection of a single solution from the ridge family. Some recent results for nonstochastic k_i and confidence intervals are also included toward the end of this chapter.

Consider the standardized normal regression model

$$y = X\beta + u = Z\gamma + u \tag{1}$$

with $X = H\Lambda^{1/2}G'$, $\gamma = G'\beta$, $Z = H\Lambda^{1/2} = XG$, from the singular value decomposition discussed in Chapter 1.

We restate the following conventional assumptions:

Assumption 1 The matrix X of explanatory variables is nonstochastic and of rank p.

Assumption 2 The disturbance vector u is distributed as multivariate normal with mean vector zero and covariance matrix σ^2 I , i.e.,

$$u \sim N(0, \sigma^2 I) . \tag{2}$$

Assumption 3 The sample size T is greater than the total number of explanatory variables p.

The OLS estimator $b = (X'X)^{-1}X'y$ of β is related to its "uncorrelated components" $c = G'b$ which estimates γ. The generalized ridge estimator is written as $b_K = (X'X + GKG')^{-1}X'y = (X'X + GKG')^{-1}X'Xb$. For diagonal K we have

$$b_K = G\Delta G'b = G\ diag(\delta_i)G'b\ , \tag{3}$$

$$= G\ diag(\delta_i c_i) = Gc_K\ .$$

where $\delta_i = \lambda_i(\lambda_i + k_i)^{-1}$ are "shrinkage fractions," and where c_K denotes a $p \times 1$ vector of $c_{Ki} = \delta_i c_i$ which are (shrinkage) ridge estimators of γ_i.

Now we minimize the MSE of b_K defined for nonstochastic biasing parameters k_i as a sum of two terms (variance and squared bias)

$$MSE(b_K) = \sigma^2 \sum_{i=1}^{p} \delta_i^2\lambda_i^{-1} + \sum_{i=1}^{p} (\delta_i - 1)^2\gamma_i^2\ . \tag{4}$$

See equation (12) in Chapter 7. Also, from (12a) or (17) in Chapter 7, the MSE of the ith uncorrelated component of c_K is

$$E(c_{ki} - \gamma_i)^2 = \sigma^2\delta_i^2\lambda_i^{-1} + (\delta_i - 1)^2\gamma_i^2\ . \tag{4a}$$

The necessary condition for a minimum of (4) requires that its derivatives with respect to δ_i be zero. We require:

$$0 = 2\sigma^2\delta_i\gamma_i^{-1} + 2(\delta_i - 1)\gamma_i^2\ ,$$

$$= \delta_i(2\sigma^2\lambda_i^{-1} + 2\gamma_i^2) - 2\gamma_i^2\ .$$

Hence the minimum MSE value of δ_i is

$$\delta_i^{MSE} = 2\gamma_i^2(2\sigma^2\lambda_i^{-1} + 2\gamma_i^2)^{-1}\ , \tag{5}$$

$$= \lambda_i(\sigma^2\gamma_i^{-2} + \lambda_i)^{-1}\ ,$$

where we cancel 2, multiply by λ_i and divide by γ_i^2, both the numerator and the denominator. The corresponding MSE optimal value of k_i is

$$k_i^{MSE} = \sigma^2\gamma_i^{-2}\ . \tag{6}$$

To verify the derivation of this from (5), note that $\delta_i^{MSE} = \lambda_i(\lambda_i + k_i^{MSE})^{-1}$ by definition of shrinkage factors. Also note that the minimum MSE values of δ_i and k_i for (4a) are the same as in (5) and (6), respectively.

Remark 1. Since MSE optimal values k_i^{MSE} and δ_i^{MSE} involve unknown parameters σ^2 and γ_i (i.e., β_i) these expressions are useless from a practical viewpoint, because if we really knew σ^2 and γ it would be silly to perform ridge regression.

The above derivation of optimal values assumes that k_i (and therefore δ_i) are nonstochastic. If we substitute any stochastic estimates for k_i^{MSE}, the optimality property cannot be shown to hold true. Many researchers feel that the theoretically optimal k_i^{MSE} are suggestive of the true optimal values. In some cases simulations are used to assess their optimality.

8.2 OPERATIONAL VALUES OF SHRINKAGE FACTORS

Recall that "MSE optimal" expressions for biasing parameters k_i and shrinkage factors δ_i involving true unknown parameters are denoted by the superscript MSE. Now, we consider estimating $k_i^{MSE} = \sigma^2 \gamma_i^{-2}$. Recall that an unbiased estimate of σ^2 is $s^2 = (y-Xb)'(y-Xb)/n$, where $n=T-p$. Since the expectation of c_i^2 is $\gamma_i^2 + \sigma^2/\lambda_i$, the unbiased estimate of γ_i^2 is $c_i^2 - s^2/\lambda_i$. In general, we write the following family of choices of k_i which may be called a "double f-class" family.

$$\hat{k}_i = (\hat{k}_{f_1,f_2})_i = \frac{f_1 s^2}{c_i^2 - f_2 s^2 \lambda_i^{-1}} \; ; \; s^2 = \frac{1}{n}\hat{u}'\hat{u}, \; \hat{u} = y - Zc \; . \tag{7}$$

For f_1=1 and f_2=0 we get one member of the family

$$(\hat{k}_{1,0})_i = s^2/c_i^2 \tag{8}$$

which is suggested by Hoerl and Kennard [1970] as their first iteration (HKFI). Recently Dwivedi et al. [1980] report some exact results, and tabulate the relative efficiency of this estimator with respect to OLS for various values of the non-centrality parameter. Further for f_1=2 and f_2=2 we get

$$(\hat{k}_{2,2})_i = 2s^2/(c_i^2 - 2s^2\lambda_i^{-1}) \tag{9}$$

based on Stein's "unbiased" estimate of MSE discussed in Vinod [1977]. Also, for f_1=2 and f_2=1 we obtain

$$(\hat{k}_{2,1})_i = 2s^2/(c_i^2 - s^2\lambda_i^{-1}) \; , \tag{10}$$

which corresponds to a theoretical upper bound of k_i motivated in Vinod [1977]. Recently, Hemmerle and Brantle [1978] study $f_1=f_2=1$ based on their minimization of an "unbiased" estimate of MSE.

8.2.1 Hemmerle's Analytic Solution to Iterative Ridge

Hoerl and Kennard's [1970] iterative ridge estimator starting with the initial choice f_1=1, f_2=0 in (8) may be conveniently denoted by

$$k_i^{(0)} = s^2/(c_i^{(0)})^2 \; ,$$

where a superscript (j), j=0,1,2,..., is introduced to number the iterates. The corresponding shrinkage factor $\delta_i = \lambda_i(\lambda_i + k_i)^{-1}$ may be similarly denoted by $\delta_i^{(0)}$:

$$\delta_i^{(0)} = \frac{\lambda_i}{\lambda_i + s^2/(c_i^{(0)})^2} = \frac{1}{1 + (s^2/\lambda_i(c_i^{(0)})^2)} = \frac{1}{1 + F_i^{-1}} , \tag{11}$$

where F_i is the traditional F statistic for testing the null hypothesis γ_i=0. The generalized ridge estimator is now $\delta_i^{(0)} c_i^{(0)}$. At the jth iteration we have

$$\delta_i^{(j+1)} = \frac{\lambda_i}{\lambda_i + s^2(\delta_i^{(j)} c_i^{(0)})^{-2}} = \frac{[\delta_i^{(j)}]^2}{[\delta_i^{(j)}]^2 + F_i^{-1}} . \tag{12}$$

Hemmerle [1975] shows by some cumbersome algebra that the converging solution δ_i^* of the sequence is analytically known to solve a cubic equation,

$$\delta_i^{*3} + F_i^{-1}\delta_i^* - \delta_i^{*2} = 0 , \tag{13}$$

based on rewriting (12) as

$$\delta_i^* = \delta_i^{*2}(\delta_i^{*2} + F_i^{-1})^{-1} . \tag{12a}$$

Note that the superscript $*$ represents the converging solution to (12). This cubic is written in terms of three factors as

$$0 = (\delta_i^* - 0)(\delta_i^* - \delta_i^{POS})(\delta_i^* - \delta_i^{NEG}) , \tag{14}$$

where

$$\delta_i^{POS} = (1/2) + (\frac{1}{4} - F_i^{-1})^{1/2} \tag{15}$$

and

$$\delta_i^{NEG} = (1/2) - (\frac{1}{4} - F_i^{-1})^{1/2} \tag{16}$$

are respectively the two solutions involving a positive and negative radicand. The reader can verify the equivalence of (13) and (14) by direct multiplication. These three solutions for δ_i given by 0, δ_i^{POS} and δ_i^{NEG} are based on Hemmerle's methods. If the radicand is negative, the solution is imaginary.

Next, we consider the stability properties of this iterative scheme by considering whether the difference $\delta_i^{(j+1)} - \delta_i^{(j)}$ is monotonically decreasing. The derivative of $[\delta_i^{(j+1)} - \delta_i^{(j)}]$ with respect to $\delta_i^{(j)}$ is negative when

$$(\delta_i^2 + F_i^{-1})2\delta_i - \delta_i^2(2\delta_i) < (\delta_i + F_i^{-1})^2 ,$$

where the superscript (j) is omitted for convenience. The inequality simplifies to the following stability (convergence) condition

$$(\delta_i^2 + F_i^{-1})^2 > 2\delta_i F_i^{-1} , \tag{17}$$

for Hemmerle's analytic solution (14) to Hoerl and Kennard's iterative ridge estimator starting with $f_1 = 1$, $f_2 = 0$ in (7) written as (12). Note that this condition depends on F_i (the F statistic for the null hypothesis $\gamma_i = 0$) which is a random variable.

Case 1. When $F_i < 4$ the radicand $(1/4 - F_i^{-1})$ is negative, and therefore the solutions δ_i^{POS} and δ_i^{NEG} are imaginary and useless as shrinkage factors. Omitting the imaginary part, the solution is 1/2 from (15) and (16), but this choice does not satisfy the condition (17). Hence $\delta_i = 0$ is the stable choice in this case.

Case 2. When $F_i > 4$, the solutions δ_i^{POS} and δ_i^{NEG} are real, but δ_i^{NEG} fails to satisfy the stability condition (17). Hence δ_i^{POS} is the recommended solution for this case.

Remark 2. Hemmerle's iteration does not permit stable solutions in the half open interval (0, 1/2], and thereby arbitrarily restricts the choice of permissible shrinkage factors δ_i.

Hemmerle [1975] suggested that the solutions $\delta_i = 0$ or $\delta_i = \delta_i^{POS}$ may lead to a poor fit (low R^2), and may involve too much bias, and he goes on to suggest some constraints on the bias. These constraints are criticized by Hocking, Speed and Lynn [1976] as being unsatisfactory.

8.3 BIAS AND MSE OF DOUBLE F-CLASS GENERALIZED RIDGE ESTIMATOR

Let us rewrite the GRE from (3) as

$$b_K = Gc_K \,, \tag{18}$$

where

$$c_K = \Delta c = diag(\delta_i)c \tag{19}$$

and $\delta_i = \lambda_i(\lambda_i + k_i)^{-1}$ are shrinkage fractions. Since $MSE(b_K) = MSE(c_K)$ we shall be considering the properties of c_K only, when the diagonal matrix K has elements given by (7) above, $\hat{k}_i = f_1 s^2(c_i^2 - f_2 s^2 \lambda_i^{-1})^{-1}$. In terms of shrinkage factors, $\delta_i = \lambda_i(\lambda_i + \hat{k}_i)^{-1}$, we write the ith element of c_K as

$$(c_{DFC})_i = (c_{f_1,f_2})_i = \delta_i c_i = (1 - \frac{f_1 s^2}{\lambda_i c_i^2 + (f_1 - f_2)s^2})c_i \,, \tag{20}$$

where DFC represents double f-class, f_1, f_2, which are arbitrary scalars. It is interesting to note that for $f_1 = 0$, the double f-class estimator reduces to the OLS estimator c_i.

Using confluent hypergeometric functions, Vinod, Ullah and Kadiyala [VUK, 1979] derive the bias and MSE expressions for all members of the double f-class family of the generalized ridge estimators defined above. From the exact bias and MSE expressions VUK [1979] derive asymptotic results for large non-centrality parameters (essentially because of small σ^2):

$$\theta_i = \lambda_i \gamma_i^2 / 2\sigma^2 \tag{21}$$

to give approximations for the bias and MSE of the c_{DFC} estimator. These approximations do not involve the confluent hypergeometric functions, and are equivalent to the results obtained by directly using Kadane's [1970] "small sigma" approximations (see explanatory note 6.2 in Chapter 6 regarding small sigma expansions).

Theorem 1 The asymptotic expansion for the exact bias of an element of the double f-class estimator up to the order σ^2(equivalently, order $1/\theta_i$) is given as

$$E(c_{DFC} - \gamma)_i = \frac{-f_1 \gamma_i}{2\theta_i} . \tag{22}$$

Theorem 2 The asymptotic expansion of the exact MSE of an element of $(c_{DFC})_i$ up to the order $\sigma^6(1/\theta_i^3)$ is given as

$$E(c_{DFC} - \gamma)_i^2 = \frac{\sigma^2}{\lambda_i} + \frac{\gamma_i^2 f_1}{4\theta_i^2 n} A_1 + \frac{\gamma_i^2 f_1}{8\theta_i^3 n} [3A_1 - 2(f_1 - f_2)A_2] , \tag{23}$$

where $A_1 = f_1(n+2) + 2n$, $A_2 = \frac{n+2}{n}[f_1(n+4)+3n]$, and $f_1 - f_2 > 0$. Recall that $n = T - p$ equals the degrees of freedom. The condition $f_1 - f_2 > 0$ is needed for the existence of the Bias and MSE (see VUK 1979).

Clearly, (23) implies that the asymptotic expansion of the MSE (c_{DFC}) is obtained by a summation of the right-hand side of (23) over the range $i = 1,...,p$. Using the definition $\theta_i = \lambda_i \gamma_i^2 / 2\sigma^2$ we may rewrite (23) as

$$MSE(c_{DFC})_i = \sigma^2 \lambda_i^{-1} + P_1 f_1 A_1 + P_2 f_1 A_1 - P_3 (f_1 - f_2) f_1 P_4 A_3 , \tag{24}$$

where $P_1 = \sigma^4 / n\lambda_i^2 \gamma_i^2$, $P_2 = 3\sigma^6 / n\lambda_i^3 \gamma_i^4$, $P_3 = 2\sigma^6 / n\lambda_i^3 \gamma_i^4$, $P_4 = (n+2)/n$ and $A_3 = f_1(n+4) + 3n$. The notation P_i is chosen to suggest positive quantities since $P_i > 0$ for $i = 1,..,4$.

This MSE is less than the $MSE(c) = \sigma^2 \lambda_i^{-1}$ for OLS provided we have

$$f_1(\theta_i + \frac{3}{2}) P_4^{-1} (A_1 / A_3) < f_1 (f_1 - f_2) , \tag{25}$$

where f_1 is not canceled from both sides because it is not assumed to be positive. Now, verify that no choice of f_1 and f_2 (two finite real numbers subject to $f_1 - f_2 > 0$) can satisfy (25) everywhere in the parameter space, which is also an implication fo the well-known "admissibility" of OLS for $p = 1$, i.e., the univariate model. For example, when θ_i is infinitely large, no finite (large and negative) f_2 can make the right-hand side of the inequality (25) large enough.

The inequality (25) can be useful in choosing f_1 and f_2 values which distinguish various members of the double f-class family of generalized ridge estimators. Since most of the generalized ridge estimators proposed in the literature are members of the DFC family, we can immediately determine the region of dominance by substituting the corresponding values of f_1 and f_2 in (25).

First, we note that any generalized ridge estimator having $f_1=f_2$ cannot dominate least squares. This follows from the fact that it will make the right-hand side of (25) equal to zero whereas the left-hand side must be strictly positive. Thus Hemmerle and Brantle's [1978] choice of $f_1=f_2=1$ or Vinod's [1977] choice $f_1=f_2=2$ cannot be recommended. Secondly, Hoerl and Kennard's [1970] first iteration (HKFI) choice $f_1=1$ and $f_2=0$ is also not recommended in large samples. For large n the term $P_4^{-1}A_1/A_3$ may be approximated by $(f_1+2)/(f_1+3)$. The condition for superiority over least squares for large n becomes

$$(\theta_i + \frac{3}{2})\frac{(f_1+2)}{(f_1+3)} < f_1-f_2>0 \ . \tag{26}$$

This will certainly not hold for $f_1\leq 1$ and $f_2=0$, even if $\theta_i=0$. If we want to keep $f_1\leq 1$, our f_2 would have to be negative. Obenchain [1975] shows that an analytic solution to component-wise minimization of Mallows' [1973] (Chapter 7, Eq. 49) C_L is close to HKFI. We have noted above that in small samples, HKFI can be recommended for low values of non-centrality according to Dwivedi et al. [1980].

From the simplified expression (26) a favorable region of the parameter space for large samples is indicated by

$$\theta_i < (f_1 - f_2)\frac{(f_1+3)}{(f_1+2)} - \frac{3}{2} \ . \tag{27}$$

If $f_1=1$ is fixed, we need f_2 to satisfy $\theta_i<-0.17-1.33f_2$. From definition (21) note that $2\theta_i=\gamma_i^2/(\sigma^2/\lambda_i)$ is the "true" unknown value of the corresponding F_i statistic, $\lambda_i c_i^2/s^2$, used in (11). For example, consider a tabulated $F_i^T(1,30) = 4.17$ value of the F statistic when the degrees of freedom for s^2 are 30. We can replace $2\theta_i$ by $F_i^T(1,30)$ and determine that

$$(-f_2) \geq (F_i^T + 0.34)/(2.66) \ . \tag{28}$$

This suggests that when n=30, using $f_1=1$, $f_2=-1.70$ will give a generalized ridge estimator whose MSE would be less than that of OLS for 95% of the region *along each dimension* of the parameter space. This may be contrasted with the properties of the Stein-rule or "ordinary" ridge estimators which combine information across parameter dimensions, and run the risk of doing worse than least squares along some dimensions.

The MSE properties of a double f-class family of "operational" generalized ridge estimators studied in this section does not require the biasing parameters k_i to be nonstochastic. Since the MSE expressions in the early literature on ridge regression are not valid for stochastic k_i, the results discussed above appear to remove certain criticisms of that literature.

An existence theorem for stochastic and distinct k_i given in Vinod [1978] results from a reinterpretation of Bhattacharya's [1966] estimator as a generalized ridge estimator. For certain (complicated and stochastic) choices of k_i it can be guaranteed that Bhattacharya's ridge estimator will reduce the MSE of OLS anywhere in the parameter space. Strawderman [1978] and Alam and Hawkes [1978] have considered the MSE properties with stochastic k from the "ordinary" rather than "generalized" family of ridge estimators. These will be discussed in the following section.

8.4 BIAS AND MSE FOR "OPERATIONAL" ORDINARY RIDGE ESTIMATOR

In the previous section we considered the "double f-class" family of "generalized" ridge estimators based on an estimator $\hat{k}_i$ from (12), of the MSE "optimal" $k_i^{\text{MSE}} = \sigma^2 \gamma_i^{-2}$. For "ordinary" ridge regression we consider a harmonic mean (HM) of k_i in (7) as

$$k_{HM} = p\left[\sum_{i=1}^{p} \hat{k}_i^{-1}\right]^{-1} = \frac{pf_1 s^2}{\sum_{i=1}^{p} c_i^2 - f_2 s^2 \sum_{i=1}^{p} \lambda_i^{-1}} . \tag{29}$$

Hoerl, Kennard and Baldwin [1975] suggested

$$k_{HKB} = ps^2(b'b)^{-1} = ps^2(c'c)^{-1} , \tag{30}$$

which is a special case of harmonic mean k_{HM} when $f_1 = 1$ and $f_2 = 0$. Farebrother's [1975] choice $k_{FA} = k_{HKB}/p$ has $f_1 = 1/p$ and $f_2 = 0$. Both of these choices involve the random vector of OLS regression coefficients b.

Lawless and Wang's [1976] choice motivated by a Bayesian formulation is

$$k_{LW} = ps^2\left[\sum_{i=1}^{p} \lambda_i c_i^2\right]^{-1} = ps^2(c' \Lambda c)^{-1} . \tag{31}$$

This becomes a special case of the harmonic mean k_{HM}, when $f_1 = 1$, and $f_2 = c'(I - \Lambda)cs^{-2}\left[\sum_{i=1}^{p} \lambda_i^{-1}\right]^{-1}$. To verify this, note that $f_2 s^2 \sum_{i=1}^{p} \lambda_i^{-1} = c'c - c'\Lambda c$, and write the denominator of k_{HM} as $c'c - c'c + c'\Lambda c$ which equals $c' \Lambda c$ in (31). This choice of f_2 also involves the random vector of OLS regression coefficients c.

Note that k_{HKB} and k_{LW} differ with respect to the matrices $c'c$ and $c' \Lambda c$. Therefore, to analyze both the HKB and LW type choices together it will be convenient to introduce the constants h_1 and h_2 and rewrite k_{HM} in (29) as

$$k_{h_1,h_2} = \hat{k} = \frac{h_1 \hat{u}'\hat{u}}{c'Wc + h_2 \hat{u}'\hat{u}} , \tag{32}$$

where $W = diag(w_i)$ is a given diagonal matrix, h_1 and h_2 are arbitrary scalars, and we have used $s^2 = \hat{u}'\hat{u}/n$. It is then clear that for $W = I$, $h_1 = p/n$ and $h_2 = 0$ equation (32) equals k_{HKB}, and for $W = \Lambda$, $h_1 = p/n$ and $h_2 = 0$ we get

k_{LW}.

Now in terms of shrinkage factors, $\delta_i = \lambda_i(\lambda_i + \hat{k})^{-1}$, we write the ith element of the ORE denoted by c_k from (19) as

$$(c_{DHC})_i = (c_{h_1,h_2})_i = \delta_i c_i = \left[1 - \frac{h_{1i}\hat{u}'\hat{u}}{c'Wc + h_{2i}\hat{u}'\hat{u}}\right] c_i \ , \tag{33}$$

where DHC represents double h-class family of estimators, and

$$h_{1i} = h_1 \lambda_i^{-1}, \ h_{2i} = h_1 \lambda_i^{-1} + h_2 \ . \tag{34}$$

When $W=I$, then (i) $h_1=p/n$, $h_2=0$ gives the HKB estimator; (ii) $h_1=1/n$, $h_2=0$ gives Farebrother's [1975] estimator. Further, when $W=\Lambda$, $h_1=p/n$, and $h_2=0$ we get the LW estimator. As an additional remark, note that if h_{1i} and h_{2i} are just any fixed constants for all $i=1,...,p$, and $W=\Lambda$ then DHC in (33) is the Stein-rule double k-class estimator discussed in Chapter 6.

Our next task is to study the Bias and MSE properties of the choices of k for ordinary ridge estimators. Some of these results are given by Ullah, Vinod, and Kadiyala [1978], Ullah and Vinod [1979] and Strawderman [1978]. They involve conditions under which the MSE of various ridge estimators is less than the MSE of OLS estimator b. The latter was denoted earlier by MSE(b) which equals the expectation of the squared Euclidean "distance" between b and the parameter β.

8.4.1 Euclidean, Mahalanobis, and Strawderman Distance

For a vector c, the OLS estimators of the parameter vector γ of the canonical model, $y=Z\gamma+u$, the squared Euclidean distance between c and γ is

$$dist\,(c, \gamma, Euclid) = (c - \gamma)'(c - \gamma) \ . \tag{35}$$

The variance of c is $\sigma^2 \Lambda^{-1}$, and therefore the "precision" of c is $\sigma^{-2}\Lambda$, the reciprocal of the variance. When a "distance" is measured in units of "precisions" by choosing a weight matrix proportional to the precisions, the corresponding distance measure becomes

$$dist\,(c, \gamma, Mahalanobis) = (c - \gamma)' \Lambda (c - \gamma) \ , \tag{36}$$

named after Professor Mahalanobis of the Indian Statistical Institute who had first used it in 1930's in one for the early applications of multivariate statistical methods. However, it should be noted that Mahalanobis' [1936] distance, which uses precision weights, is not identical with the above expression, because we are not considering a distance between two populations.

An additional "distance" measure is from Strawderman [1978], where a remark similar to the following was first made. An adaptive ordinary ridge estimator can be guaranteed to dominate OLS if weights proportional to squared precision, Λ^2, are used in a distance concept. We write

$$dist\,(c,\gamma,Strawderman) = (c-\gamma)'\Lambda^2(c-\gamma)\,. \tag{37}$$

Thus for shrinkage estimators $c_{ki} = \delta_i c_i$ we can define three analogous distance concepts. The expectations of these distances are "weighted" MSE's or WMSE's with diagonal weights which are nonstochastic quantities. For the case where k is nonstochastic, the existence theorem and some MSE properties are already studied in the previous chapter. Below, we analyze the bias and other properties *when k is stochastic,* as in (32). Although the expressions for DHC estimators differ for $W=I$ and $W=\Lambda$ only slightly, their MSE properties are sufficiently different. The following results can be verified using Kadane's [1970, 1971] "small σ" expansion as discussed in explanatory note 6.2 in Chapter 6.

Before stating the results we introduce the following notations:

$$w_{\min} = min\,(w_1, \ldots, w_p), \tag{38}$$

where w_i are elements of W in equation (32). Further, denote

$$q_{\min} = \min(\lambda_1^2 d_1^{-1}, \ldots, \lambda_p^2 d_p^{-1})\,, \tag{39}$$

where $d_1,..,d_p$ are the diagonal elements of the weight matrix D in WMSE. Also, $\lambda_1 \geq \cdots \geq \lambda_p$ such that $\lambda_{\min} = min\,(\lambda_1, \ldots, \lambda_p) = \lambda_p$, and

$$\theta = \frac{\gamma' W \gamma}{2\sigma^2}$$

is a (weighted) noncentrality parameter. Now, we present the results.

Theorem 3 The asymptotic expansion of the bias of the double h-class ORE estimator in (33) up to order σ^2 (i.e. order $1/\theta$) is

$$E\,(c_{DHC}-\gamma)_i = -\frac{h_1 p}{2\theta}\lambda_i^{-1}\gamma_i\,, \tag{40}$$

when $h_1 > 0$.

Theorem 4 The asymptotic expansion of the MSE of the double h-class ORE in (33) for $h_1 > 0$, up to order $\sigma^4(1/\theta^2)$ is given by

$$E\,(c_{DHC}-\gamma)_i^2 = \frac{\sigma^2}{\lambda_i} + \frac{nh_{1i}}{4\theta^2}\left[\alpha_i^2\{4\frac{w_i}{\lambda_i} + h_{1i}(n+2)\} - 2\frac{\alpha' W\alpha}{\lambda_i}\right], \tag{41}$$

where $h_{1i} = h_1 \lambda_i^{-1}$ and w_i is the ith diagonal element of W.

Using (41),the MSE of DHC estimators can be written as

$$E\,(c_{DHC}-\gamma)'(c_{DHC}-\gamma) = \sigma^2 Tr\,\Lambda^{-1} + \frac{nh_1}{4\theta^2}\,[4\alpha'\Lambda^{-2}W\alpha + h_1(n+2)\alpha'\Lambda^{-2}\alpha$$

$$-\,2\alpha' W\alpha Tr\,\Lambda^{-2}] \tag{42}$$

or alternatively as

$$E(c_{DHC}-\gamma)'(c_{DHC}-\gamma)=\sigma^2 Tr\,\Lambda^{-1}+\frac{nh_1}{4\theta^2}\alpha'\Lambda^{-2}\alpha\Big[h_1(n+2)$$

$$-2\frac{\alpha'\Lambda^{-2}W\alpha}{\alpha'\Lambda^{-2}\alpha}\{\frac{\alpha'W\alpha}{\alpha'\Lambda^{-2}W\alpha}Tr\,\Lambda^{-2}-2\}\Big].$$

Further, the WMSE with weight matrix D is as follows:

$$E(c_{DHC}-\gamma)'D(c_{DHC}=\gamma)=\sigma^2 Tr\,\Lambda^{-1}D+\frac{nh_1}{4\theta^2}[4\alpha'D\Lambda^{-2}W\alpha+h_1(n+2)\alpha'D\Lambda^{-2}\alpha$$

$$-2\alpha'W\alpha\; Tr\,\Lambda^{-2}D] \tag{44}$$

$$=\sigma^2 Tr\,\Lambda^{-1}D+\frac{nh_1}{4\theta^2}\alpha'D\Lambda^{-2}\alpha\,[h_1(n+2)$$

$$-2\frac{\alpha'D\Lambda^{-2}W\alpha}{\alpha'D\Lambda^{-2}\alpha}\{\frac{\alpha'W\alpha Tr\,\Lambda^{-2}D}{\alpha'D\Lambda^{-2}W\alpha}-2\}]\,. \tag{45}$$

We note that $\min_\alpha$ of $\frac{\alpha'A\alpha}{\alpha'B\alpha}$ is the minimum value of μ in the determinant of $A-\mu B$, i.e. $|A-\mu B|$, equal to zero, see Rao [1973, p. 74]. Using this result it can easily be verified that the second term on the right of (45) is negative, i.e.,

$$WMSE(c_{DHC})-WMSE(c)<0$$

for

$$0<h_1\le\frac{2w_{\min}}{n+2}(d^*-2),\ d^*=q_{\min}(Tr\,\Lambda^{-2}D)>2\,, \tag{46}$$

where $w_{\min}$ and $q_{\min}$ are as defined in (38) and (39), respectively. Note that h_2 does not appear in the approximate MSE and therefore the above results are true for any $h_2\ge 0$. The positiveness of h_2 is required for the existence of the moments of the DHC estimator. The condition (46), under the small-σ assumption, is true for all values of γ. Thus the set of DHC estimators given by (46) are better than the OLS estimator in the WMSE sense. In the special case of $D=I$, it has been shown by Alam and Hawkes [1978] that the exact MSE of the double h-class DHC estimator dominates the OLS under the (approximate) condition (46) after substituting $d_i=1$ into it. In an Explanatory Note 8-1 at the end of this chapter we derive the exact conditions for dominance for any D and show that they are same as (46). We also include a more general class of ORE in that Explanatory Note.

The condition $d^*>2$ from (46) is necessary for the WMSE advantage in using the DHC estimator. Since d^* depends on the eigenvalues of $X'X$, the implication is that one should check up whether or not the data satisfy this condition before using any of these estimators. The forms of (46) for various ridge estimators which correspond to the alternative choices of W and D are summarized in Table 8.1. Since these ranges involve only the observable quantities, they are completely known for any given regression problem. In computing these (Minimax) ranges we have not assumed that the biasing parameters k_i of ridge estimators are non-stochastic. Thus recent literature has given us powerful reasons for using ridge regression, which go well beyond the early existence proofs, heuristics, and simulations. This literature is continuing to advance at a rapid pace.

Each entry of Table 8.1, in some sense, explains the implication of the condition $d^* > 2$. It is implicit in this that $p > 2$. Thus one need not even look at Table 8.1 for models having only one or two regressors since no member of the DHC family will dominate OLS for $p=1$ or $p=2$. The bottom left term of Table 8.1 suggests that if we accept Strawderman's distance concept and identity weight matrix, the structure of eigenvalues of $X'X$ and the existence of multicollinearity does not matter for DHC to dominate OLS. For the other entries in the table, the range is narrow for multicollinear data having $\lambda_p \to 0$.

In some sense we confirm Casella's [1980, p. 1052] remark that there is an "incompatibility between minimaxity and the conditioning problem." Vinod [1978i, p. 731] had also pointed out the potential incompatibility of "declining deltas" and Bhattacharya's minimax rule. From a practical viewpoint we feel that the reader should recognize that minimaxity is a very pessimistic (the worst that can happen about the unknown parameter will happen, see Section 2.8) criterion. It is often appropriate to use side computations and preliminary tests to rule out unfavorable values of the parameters. One can then use larger biasing parameters k than indicated by Table 8.1, when severe multicollinearity is present. The "acceptable" range of values of stochastic k for which ORE dominate OLS can be narrower than a similar range for non-stochastic k based on equation (22) of Chapter 7. Certain "self-consistency" checks proposed by Causey [1980] are not satisfied by minimax rules, and she does not favor Strawderman's minimax rule for multicollinear data.

8.5 MSE MATRIX FOR GENERALIZED RIDGE INVOLVING A COMBINATION OF OLS AND RESTRICTED LEAST SQUARES ESTIMATORS

Unlike the material in previous sections, this section is concerned with some of the recent results for non-stochastic k based on Farebrother [1980]. For the model $y = X\beta + u$, $Eu=0$, $Euu' = \sigma^2 I_T$ recall from equation (9) of Chapter 3 that the restricted least squares estimator is given by

$$b_R = b - VR'[RVR']^{-1}\,(Rb-r) \tag{47}$$

where $V = (X'X)^{-1}$, $b = VX'y$, and $R\beta = r$ gives m "consistent" linear constraints.

TABLE 8.1

Range of h_1

D \ W	I	Λ
I	$\bar{m}(\lambda_p^2 \sum_{i=1}^{p} \lambda_i^{-2} - 2)$	$\bar{m}\lambda_p(\lambda_p^2 \sum_{i=1}^{p} \lambda_i^{-2} - 2)$
Λ	$\bar{m}(\lambda_p \sum_{i=1}^{p} \lambda_i^{-1} - 2)$	$\bar{m}\lambda_p(\lambda_p \sum_{i=1}^{p} \lambda_i^{-1} - 2)$
Λ^2	$\bar{m}(p-2)$	$\bar{m}\lambda_p(p-2)$

Notes to Table 8.1

1) h_1 in (46) can be written as $h_1 = \bar{m} w_{\min}(d^* - 2)$, where $0 < \bar{m} < \frac{2}{n+2}$. Since $\bar{m}$ lies in this range, this Table gives us a range. Note that $d^* = \min(\lambda_1^2 d_1^{-1}, \ldots, \lambda_p^2 d_p^{-1})$ Trace $(\Lambda^{-2} D)$.

2) $W = I$ represents HKB type DHC estimators, and $W = \Lambda$ corresponds to the LW type DHC estimators. Recall that W appears in the definition of the double h-class estimator given in (32) as $(c'Wc)$.

3) $D = I$ represents equal weights, $D = \Lambda$ represents weights in Mahalanobis' distance, and $D = \Lambda^2$ represents weights in Strawderman's distance. Recall that weights D are present in the definition of the weighted MSE in equation (44).

Farebrother [1980] proposes a combination of b and b_R defined by a ridge-like parameter k

$$b(k) = (X'X + kR'R)^{-1}(X'y + kR'r), \tag{48}$$

$$= (X'X + kR'R)^{-1}(X'Xb + kR'Rb_R), \tag{49}$$

where $k=0$ implies that $b(k) = b$ and $k = \infty$ implies that $b(k) = b_R$. As k increases from 0 to ∞, greater weight is given to the restricted estimator. This is a generalization of ORE defined by $(X'X + kI)^{-1}X'y$ which is obtained when $R = I_p$ and $r = 0$ in (48).

Since X, k and R contain non-stochastic quantities, we may write the expectation of $b(k)$ as follows. Denoting $W = (X'X + kR'R)^{-1}$ we have

$$Eb(k) = EW(X'Xb + kR'Rb_R) \tag{50}$$

$$= W(X'X\beta + kR'R[\beta - VR'(RVR')^{-1}(R\beta - \gamma)])$$

$$= W[X'X\beta + kR'R\beta - kR'(R\beta - \gamma)]$$

$$= \beta - kWR'\phi$$

where $\phi = R\beta - \gamma$. Hence $b(k) - Eb(k) = W(X'u)$. Now the MSE matrix of $b(k)$ is given by (writing $b(k) - \beta = b(k) - Eb(k) + Eb(k) - \beta$)

$$E(b(k) - \beta)(b(k) - \beta)' = EW[X'uu'X + k^2R'\phi\phi'R]W$$

$$= W[\sigma^2 X'X + k^2R'\phi\phi'R]W. \tag{51}$$

This gives the MtxMSE$[b(k)]$. For $k=0$ we know that MtxMSE$[b(0)] = \sigma^2(X'X)^{-1} = \sigma^2 V$, for OLS. We write this in a form which is suitable for comparison with (51) as: MtxMSE$(b) = \sigma^2 WW^{-1}VW^{-1}W$, and consider

$$W^{-1}VW^{-1} = (X'X + kR'R)V(X'X + R'R)$$

$$= (X'X + kR'R)(I_p + kVR'R)$$

$$= X'X + 2kR'R + k^2R'RVR'R.$$

$$\text{MtxMSE}(b) = W[\sigma^2 X'X + 2\sigma^2 kR'R + \sigma^2 k^2 R'RVR'R]W \tag{52}$$

When MtxMSE$[b(k)]$ of (51) is subtracted from this $\sigma^2 X'X$ cancels, the remaining terms can be written as

$$WR'[2\sigma^2 kI_m + \sigma^2k^2RVR' - \phi\phi']RW' \tag{53}$$

Hence the necessary and sufficient (iff) condition for superiority (in MtxMSE sense) of $b(k)$ over OLS is given by studying the eigenvalues of the specially patterned matrix in (53). Farebrother [1980] shows it to be

$$\phi'[\frac{2}{k}I_m + RVR']\phi \le \sigma^2, \tag{54}$$

which is based on the following algebraic Lemma.

Lemma (Farebrother) If A is $m \times m$ symmetric positive definite matrix F is $n \times m$ matrix of rank m, b is $m \times 1$ vector and d is a scalar, then $F[dA - bb']F'$ has $n-m$ zero eigenvalues, $m-1$ eigenvalues whose sign is given by d and one eigenvalue whose sign is given by $d - b'A^{-1}b$.

The proof given by Farebrother [1980] is based on Sylvester's law of inertia.

Next we can compare MtxMSE$(b(k))$ from (51) with MtxMSE of the restricted estimator in equation (26) of Chapter 3.

$$\text{MtxMSE}(b_R) = \sigma^2 V - \sigma^2 PQP' + P\phi\phi'P' , \tag{55}$$

where $P = VR'Q^{-1}$ and $Q = RVR'$ similar to Chapter 3.

It can be shown that the necessary and sufficient condition for MSE matrix superiority of the combined (ridge type) estimator $b(k)$ over b_R is given by

$$k^2[(\phi'Q\phi)(\phi'Q^{-1}\phi) - (\phi'\phi)^2] + 2k\sigma^2\phi'\phi + \sigma^2\phi'Q^{-1}\phi \leq \sigma^4 \tag{56}$$

This is a generalization of Toro-Vizcarrondo and Wallace's [1968] result (for non-zero k) given in equation (30) of Chapter 3, which says that MtxMSE(b) "$\geq$" MtxMSE(b_R) iff

$$\phi'Q^{-1}\phi \leq \sigma^2 , \tag{56a}$$

and for which a statistical test is available. This test does not appear to be popular among practitioners, perhaps because the experience with this test is that it is too strong. It is unlikely that the experience for similar tests based on (54) and (56) would be much different. A simple method of weakening the test is to use a high level of significance.

Confidence Intervals for Ridge Parameters

Bayesian viewpoint can be used to construct classical confidence intervals for ridge regression by integrating over the posterior distribution. Obenchain [1977] suggests that the OLS confidence intervals from normal regression theory are appropriate for ridge regression. If ridge regression is successful in reducing the MSE of b, we would hope to have narrower confidence intervals centered near the ridge solution, rather than at the OLS solution b which may be unstable and may have unrealistic signs.

Vinod [1977, 1980b] suggests a confidence interval based on Stein's [1974] results modified to always have positive estimates of MSE, and extended to a study of the regression problem. Vinod works with $c_i^* = \sigma^{-1}\lambda_i^{1/2}c_i^0 \sim N(\gamma_i^*,1)$, where $\gamma_i^* = \sigma^{-1}\lambda_i^{1/2}\gamma_i$. Since c_i^* are unit normal variables from the canonical model in (1) Stein's [1974] suggestion becomes applicable with minor modifications.

If $x \sim N(\xi,1)$, the MSE of the shrunken estimator δx is

$$E(\delta x-\xi)^2 = \delta^2 + (\delta-1)^2\xi^2,$$

where an unbiased estimate of ξ^2 is x^2-1. Considering its positive part, we have the following estimate of the right hand side

$$\max\left[\delta^2,(\delta-1)^2x^2 + 2\delta-1\right] .$$

Next, consider the expectation of the squared difference between $(\delta x-\xi)^2$ and its unbiased estimate

$$E\left[\{\delta^2 + (\delta-1)^2(x^2-1)\} - (\delta x-\xi)^2\right]^2 = 8\delta^2 - 8\delta + 2 + 4(\delta-1)^2\xi^2$$

$$= 2(2\delta-1)^2 + 4(\delta-1)^2\xi^2$$

which can be estimated by replacing the ξ^2 on the right hand side by max $[0,x^2-1]$. Stein [1974] proposes that a multivariate analogue of

$$\left[(\delta-1)^2x^2 + 2\delta-1 - (\delta x-\xi)^2\right]^2 / \left[4\delta^2 - 2 + 4(\delta-2)^2x^2\right]$$

is a central Chi-square variable with one degree of freedom. This approximation is exact for $\delta = 1$ but it is obviously poor for $\delta = 0$ or $\delta = \frac{1}{2}$. Hence Vinod [1980b] makes certain modifications to obtain a more meaningful approximation given below.

The upper and lower 95% bounds on c_i is given by

$$c_i^{UP} = \delta_i c_i + s\lambda_i^{-1/2}(U^* + 2.0095\, S^*)^{1/2} \text{ ,and}$$

$$c_i^{LO} = \delta_i c_i - s\lambda_i^{-1/2}(U^* + 2.0095\, S^*)^{1/2} \text{ ,} \qquad (57)$$

where

$$s^2 = (y-Xb)'(y-Xb)(T-p-1)^{-1}$$

$$U^* = \max\left[\delta_i^2, (\delta_i-1)^2 s^{-2}\lambda_i c_i^2 + 2\delta_i - 1\right]$$

$$S^* = \left[\max\left(1, 2(2\delta_i-1)^2, 4\delta_i^2 - 2 + 4(\delta_i-1)^2 s^{-2}\lambda_i c_i^2\right)\right]^{1/2} .$$

Hence the 95% confidence bounds on b_i are given by

$$b_i^{UP} = \sum_{j=1}^{p} \max\left(g_{ij}c_j^{UP}, g_{ij}c_j^{LO}\right) ,$$

$$b_i^{LO} = \sum_{j=1}^{p} \min\left(g_{ij}c_j^{UP}, g_{ij}c_j^{LO}\right) , \qquad (57a)$$

where g_{ij} represents the $(i,j)^{th}$ element of the matrix of eigenvectors of $X'X$ consistent with the notation in (1) and where c_i^{UP} and c_i^{LO} are from (57). These confidence intervals are conservative and the actual confidence level may be larger, especially for large values of p. More research is needed on this topic. In the example and simulations given in Vinod [1980b], (57a) yields narrower and more meaningful confidence intervals than OLS. These intervals are centered at the ridge solution. Stein's [1974] method is applicable to the ridge problems having stochastic k, except that the computations become somewhat complicated.

Explanatory Note 8.1

Exact Dominance of Operational Ordinary Ridge Estimators

Let us write the linear regression model as

$$y = X\beta + u = H\Lambda^{1/2}G'\beta + u = Z\gamma + u,$$

where X is a matrix of p regressors which is replaced by its "singular value decomposition," see equation (1), and where $Z = H\Lambda^{1/2}$ and $\gamma = G'\beta$. Notice that $Z'Z = \Lambda$ and $X'X = G\Lambda G'$ because $H'H=I$ and $G'G=I$. Thus we can write

$$b = (X'X)^{-1}X'y = Gc;\; c = \Lambda^{-1/2}H'y = \Lambda^{-1}Z'y. \tag{58}$$

Let us consider

$$\hat{\beta} = [X'X + \hat{k}A]^{-1}X'y$$

$$= [I + \hat{k}(X'X)^{-1}A]^{-1}b,$$

where A is a known positive definite matrix and

$$\hat{k} = \frac{h_1 s}{b'Pb + h_2 s},\; s = \hat{u}'\hat{u} = (y-Xb)'(y-Xb). \tag{59}$$

The matrix P in (48) is a known positive definite matrix. Now note from Rao [1973, page 41] that if the matrices $X'X$ and A commute, we can find an orthogonal matrix G such that

$$G'AG = Diag.[\eta_1 \cdots \eta_p] = \nabla \tag{60}$$

$$G'X'XG = Diag.[\lambda_1 \cdots \lambda_p] = \Lambda \tag{61}$$

and therefore, denoting $r_i = \lambda_i^{-1}\eta_i$ for $i=1,...,p$ as the eigenvalues of $(X'X)^{-1}A$,

$$G'(X'X)^{-1}AG = Diag.[\lambda_1^{-1}\eta_1 \cdots \lambda_p^{-1}\eta_p] = Diag.[r_1...r_p] = R. \tag{62}$$

Thus we can write $\hat{\beta}$, after algebraic manipulation, as

$$\hat{\beta} = [GG' + \hat{k}GRG']^{-1}Gc \tag{63}$$

$$= G\hat{\gamma}$$

where

$$\hat{\gamma} = [I + \hat{k}R]^{-1}c \tag{64}$$

and

$$\hat{k} = \frac{h_1 s}{c'Wc + h_2 s};\; W = G'PG, \tag{65}$$

as given in (32), except that W need not be a diagonal matrix in (65).

Observe that $E(\hat{\beta}-\beta)'(\hat{\beta}-\beta) = E(\hat{\gamma}-\gamma)'(\hat{\gamma}-\gamma)$.

Consider the i-th element of $\hat{\gamma}$ and write it as

$$\hat{\gamma}_i - \gamma_i = c_i - \gamma_i - h_1 r_i c_i \left(\frac{s}{g_i}\right); \tag{66}$$

where $r_i = \lambda_i^{-1}\eta_i$ from (62) and

$$g_i = c'Wc + h_{2i}s;\ h_{2i} = h_1 r_i + h_2 . \tag{67}$$

Notice that for $A=I$ in $\hat{\beta}$, $r_i = \lambda_i^{-1}$ from (62). In this special case the estimator $\hat{\gamma}_i$ in (66) is identical with the double h-class estimators in (33). Thus, here we are considering a more general class of ordinary ridge estimators (ORE).

Using (66), we can obtain

$$E(\hat{\gamma}_i-\gamma_i)^2 = E(c_i-\gamma_i)^2 + h_1^2 r_i^2 E(sc_i g_i^{-1})^2 - 2h_1 r_i E(sc_i(c_i-\gamma_i)g_i^{-1}) \tag{68}$$

$$= \sigma^2\lambda_i^{-1} + h_1^2 r_i^2 E(sc_i g_i^{-1})^2 - 2h_1 r_i \sigma^2 \lambda_i^{-1} E[sg_i^{-1}$$

$$- 2sg_i^{-2}(c_i^2 w_{ii} + \sum_{j\neq i} c_i c_j w_{ij})]$$

$$= \sigma^2\lambda_i^{-1} + h_1\sigma^2 E[g_i^{-2}\{h_1 r_i^2 c_i^2 (s\,\sigma^{-1})^2 + 4sr_i\lambda_i^{-1}(c_i^2 w_{ii} + \sum_{j\neq i} c_i c_j w_{ij})\}$$

$$- 2r_i\lambda_i^{-1}g_i^{-1}s],$$

where the third term in the second equality has been obtained through integrating by parts with respect to c_i. Using (61) and (62), note that

$$\sum_{i=1}^{p} r_i^2 c_i^2 = c'R^2 c = c'\nabla^2\Lambda^{-2}c \tag{69}$$

and

$$\sum_{i=1}^{p} r_i (c_i^2 w_{ii} \sum_{j\neq i} c_i c_j w_{ij}) = c'\nabla W \Lambda^{-2} c . \tag{70}$$

Further for $h_2 \geq 0$, let

$$g_i = c'Wc + h_{2i}s \geq c'Wc + h_1 r_i s \geq c'Wc + h_1 r_* s = g , \tag{71}$$

where

$$r_* = \frac{\eta_p}{\lambda_1}; \tag{72}$$

$\lambda_1 \geq \cdots \geq \lambda_p$ are the eigenvalues of $X'X$ and $\eta_1 \geq \cdots \geq \eta_p$ are the eigenvalues of the matrix A in (60).

Thus from (68)

$$E(\hat{\gamma}_i - \gamma_i)^2 - E(c_i - \gamma_i)^2 < 0 \tag{73}$$

if

$$h_1\sigma^2E\ [\{h_1r_i^2c_i^2(s\,\sigma^{-1})^2 + 4sr_i\lambda_i^{-1}(c_i^2w_{ii} + \sum_{j\neq i}c_ic_jw_{ij})\}g^{-1} - 2r_i\lambda_i^{-1}s]g^{-1} < 0\,. \tag{74}$$

Summing over i=1,...,p, representing $w_1 \geq w_2 \cdots \geq w_p$ as eigenvalues of W, and employing

$$\frac{c'\nabla^2\Lambda^{-2}c}{g} < \frac{c'\nabla^2\Lambda^{-1}c}{c'Wc} < \frac{\eta_1^2}{\lambda_p^2w_p} \tag{75}$$

$$\frac{c'\nabla W\Lambda^{-2}c}{g} < \frac{c'\nabla W\Lambda^{-2}c}{c'Wc} < \frac{\eta_1}{\lambda_p^2}$$

it follows from (73) and (74) that

$$E(\hat{\gamma}-\gamma)'(\hat{\gamma}-\gamma) - E(c-\gamma)'(c-\gamma) < 0 \tag{76}$$

when

$$h_1\sigma^2E(g^{-1})[\eta_1\lambda_p^{-2}\{h_1\eta_1w_p^{-1}(n+2) + 4\} - 2\sum_{i=1}^{p}\eta_i\lambda_i^{-2}] < 0, \tag{77}$$

where we use $Es = \sigma^2n$, $Es^2 = n(n+2)\sigma^4$, $n = T-p$ and the fact that c and s are independently distributed. The inequality (77) holds when

$$0 < h_1 < \frac{2w_p(d-2)}{\eta_1(n+2)}\ ;\ d = \lambda_p^2\eta_1^{-1}Tr\,\Lambda^{-2}\nabla > 2\,. \tag{78}$$

Notice that the eigenvalues of the matrix $W = G'PG$ in (65) is the same as the eigenvalues of the matrix P. In a special case when A=I so that $\eta_i = 1$, the condition (78) reduces to

$$0 < h_1 < \frac{2w_p}{n+2}(\lambda_p^2Tr\,\Lambda^{-2}-2), \tag{79}$$

which is given in Alam and Hawkes [1978].

Thus, it is clear that the double h-class estimators in (33) dominate the OLS estimator, under MSE error criterion, when $h_2 \geq 0$ and h_1 is given by (79). This condition based on exact MSE is identical with the condition based on the approximate MSE as given in (46) [substitute D=I].

If we consider the loss function $(\hat{\gamma}-\gamma)'D(\hat{\gamma}-\gamma)$ where D is a diagonal positive definite matrix, then it is straightforward to extend (76) and (78) respectively, as

$$E(\hat{\gamma}-\gamma)'D(\hat{\gamma}-\gamma) - E(a-\gamma)'D(a-\gamma) < 0$$

for $h_2 \geq 0$ and

$$0 < h_1 < \frac{2w_p(d^0-2)}{\eta_1(n+2)}; \quad d^0 = \eta_1^{-1} q_{\min} Tr\,\Lambda^{-2} \nabla D > 2, \tag{80}$$

where $q_{\min} =$ minimum cigenvalue of $\Lambda^2 D^{-1}$. Again when $A=I$ such that η's $= 1$, the exact condition (80) is identical with the condition in (46) based on small-σ approximate WMSE.

It is obvious that the double h-class estimators $\hat{\gamma}$ are a class of minimax estimators under the condition (80).

EXERCISES

8.1 Show that the k_i^{MSE} of (6) and δ_i^{MSE} of (5) are MSE optimal, but useless. If true γ_i were known, show that it is possible to obtain MSE $\equiv 0$ by using a stochastic k_i. (Hint: see Vinod [1978] letter.)

8.2 Describe Hoerl and Kennard's first iteration estimator (HKFI). Is this optimal? Find an analytic solution to the iterative process. Discuss the MSE properties of HKFI estimator.

8.3 Show the relationship between Double h-class, Double f-class and Double k-class estimators, and list their special cases.

8.4 Show that Strawderman's distance measure allows for the stability goal in using declining deltas.

8.5 In the notation of Section 8.5 show that $(X'X + kR'R)VR' = R'(I_m + kRVR')$ is a definitional identity, while MtxMSE(b_R) is comparable to (51). By using the following identity

$$(A + BDB')^{-1} = A^{-1} - A^{-1}B(B'A^{-1}B + D^{-1})^{-1}B'A^{-1}$$

show that

$$(\sigma^2 Q^{-1} + k^2 \phi\phi')^{-1} = \frac{(\sigma^2 + k^2\phi' Q \phi)Q - k^2 Q\phi\phi' Q}{\sigma^2(\sigma^2 + k^2\phi' Q\phi)} .$$

Using these show that MtxMSE(b_R) $-$ MtxMSE$[b(k)]$ can be written as $WR'BRW$ where

$$B = \sigma^2 Q^{-1} + k^2\phi'\phi - (Q^{-1} + kI_m)\phi\phi'(Q^{-1} + kI_m) .$$

Use the Lemma of Section 8.5 to consider the smallest eigenvalue of B and write the iff condition for it to be positive. Hence derive (56) in the text.

8.6 Construct a simulation study to compare Vinod's confidence intervals (57a) with Bayesian posterior intervals and OLS intervals.

9

Estimation of Polynomial Distributed Lag Models

9.1 INTRODUCTION

Consider the finite "distributed lag" model

$$y_t = \beta_0 x_t + \beta_1 x_{t-1} + \ldots + \beta_p x_{t-p} + u_t \tag{1}$$

$$= \sum_{i=0}^{p} \beta_i x_{t-i} + u_t, \quad t = p+1,\ldots,T .$$

This model implies that the current value of y_t depends not just on x_t but also on some past values, (say p in number) of x_t. The coefficients β_i are called lag weights. The problems with the direct ordinary least squares (OLS) estimation of (1) are: (a) we lose p degrees of freedom as only $T - p$ observations can be used. If p is large this amounts to a substantial loss of observations. (b) often there is a high degree of multicollinearity among the x's and this leads to problems discussed in earlier chapters. In the past, many solutions have been suggested to tackle these two problems. Almost all of them lead to introducing first some prior information about the behavior of the β's in (1) before its estimation. Broadly, these two sources of prior information can be classified as (i) nonstochastic smoothness prior (ii) stochastic smoothness prior.

Irving Fisher [1937] was the first econometrician to introduce "nonstochastic smoothness prior" information of the following type:

$$\beta_i = (p+1-i)\alpha \quad , \quad 0 \le i \le p \tag{2}$$

$$= 0 \qquad , \quad i > p$$

where α is any unknown parameter. This can be called a linear lag specification, and (1) along with this can be called a linear distributed lag model.

Substituting (2) in (1) gives

$$y_t = \left[\sum_{i=0}^{p} (p+1-i)x_{t-i} \right] \alpha + u_t \tag{3}$$

$$= z_t \alpha + u_t \ .$$

Thus the OLS estimate of α can be obtained from this equation, and then using (2), the estimate of β_i can be obtained.

A generalization of the linear nonstochastic prior on β_i can be written as

$$\beta_i = \alpha_0 + \alpha_1 i + \alpha_2 i^2 + \ldots + \alpha_r i^r \ \ , \ p \geq r \geq 0 \tag{4}$$

which is a polynomial of the rth degree. This structure on lag weights β_i was proposed by Almon [1965] and therefore is known as the Almon polynomial lag. Again, substituting (4) in (1) we can get estimates of the α's and then using (4) we can obtain the estimates of β_i. Sometimes estimation is done under the end point restrictions $\beta_{-1} = 0 = \beta_{p+1}$.

A major problem with the estimation of an Almon distributed lag model is with respect to the choice of the lag length p and the degree of polynomial r. Any misspecification with respect to p and r would affect the efficiency of estimates, Trivedi and Pagan [1979]. A further problem, both with Fisher and Almon nonstochastic smoothness priors, is that they imply a "strong prior" because the shapes of the estimated lag structures are dictated by the stringency of the above specification. Often these shapes have very little theoretical justification and offer little more than a method of reducing the number of parameters to be estimated and tackling the collinearity effects implied by (1). An alternative to this is the stochastic smoothness prior, say $\beta_i = \alpha_0 + \alpha_1 i + \ldots + \alpha_r i^r + v_i$ where v_i is stochastic such that $Ev_i = 0$, $Ev_i^2 = \sigma_v^2$ and $Ev_i v_j = 0$, $i \neq j$. This imposes a weaker restriction in that the degree of "smoothness" of the lag structure is specified rather than its exact shape. Also, this stochastic prior is more likely to be encountered by a practitioner than the nonstochastic priors. There are, however, some other methods, e.g. ridge, which estimate the model (1) under some restriction on β_i but with no smoothness prior. In the following sections we deal with the estimation of (1) under nonstochastic smoothness priors, stochastic smoothness priors, and without smoothness priors (that is, ridge regression). In the literature, the "nonstochastic smoothness priors" cases has been termed as imposing *strong* restriction on the parameters β_i whereas the latter two cases as imposing *weak* or no restrictions on β_i.

9.2 ESTIMATION UNDER NONSTOCHASTIC SMOOTHNESS PRIORS

Consider the distributed lag model in (1) and write it in matrix notation as

$$y = X\beta + u \ , \tag{5}$$

where

$$y = \begin{bmatrix} y_{p+1} \\ \cdot \\ \cdot \\ \cdot \\ y_T \end{bmatrix}, \ \beta = \begin{bmatrix} \beta_0 \\ \cdot \\ \cdot \\ \cdot \\ \beta_p \end{bmatrix}, \ X = \begin{bmatrix} x_{p+1} & \cdots & x_1 \\ \cdot & & \cdot \\ \cdot & & \cdot \\ \cdot & & \cdot \\ x_T & \cdots & x_{T-p} \end{bmatrix}, \ u = \begin{bmatrix} u_{p+1} \\ \cdot \\ \cdot \\ \cdot \\ u_T \end{bmatrix}$$

The elements of u are normally distributed with zero mean and constant variance σ_u^2, i.e., $N\ (0,\ \sigma_u^2)$. In the Almon method we assume that the β_i's lie on an rth degree polynomial given in (4). This can be written in matrix notation as

$$\beta = A\alpha \ , \tag{6}$$

where β is given before, and

$$A = \begin{bmatrix} 1 & 0 \ldots & 0 \\ 1 & 1 & 1 \\ \cdot & \cdot & \cdot \\ \cdot & \cdot & \cdot \\ \cdot & \cdot & \cdot \\ 1 & p & p^r \end{bmatrix}, \quad \alpha = \begin{bmatrix} \alpha_0 \\ \cdot \\ \cdot \\ \cdot \\ \alpha_r \end{bmatrix}$$

are a $(p+1)\times(r+1)$ matrix and a $(r+1)\times1$ vector, respectively. The ranks of matrices X and A are assumed to be $(p+1) < (T-p)$ and $(r+1) < (p+1)$ respectively. If $r < p$ then the rank of A is $r + 1$.

In the Almon estimation method we estimate β in (5) under the nonstochastic prior on β given by (6). There are two ways of doing this estimation.

Substitute (6) in (5) and write

$$y = XA\alpha + u \ . \tag{7}$$

Then the OLS estimator of α is $a = (A'X'XA)^{-1}A'X'y$. The Almon estimator of β is then simply

$$\hat{\beta} = Aa \ . \tag{7a}$$

If $\beta = A\alpha$ is true, then $E\hat{\beta} = \beta$ and $V(\hat{\beta}) = AV(a)A' = \sigma_u^2 A(A'X'XA)^{-1}A'$, i.e., $\hat{\beta}$ is BLUE.

The form of the Almon estimator in (7) is useful for calculations. An alternative interpretation of $\hat{\beta}$ is in terms of the restricted least squares (RLS) estimator. Since the columns of A are linearly independent we can define an idempotent matrix

$$M = I - A(A'A)^{-1}A \tag{8}$$

such that $\beta = A\alpha$ implies the set of restrictions

$$M\beta = 0 .$$

The RLS estimator is

$$\hat{\beta}_R = b - (X'X)^{-1}M'[M(X'X)^{-1}M]^{+}Mb \ , \tag{8a}$$

where the superscript + represents a generalized inverse.

Note that M can be written following Terasvirta [1976] as

$$M = I - A(A'A)^{-1}A' = R'(RR')^{-1}R = M^2, \tag{9}$$

where R is a $(p-r)\times(p+1)$ known matrix involving binomial type coefficients given by the following

$$\begin{vmatrix} (-1)^0\begin{bmatrix} r+1 \\ 0 \end{bmatrix} & (-1)^1\begin{bmatrix} r+1 \\ 1 \end{bmatrix} & \cdots & (-1)^{r+1}\begin{bmatrix} r+1 \\ r+1 \end{bmatrix} & 0 & \cdots & 0 \\ 0 & (-1)^0\begin{bmatrix} r+1 \\ 0 \end{bmatrix} & \cdots & (-1)^r\begin{bmatrix} r+1 \\ r \end{bmatrix} & (-1)^{r+1}\begin{bmatrix} r+1 \\ r+1 \end{bmatrix} & \cdots & 0 \\ \vdots & & & & & & \vdots \\ 0 & 0 & \cdots & \cdots & \cdots & & (-1)^{r+1}\begin{bmatrix} r+1 \\ r+1 \end{bmatrix} \end{vmatrix}$$

This matrix R can be generated by noting that if β_i follows an rth degree polynomial, then $\Delta^{r+1}\beta_i = 0$ where $\Delta\beta_i = \beta_i - \beta_{i-1}$. One can verify this for $r = 2, 3$, say. This definition satisfies $RA = 0$. Now the restrictions $M\beta = 0$ also imply that

$$R\beta = 0 = RA\alpha. \tag{10}$$

Thus, to get Almon's estimator we minimize $(y - X\beta)'(y - X\beta)$ subject to $R\beta = 0$. The resulting estimator (see Eq. (9) of Chapter 3)

$$\hat{\beta}_R = b - (X'X)^{-1}R'[R(X'X)^{-1}R']^{-1}Rb, \tag{11}$$

where $b = (X'X)^{-1}X'y$ is the OLS estimator and subscript R represents restricted. This expression can also be shown to be identical with the expression in (7a) and (8a). Note that $\hat{\beta}_R$ will be biased if $R\beta = 0$ is not true.

From equation (30) of Chapter 3, $\hat{\beta}_R$ will be better than the OLS estimator b, that is, *Mtx*MSE(b) "$\geq$" *Mtx*MSE$(\hat{\beta}_R)$ if and only if $\beta'[R(X'X)^{-1}R']^{-1}\beta/\sigma_u^2 \leq 1$. A test for this condition can be carried out by using Toro Vizcarrondo and Wallace's test (see Section 3.3.1 in Chapter 3).

9.3 ESTIMATION UNDER STOCHASTIC SMOOTHNESS PRIORS

The Almon estimator is BLUE only if the specification form $\beta = A\alpha$ is true and also if the true values of p and r are known. Any misspecification with respect to p and r would result in the inefficiency of Almon's estimator.

However, even if the lag length p is correctly specified, the lag weights could not be expected to lie exactly on an rth order polynomial. A slightly more flexible approach is to assume that the prior information is uncertain and of the form

$$\beta = A\alpha + v \,, \tag{12}$$

where $v \sim N\left(0, \sigma_v^2 I\right)$. Since $RA\alpha = 0$, (12) implies that the stochastic restrictions on β are

$$R\beta = Rv \,, \tag{13}$$

where $Rv \sim N\left(0,\ \sigma_v^2 RR'\right)$.

Now we can estimate $y = X\beta + u$ under the stochastic prior $R\beta = Rv$. This can be done by minimizing

$$S = \frac{(y - X\beta)'(y - X\beta)}{\sigma_u^2} + \frac{(R\beta)'(RR')^{-1}(R\beta)}{\sigma_v^2}$$

with respect to β. The resulting estimator, which is called the mixed estimator in the terminology of Theil-Goldberger and has been discussed in Chapter 3, is (See eq. (48) and Section 3.4).

$$\hat{\beta}_M = \left(\frac{X'X}{\sigma_u^2} + \frac{R'(RR')^{-1}R}{\sigma_v^2}\right)^{-1} \left(\frac{X'y}{\sigma_u^2}\right) \tag{14}$$

$$= (X'X + \lambda M)^{-1} X'y,$$

where $\lambda = \sigma_u^2/\sigma_v^2$, and M is the idempotent matrix used before in (8).

If we assume that $Rv \sim N(0,\ \sigma_v^2 I)$, where the covariance matrix is different from (13) then the estimator of β under $R\beta = Rv$ as proposed by Shiller [1973] is

$$\hat{\beta}_{SH} = (X'X + \lambda\, R'R)^{-1} X'y \,, \tag{15}$$

where SH represents Shiller. The difference between $\hat{\beta}_M$ and $\hat{\beta}_{SH}$ arises because Shiller begins with $\beta = A\alpha$ which implies $R\beta = 0$, and then uses a stochastic vector, say v, to specify stochastic restrictions as $R\beta = v$. Thus it makes a difference whether we make $\beta = A\alpha$ or $R\beta = 0$ stochastic.

The mixed estimator $\hat{\beta}_M$ offers some immediate benefits. First, the notion of a weaker prior inherent in the specification of $R\beta = Rv$ is highlighted. Specifically, as $\lambda \to 0$ $(\sigma_v^2 \to \infty)$ $\hat{\beta}_M$ approaches the OLS estimator $b = (X'X)^{-1}X'y$. In contrast, as $\lambda \to \infty (\sigma_v^2 \to 0)$ $\hat{\beta}_M$ approaches the Almon estimator in (8a). This can be checked as follows: $\hat{\beta}_M = (X'X + \lambda M)^{-1}X'y =$ $(X'X + \lambda M^2)^{-1}X'y =$ $(X'X)^{-1}[I - M(M'(X'X)^{-1}M + \frac{1}{\lambda}I)^{-1}M'(X'X^{-1}]X'y$. As $\sigma_v^2 \to 0$, $1/\lambda \to 0$ and so $\hat{\beta}_M = \hat{\beta}_R$. Thus the OLS estimation of β

coincides with the prior knowledge, assumption, or belief that any lag structure is possible, whereas the Almon estimator implies the belief that the lag structure *must* be of the polynomial form. The $\hat{\beta}_M$ is then the intermediate case, λ reflecting to what extent it is felt that the smoothness prior should be allowed to modify ordinary least squares estimation.

The estimator $\hat{\beta}_M$ is BLUE as long as the stochastic restrictions are unbiased, i.e., $ERv = 0$. The variance covariance matrix of $\hat{\beta}_R$ is (see equation (55) in Chapter 3).

$$V(\hat{\beta}_M) = \sigma_u^2(X'X + \lambda M)^{-1} .$$

However, if $ERv \neq 0$, that is the stochastic restrictions are not true, then $\hat{\beta}_M$ is biased. Such bias may result from a misspecification of the degree of smoothness inherent in the true lag structure. For example, suppose the true r was 2 whereas we used $r = 1$. The question normally of interest in this case is whether $\hat{\beta}_M$ will be better than the OLS estimator under the MtxMSE criterion. This question has been analyzed in Section 3.4.1 of Chapter 3 in the context of a general mixed estimator.

From equations (61) and (62) in Chapter 3 we can write the bias and MtxMSE of $\hat{\beta}_M$ as

$$Bias(\hat{\beta}_M) = -\lambda(X'X + \lambda M)^{-1}M\beta$$

$$MtxMSE(\hat{\beta}_M) = (X'X + \lambda M)^{-1}[\sigma_u^2 + \lambda^2 M\beta\beta' M'(X'X + \lambda M)^{-1}] .$$

Further, from equation (73) in Chapter 3, the $MtxMSE(b) - MtxMSE(\hat{\beta}_M)$ is a positive definite matrix if

$$\frac{\beta' M[M(X'X)^{-1}M' + \lambda^{-1}I]^{-1}M\beta}{2\sigma_u^2} \leq 1/2 .$$

The test statistic for this condition is the modified Toro-Vizcarrondo and Wallace's noncentral F test statistic given in equation (72) of Chapter 3. All the above results can be written for Shiller's estimator in (15) by substituting R for M.

9.3.1 Bayesian Interpretation of Stochastic Smoothness Priors

An efficient and meaningful way to handle stochastic information is by Bayesian methods. Lindley's [1971] or Lindley-Smith's [1972] hierarchical priors of hyperparameters are particularly relevant in this situation, which has been discussed earlier in Chapter 3. In the current set up our data information is

$$y \sim N(X\beta, \sigma_u^2 I)$$

while the lag weight vector β has a normal prior

$$\beta \sim N(A\alpha,\ \sigma_v^2 I),\quad \sigma_v^2 = \sigma_\beta^2 .$$

Thus the posterior mean, conditional on σ_v^2 and α, is given by

$$\hat{\beta}_B = \left(\frac{X'X}{\sigma_u^2} + \frac{1}{\sigma_v^2}\right)^{-1} \left(\frac{X'y}{\sigma_u^2} + \frac{A\alpha}{\sigma_v^2}\right), \tag{16}$$

when B represents Bayes' estimator of β under a quadratic loss and the normal prior. If, however, we consider a diffuse prior on β, then $\beta_B = (X'X)^{-1}X'y = b$, which is the OLS estimator.

Now, since α is rarely known, one can go one step further by assuming the hyperparameter α to be normal. Then we have Lindley and Smith's Bayesian model

$$y \sim N(X\beta,\ \sigma_u^2 I)$$

$$\beta \sim N(A\alpha,\ \sigma_v^2 I)$$

$$\alpha \sim N(\bar{\alpha},\ \sigma_\alpha^2 I).$$

The posterior density of β in this case is $N(\beta^*, D_0)$ where $\beta^* = D_0 d_0$ and where

$$D_0^{-1} = \left[\frac{X'X}{\sigma_u^2} + \frac{I}{\sigma_v^2} - \frac{A}{\sigma_v^2}\left(\frac{A'A}{\sigma_v^2} + \frac{I}{\sigma_\alpha^2}\right)^{-1} \frac{A'}{\sigma_v^2}\right]$$

$$d_0 = \left[\frac{X'y}{\sigma_u^2} + \left(\frac{I}{\sigma_v^2} - \frac{A}{\sigma_v^2}\left[\frac{A'A}{\sigma_v^2} + \frac{I}{\sigma_\alpha^2}\right]^{-1} \frac{A}{\sigma_v^2}\right) A\bar{\alpha}\right.$$

If, however, we use a diffuse prior for α, i.e., $\sigma_\alpha^2 \to \infty$, then β^* simplifies (Hint: eq. (9) and $A(A'A)^{-1}A'A(A'A)^{-1}A' = A(A'A)^{-1}A'$ are used) to

$$\hat{\beta}_L = \beta^* = (X'X+\lambda M)^{-1}X'y \tag{17}$$

$$= (I+\lambda(X'X)-1M)-1b \qquad , \lambda = \frac{\sigma_u^2}{\sigma_v^2},$$

where $b = (X'X)^{-1}X'y$ is the OLS estimator, $M = I-A(A'A)^{-1}A'$ and the subscript L refers to Lindley. Using (9) we note that β^* is the same as the mixed estimator in (14), though their interpretations are different.

An alternative form of (17) can be written as

$$\beta^* = (X'X+\lambda I)^{-1}(X'Xb+\lambda A\alpha^*) \tag{18}$$

$$= A\alpha^* + [I+\lambda(X'X)^{-1}]^{-1}(b-A\alpha^*),$$

where

$$\alpha^* = \left[A'\left((X'X)^{-1} + \lambda^{-1}I\right]^{-1} A\right]^{-1} A'\left[(X'X)^{-1} + \lambda^{-1}I\right]^{-1} b \tag{19}$$

is the posterior mean of α. The (18) and (19) are direct outcomes of equations (105) and (104), respectively, in Chapter 3 by substituting $\Omega_o = \lambda^{-1}I$ in them. It is not difficult to verify from (18) and (19) that

$$\alpha^* = (A'A)^{-1}A'\beta^* \tag{20}$$

which is essentially the least squares estimator in the model $\beta^* = A\alpha + V$.

The posterior mean (Bayes estimator with a diffuse prior on α) β^* appears as the weighted matrix combination of the estimators b and $A\alpha^*$. Also, it shrinks the OLS estimator b towards $A\alpha^*$, which is the posterior mean of $A\alpha$. Comparing (18) with (16) we note that while Bayes estimator with a known α shrinks the OLS towards $A\alpha$, Lindley's Bayesian estimator with a diffuse prior on α shrinks it towards $A\alpha^*$. The estimator β^* thus appears as if we are using β_B defined in (16) with α replaced by α^*.

Regarding λ in (17) Lindley [1971] and Lindley and Smith [1972] suggested an iterative procedure which essentially consists of starting with the OLS estimate of β (λ=0) and then taking the estimates of σ_u^2 and σ_v^2, respectively, as $(y-Xb)'(y-Xb)/n_1$ and $(b-Aa)'(b-Aa)/n_2$, where $a = (A'A)^{-1}A'b$ for λ=0 (see (20)), and where n_1 and n_2 are scalars. Using these estimates of σ_u^2 and σ_v^2 we can formulate λ as

$$\hat{\lambda} = \frac{gs}{b'Mb}, \quad s = (y-Xb)'(y-Xb),$$

where g is an arbitrary scaler, and $M=I-A(A'A)^{-1}A'$ as before.

9.3.2 Operational Estimators

Substituting $\hat{\lambda}$ in (17) we can write the following Lindley-type estimator of β:

$$\tilde{\beta}_L = [I + \hat{\lambda}(X'X)^{-1}M]^{-1}b \tag{21}$$

$$= [I + \frac{gs}{b'Mb}(X'X)^{-1}M]^{-1}b.$$

Since $\alpha^* = a = (A'A)^{-1}A'b$ for λ=0, the alternative form of $\tilde{\beta}_L$ from (18) will indicate that $\hat{\beta}_L$ shrinks the OLS estimator b towards $A\hat{\alpha}$. Thus, it is not surprising that $\tilde{\beta}_L$, when plotted will lie on a smooth polynomial (see Ullah and Raj [1980]). Notice that $\hat{\lambda} = \lambda \to \infty$ would give Almon's estimator as indicated before.

One can also write an operational version of Shiller's estimator in (15) by considering $\lambda = gs/b'R'Rb$. This gives

$$\tilde{\beta}_{SH} = [I + \frac{gs}{b'R'Rb} (X'X)^{-1}R'R]^{-1}b. \tag{22}$$

Notice that $\tilde{\beta}_L$ of (21) contains the matrix M in place of $R'R$ in (22) for $\tilde{\beta}_{SH}$. Since $M=M^2$, but $R'R \neq (R'R)^2$, $\tilde{\beta}_{SH}$ may not be written in the form of (18).

A more general form of the estimator $\tilde{\beta}_L$ for the Nonnormal case will be discussed in Chapter 13. We analyze here only $\tilde{\beta}_L$ and $\tilde{\beta}_{SH}$. Their iterative versions will be taken up later in this section.

The approximate bias, to order σ^2, and MtxMSE, to order σ^4, of $\tilde{\beta}_L$ are:

$$E(\tilde{\beta}_L - \beta) = -\frac{c\,\sigma^2 n}{\beta' M \beta} (X'X)^{-1} M \beta, \tag{23}$$

$$MtxMSE(\tilde{\beta}_L) = \sigma^2 (X'X)^{-1} + \frac{\sigma^4 gn(n+2)}{\beta' M \beta}\left[\frac{g}{\beta' M \beta}(X'X)^{-1} M \beta \beta' M (X'X)^{-1} - \frac{2}{n+2}\left\{N - \frac{H}{\beta' M \beta}\right\}\right], \tag{24}$$

where $n = T-2p-1$, $N = (X'X)^{-1}M(X'X)^{-1}$, and $H = 2(X'X)^{-1} M\beta\beta' M(X'X)^{-1}$. Further, for a given positive definite matrix Q, we have $E(\tilde{\beta}_L-\beta)'Q(\tilde{\beta}_L-\beta) = Tr[MtxMSE(\tilde{\beta}_L)Q] = WMSE(\tilde{\beta}_L)$, to order σ^4 as:

$$WMSE(\tilde{\beta}_L) = \sigma^2 Tr(X'X)^{-1}\alpha + \sigma^4 gn(n+2)\frac{\beta' M (X'X)^{-1} Q (X'X)^{-1} M \beta}{(\beta' M \beta)^2}\left[g - \frac{2}{n+2}\frac{\beta' M \beta}{\beta' M (X'X)^{-1} Q (X'X)^{-1} M \beta}\left\{Tr\,(X'X)^{-1}M(X'X)^{-1}Q - 2\frac{\beta' M (X'X)^{-1} Q (X'X)^{-1} M \beta}{\beta' M \beta}\right\}\right] \tag{25}$$

The results in (23) and (25) can be obtained by directly substituting $c=g$, $\gamma_2=0$, $B=M$ and $D=(X'X)^{-1}M$ in the equation (16) and (18) of the Chapter 13. With the same substitutions in the equation (53) of Chapter 13 one can obtain the MtxMSE as given in (24). These results are also derived in Ullah [1980a].

The approximate bias, MtxMSE and WMSE of the Shiller-type estimator $\tilde{\beta}_{SH}$ can be obtained by simply substituting $M = R'R$ in (23) to (25).

Now we note that if $\mu_1 \geq \mu_2 \cdots \geq \mu_p$ are the roots of the determinantal equation $|E-\mu F|=0$, where E is a $p\times p$ symmetric matrix, and F is a positive definite matrix, then $\mu_p = \min_\beta(\beta' E\beta/\beta' F\beta)$ and $\mu_1 = \max_\beta(\beta' E\beta/\beta' F\beta)$. Using this we can verify that $\tilde{\beta}_L$ has a smaller WMSE than b, i.e.,

$$WMSE(\tilde{\beta}_L) - WMSE(b) \leq 0 \tag{26}$$

when

$$0 < g \leq \frac{2}{n+2}(d-2); \quad d = \frac{1}{\mu_1} Tr(X'X)^{-1}M(X'X)^{-1}Q > 2, \tag{27}$$

where μ_1 is the maximum eigenvalue of $G'(X'X)^{-1}Q(X'X)^{-1}G$, where G is a $p\times(p-r)$ matrix of orthonormal vectors satisfying $M = GG' = M^2$. If we take $Q=I$ then the condition of dominance (27) becomes $0 < g \leq \frac{2}{n+2}(d-2)$ and $d = \mu_1^{-1}TrM(X'X)^{-2} > 2$ where $\mu_1 \geq \cdots \geq \mu_p$ are the eigenvalues of the data matrix $G'(X'X)^{-2}G$. However, if we take $Q = (X'X)^2$ then

$$0 < g \leq \frac{2}{n+2}(p-r-2), \quad p-r > 2. \tag{28}$$

This is because for $Q=(X'X)^2$, $\mu_1=1$ and $d=Tr(X'X)^{-1}M(X'X)^{-1}(X'X)^2 = TrM = Tr[I-A(A'A)^{-1}A'] = p+1-(r+1) = p-r$ by using (9).

The result in (28) implies that the Lindley-type estimator, under the WMSE criterion with $Q=(X'X)^2$, does better than the OLS for all β when the length of the lag minus the degree of the polynomial is at least 2. For the WMSE with $Q \neq (X'X)^2$, the condition $p-r > 2$ is necessary but not sufficient to have $d > 2$ in (27). It is interesting to observe that when $r=0$, the conditions (27) and (28) will be similar to the ones discussed in the Chapter 8 for the operational ordinary ridge estimator. This is obvious because when $r=0$, i.e., when there is no additional structure on β, M will become I and $\tilde{\beta}_L$ will reduce to an operational ordinary ridge estimator.

It is also clear from (26) that the estimator $\tilde{\beta}_L$ is minimax, for small σ, under the condition (27). If one considers the MtxMSE criterion and compares MtxMSE $(\tilde{\beta}_L)$ with the MtxMSE (b), then it is not possible to get a condition on g under which $\tilde{\beta}_L$ will be better than the OLS for all values of β. This can be verified from (24), and it is consistent with the well-known result of the admissibility of the OLS in one and two dimensions (Ullah [1980a] and Chapter 6 have similar results).

Since Shiller-type estimator differs from Lindley-type estimator only with respect to the matrices $R'R$ (in the former) and M (in the latter), the conditions of dominance on g for the estimator $\tilde{\beta}_{SH}$ can be written from (27) as

$$0 < g_{SH} \leq \frac{2}{n+2}(d_{SH}-2); \; d_{SH} = \frac{1}{\mu_1} Tr(X'X)^{-1}R'R(X'X)^{-1}Q > 2, \tag{29}$$

where g_{SH} and d_{SH} represent g and d values in Shiller's case.

9.3.3 Iterative Operational Estimators

Let us call the estimator $\tilde{\beta}_L$ in (21) as the first step estimator and rewrite it as

$$\tilde{\beta}_L = \beta^{*(1)} = \left[I + \lambda^{(o)}(X'X)^{-1}M\right]^{-1} b,$$

where

$$\lambda^{(o)} = \frac{g(y-X\beta^{*(o)})'(y-X\beta^{*(o)})}{\beta^{*(o)\prime}M\beta^{*(o)}} = \frac{gs^{(o)}}{\beta^{*(o)\prime}M\beta^{*(o)}} = \hat{\lambda}$$

and $\beta^{*(o)}=b$ is the initial estimator β^* in (17) for $\lambda=0$. The estimator $\beta^{*(1)}$ can then be employed to provide another estimator of λ and the process can be continued.

Suppose $\beta^{*(m+1)}$ denotes the estimator of β at the $(m+1)$-th iteration. Then we have Lindley-type iterative estimator

$$\beta^{*(m+1)} = \left[I + \lambda^{(m)}(X'X)^{-1}M\right]^{-1} b \quad , \quad \lambda^{(m)} = \frac{gs^{(m)}}{\beta^{*(m)\prime}M\beta^{*(m)}} \tag{30}$$

where for $m=0,1,...$ we can write

$$s^{(m)} = s^{(o)} + (b-\beta^{*(m)})'X'X(b-\beta^{*(m)}) \tag{31}$$

$$\beta^{*(m)\prime}M\beta^{*(m)} = b'Mb + (b-\beta^{*(m)})'M(b-\beta^{*(m)})-2b'M(b-\beta^{*(m)}).$$

[Hint: $y-X\hat{\beta} = y-Xb + X(b-\hat{\beta})$ and $\hat{\beta}'M\hat{\beta} = (\hat{\beta}-b+b)'M(\hat{\beta}-b+b)$ for any β]. When $M = I-\iota(\iota'\iota)^{-1}\iota;\iota$ is a column vector of unit elements, then (30) is essentially Lindley and Smith's [1972, p. 17] iterative model estimation under the diffuse prior about σ_u^2 and independent, inverse- χ^2 prior about σ_v^2.

It has been shown in Ullah [1980a] that, up to the order σ^4; MtxMSE and WMSE of $\beta^{*(m+1)}$ is identical with that of the first stage estimator $\beta^{*(1)} = \tilde{\beta}_L$ given in (24) and (25), respectively. This can be easily verified by noting that $(b-\beta^{*(m)})$ for any $m>0$ is at least of order σ^2 [see, e.q., Explanatory Note 13.1 in Chapter 13]. This result implies that the efficiency of the iterative estimator is the same as that of the first stage estimator.

Alternatively, Ullah [1980a] considers the following iterative estimator

$$\beta_U^{*(m+1)} = \left[I + \lambda^{(m)}(X'X)^{-1}M\right]^{-1} \beta^{*(m)}, \tag{32}$$

where U refers to Ullah. This iterative estimator is better than the OLS, under the WMSE criterion, when $0 < g < 2(m+1)^{-1}(n+2)^{-1}(d-2)$ where $d>2$ is as given before. Also WMSE$(\beta_U^{*(m+1)})$ $-$ WMSE$(\beta_U^{*(m)}) \leq 0$ for $0 < g \leq \frac{2(d-2)}{(2m+1)(n+2)}$. These results show that the estimator in (32), at the $(m+1)-th$ iteration will have smaller WMSE compared to that at the m-th iteration and also it dominates the OLS estimator so long as $0 < g \leq \frac{2(d-2)}{(2m+1)(n+2)}$. In a special case where $M = X'X$; the estimator (32) is algebraically similar to the Vinod-type iterative estimator discussed in Section 6.4.2 of Chapter 6.

9.4 RIDGE ESTIMATION WITHOUT SMOOTHNESS PRIORS

The model (1), without smoothness priors, can be estimated by applying the OLS estimation. However, since the explanatory variable in practice is an autocorrelated time series the columns of X will be expected to be highly correlated. A simple solution to the problem of multicollinearity requires that we use the ridge estimators given in Chapter 7. The ordinary and generalized ridge estimators, without using smoothness priors, are respectively

$$\hat{\beta}_k = (X'X + kI)^{-1}X'y \text{ , and}$$

$$\hat{\beta}_K = (X'X + GKG')^{-1}X'y,$$

where K is a matrix of constants or biasing parameters and G is the matrix of eigenvectors of $X'X$ such that $G'G = I$. For $K = kI$ we have the ordinary ridge estimator. Various interpretations and derivations of ridge estimators and the method of selecting K have been discussed in Chapters 7 and 8.

Recall that in the Bayesian interpretation of the generalized ridge estimator we assume the prior of $\beta \sim N(0, \sigma^2 GK^{-1}G')$. Thus, although the ridge estimator is particularly useful in breaking multicollinearity, the implied prior assumption that all lag weights have the same mean may contradict our prior knowledge about the lag distributions for most of the economic variables. For example, when $K = kI$, the "ordinary" ridge estimator is the Bayes' estimator in (16) with $k = \sigma_u^2/\sigma_v^2$ and $A\alpha = 0$. Thus the prior knowledge that the mean of β is $A\alpha$ is not used. An alternative approach to this problem is to develop a restricted ridge estimator. This may be based on the methods discussed in Chapter 3, and is given in the following section.

9.5 RIDGE ESTIMATION WITH NONSTOCHASTIC SMOOTHNESS PRIORS

When the choice of β in $y = X\beta + u$ is constrained to satisfy $\beta = A\alpha$, the results of the previous section may no longer be directly applicable. Constraints which are implicit in using the ridge estimators may not be consistent with the prior constraints of the type $\beta = A\alpha$. We therefore develop below a restricted ridge estimator of β.

Let us write $y = X\beta + u$ as $y = XA\alpha + u$. The OLS estimator of α is $a = (A'X'XA)^{-1}A'X'y$. The ordinary and generalized ridge estimators of α are

$$\hat{\alpha}_k = (A'X'XA + kI)^{-1}A'X'y$$

$$\hat{\alpha}_K = (A'X'XA + GKG')^{-1}A'X'y \text{ ,}$$

where G is now a matrix of eigenvectors of $A'X'XA$ such that $G'G = I$.

The matrix mean squared error of $\hat{\alpha}_k$ is

$$MtxMSE(\hat{\alpha}_k) = E(\hat{\alpha}_k - \alpha)(\hat{\alpha}_k - \alpha)'$$

$$= (A'X'XA + kI)^{-1}(A'X'XA + k^2\alpha\alpha')(A'X'XA + kI)^{-1}.$$

The $MtxMSE(a) - MtxMSE(\hat{\alpha}_k)$ = a positive definite matrix if and only if

$$\alpha' [(A'X'XA)^{-1} + 2k^{-1}I]^{-1}\alpha \leq 1.$$

Similarly $MtxMSE(a) - MtxMSE(\hat{\alpha}_K)$ = a positive definite matrix if and only if

$$\alpha'[(A'X'XA)^{-1} + 2K^{-1}]^{-1}\alpha \leq 1.$$

The restricted ridge estimators of β are obtained from $\beta = A\alpha$. Thus $\hat{\beta}_K = A\hat{\alpha}_K$ is the restricted generalized ridge estimator. For $K = kI$ we get the restricted ordinary ridge estimator. Thus, we note that ridge estimation with smoothness priors is possible, provided we first work with α parameters.

9.6 FINAL REMARKS

There are other forms of distributed lag estimators discussed in Econometrics textbooks, e.g., Maddala [1977, Chapter 16]. The engineering literature refers to this as the so-called *impulse response* function, which measures the response of an initial impulse over a certain time period. These are usually described by a set of weights associated with polynomial in a backward shift operator, Box and Jenkins [1976, p. 378]. The order of the polynomial is usually determined by "looking at" cross correlations between "prewhitened" (made normal) input and the correspondingly transformed output, to obtain the best fit in some sense. For an interesting application outside the engineering field the reader is referred to Maloney and Ireland [1980].

The statistical estimation of various forms of lag distributions, such as Koyck, Pascal, Jorgenson, etc. involve certain multicollinearity problems, and various ridge-type or Stein-rule type methods may be useful. Theoretical expressions for the MSE of these estimators when they contain lagged dependent variable (e.g. y_{t-1} or y_{t-4}) on the right hand side of the regression equation, and when they are further shrunken by ridge-type methods have not been published in the literature so far.

EXERCISES

9.1 Consider the data on capital expenditures (Y) and appropriations (X) from Almon [1965]. These data are reported in Table 9.1 for the years 1953-1967 on a quarterly basis. These are 60 quarterly observations numbered 1 to 60.

TABLE 9.1

N	Y	X	N	Y	X	N	Y	X
1	2072	1660	21	2697	1511	41	2601	2629
2	2077	1926	22	2338	1631	42	2648	3133
3	2078	2181	23	2140	1990	43	2840	3449
4	2043	1897	24	2012	1993	44	2937	3764
5	2062	1695	25	2071	2520	45	3136	3983
6	2067	1705	26	2192	2804	46	3299	4381
7	1964	1731	27	2240	2919	47	3514	4786
8	1981	2151	28	2421	3024	48	3815	4094
9	1914	2556	29	2639	2725	49	4093	4870
10	1991	3152	30	2733	2321	50	4262	5344
11	2129	3763	31	2721	2131	51	4531	5433
12	2309	3903	32	2640	2552	52	4825	5911
13	2614	3912	33	2513	2234	53	5160	6109
14	2896	3571	34	2448	2282	54	5319	6542
15	3058	3199	35	2429	2533	55	5574	5785
16	3309	3262	36	2516	2517	56	5749	5707
17	3446	3476	37	2534	2772	57	5715	5412
18	3466	2993	38	2494	2380	58	5637	5465
19	3435	2262	39	2596	2568	59	5383	5550
20	3183	2011	40	2572	2944	60	5467	5465

Calculate (i) Shiller type estimator (ii) Lindley-type estimator and (iii) iterative estimators by using the above data. Do your calculations for the upper bound of g in each case, and for $Q = I$ as well as $Q = (X'X)^2$. Consider various choices of the degree of polynomial and lag length for both $Q = I$, and $Q = (X'X)^2$. Make sure that the dominance over the least squares is assured. Plot the estimates and interpret your results.

9.2 Consider Almon's distributed lag model $y = X\beta + u$ where $\beta = A\alpha$, show that

(i) $\alpha = A^+\beta$ and therefore $(I - A(A'A)^{-1}A')\alpha = 0$; A^+ is the generalized inverse of A.

(ii) If the degree of polynomial is correct, but the true lag length is understated, then Almon's polynomial distributed lag estimator will be biased.

(iii) If the assumed lag length is correct, but the degree of polynomial is of an order higher than the true polynomial, then Almon's estimator is unbiased but inefficient.

9.3 Show that the mixed estimator is biased when the stochastic restrictions are not unbiased. Obtain the MtxMSE of the mixed estimator and suggest a test for its dominance over the least squares estimator (reference: Chapter 3).

9.4 For the data structure of Exercise 9.1 choose a true lag structure, simulate the various estimators and compare their MSE's.

10

Multiple Sets of Regression Equations

10.1 INTRODUCTION

In the earlier chapters we were concerned with the analysis of a single equation regression model: for example the demand equation for a commodity, a consumption function, or an investment function. In this chapter we shall deal with the multi-regression model. Such a model was proposed in Statistics by Zellner [1962], and is needed, for example, when we want to analyze a set of demand equations for different commodities, or a set of consumption functions.

The general M equations model, in this context, can be written as

$$y_1 = X_1\beta_1 + u_1 , \tag{1}$$
$$\vdots$$
$$y_M = X_M\beta_M + u_M ,$$

where y_i and u_i are $T \times 1$ random vectors, X_i is a $T \times p_i$ matrix, and β_i is a $p_i \times 1$ vector, with $i = 1, ..., M$. Thus M is the number of equations, T is the number of time periods and p_i is the number of regressors in the i-th equation. The error vectors u_i and u_j, $i, j = 1, ..., M$, are assumed to be correlated only at the same time point. Within each equation the disturbance vector is assumed to be homoscedastic.

The temporal cross-section model, frequently used in applied statistics, can be seen as a special case of model (1). In such a model M represents the number of cross sectional units for example, firms, households, or geographical areas, and T represents time points. The temporal cross-section model is normally an outcome of a single equation regression model with both cross section and time series data. The error components model, in this context, has been extensively analyzed by Balestra and Nerlove [1966], Nerlove [1971a,b], Wallace and Hussain [1969], Maddala [1971], Mundlak [1978] and Avery [1977].

This model is in the form of (1) with $p_i = p$, $\beta_i = \beta$, for $i = 1, ..., M$, and the t-th element of the vector u_i as

$$u_{it} = \epsilon_i + \eta_t + v_{it}, \tag{2}$$

where ϵ_i and η_t are systematic effects specific to the i-th cross sectional unit and t-th period, $t = 1, ..., T$, respectively. These effects may be treated as fixed as in Mundlak [1961], or as random, as in Balestra and Nerlove [1966]. The v_{it} in (2) is the disturbance term which represents both the cross section and time effects.

Another related model is the random coefficient model of Swamy [1970, 1971]. In this model $p_i = p$ and $\beta_i = \beta + v_i$, $i = 1, ..., M$, where $E(v_i v_j') = \delta_{ij}\Delta$; $\delta_{ij} = 1$ if $j = i$ and 0 if $j \neq i$. Further, Δ is the contemporaneous variance matrix of the v_i. He also assumes $E(u_i) = 0$ and $E(u_i u_j') = \delta_{ij}\sigma_i^2 I$, and that β_i and u_j are independent.

10.2 SEEMINGLY UNRELATED REGRESSIONS (SUR) MODEL

Zellner [1962] calls the model (1) a SUR model. The assumptions about the error vectors in his model are

$$\begin{aligned} Eu_i &= 0 \\ Eu_i u_j' &= \sigma_{ij} I, \ (i, j = 1, ..., M), \end{aligned} \tag{3}$$

where I is a $T \times T$ identity matrix and σ_{ij} is the covariance between u_{it} and u_{jt}.

One way to estimate the model (1) is to use the ordinary least squares (OLS) estimator for each equation. For the i-th equation, this estimator will be a solution of the normal equations

$$\frac{X_i'X_i}{\sigma_{ii}} b_i = \frac{X_i'y_i}{\sigma_{ii}}. \tag{4}$$

This gives the OLS estimator

$$b_i = (X_i'X_i)^{-1} X_i'y_i . \tag{5}$$

An alternative is to use the ridge and Stein type estimators for the i-th equation. As discussed in Chapter 7, if $y_i \sim N(X_i\beta_i, \sigma_{ii}I)$, and the prior belief about $\beta_i \sim N(0, \sigma_{ii}G_i K_i^{-1} G_i')$, then the mean of the posterior density of β_i becomes the generalized ridge estimator (also see 3.6.1 of Chapter 3). This is given by

$$b_{K_i} = (X_i'X_i + G_i K_i G_i')^{-1} X_i'y_i , \tag{6}$$

or

$$\frac{(X_i'X_i + G_iK_iG_i')}{\sigma_{ii}} b_{K_i} = \frac{X_i'y_i}{\sigma_{ii}} ,$$

where G_i is the matrix of eigenvectors of $X_i'X_i$ and K_i is the matrix of ridge biasing constants. If $K_i = k_iI$ then

$$b_{k_i} = (X_i'X_i + k_iI)^{-1} X_i'y_i \tag{7}$$

is the "ordinary" ridge estimator. The determination of K_i has been discussed in Chapters 7 and 8. If the prior β_i is $\sim N(0, k_i\sigma_{ii}(X_i'X_i)^{-1})$ then the posterior mean will be a Stein-type estimator.

Note that both the OLS and ridge type estimators ignore the information across the equations given by the covariance of disturbance vectors u_i and u_j. Thus these single equation estimators are inefficient. A way out is to estimate all the M equations jointly. For this, it will be convenient to write the M equations in a compact form as

$$y = X\beta + u , \tag{8}$$

where the new meaning of these familiar symbols for the purpose of this chapter is as follows. We have

$$y = \begin{bmatrix} y_1 \\ \cdot \\ \cdot \\ \cdot \\ y_M \end{bmatrix}, \quad u = \begin{bmatrix} u_1 \\ \cdot \\ \cdot \\ \cdot \\ u_M \end{bmatrix}, \text{ and } \beta = \begin{bmatrix} \beta_1 \\ \cdot \\ \cdot \\ \cdot \\ \beta_M \end{bmatrix}$$

are, respectively, $MT \times 1$, $MT \times 1$, and $p \times 1$ ($p = \sum_{i=1}^{M} p_i$) vectors, and

$$X = \begin{vmatrix} X_1 & \cdot & \cdot & \cdot & 0 \\ \cdot & \cdot & & & \\ \cdot & & \cdot & & \\ \cdot & & & \cdot & \\ 0 & & & & X_M \end{vmatrix}$$

is an $MT \times p$ matrix. The disturbance vector u has zero mean and the following variance covariance matrix

$$V(u) = E(uu') = \Sigma = \begin{bmatrix} \sigma_{11} & \cdots & \sigma_{1M} \\ \vdots & & \\ \sigma_{M1} & \cdots & \sigma_{MM} \end{bmatrix} \otimes I \ , \tag{9}$$

where $\otimes$ represents Kronecker product and I is the $T \times T$ identity matrix. The matrix Σ is positive definite, and therefore we can obtain a nonsingular matrix P such that $P'\Sigma P = I$, or $\Sigma^{-1} = PP'$. The transformed M equations model can then be written as

$$P'y = P'X\,\beta + P'u \ , \tag{10}$$

where $P'u$ is such that $E(P'u) = 0$ and $V(P'u) = I$.

If we apply the OLS estimation technique to the transformed model we get the estimator of β as

$$\hat{\beta} = [(P'X)'(P'X)]^{-1}(P'X)'P'y$$

$$= (X'\Sigma^{-1}X)^{-1}X'\Sigma^{-1}y \ . \tag{11}$$

The estimator $\hat{\beta}$ is called Aitken's generalized least squared (GLS) and it is known to be "best linear unbiased." The mean vector and variance covariance matrix of $\hat{\beta}$ are

$$E\hat{\beta}\,b = \beta$$

$$V(\hat{\beta}) = E(\hat{\beta}-\beta)(\hat{\beta}-\beta)' = (X'\Sigma^{-1}X)^{-1} \ . \tag{12}$$

If we have $X_1 = X_2 = \cdots = X_M$, and/or $\sigma_{ij} = 0$ for $j \neq i$ then the GLS estimator $\hat{\beta}_i$ of the i-th equation will be equal to the OLS estimator $b_i = (X_i'X_i)^{-1}X_i'y_i$. Thus in these special situations there is no gain due to the joint estimation. This is so because whenever $X_1 = \cdots = X_M$ and/or $\sigma_{ij} = 0$ there is no additional information relevant to a particular equation contained in the remaining M-1 equations. For a more general condition under which $\hat{\beta}_i$ and b_i will be identical the reader should see Srivastava and Dwivedi [1979], Dwivedi and Srivastava [1978a] and Schmidt [1978].

Next, we derive the ridge type Bayesian estimators. Consider $P'y \sim N(P'X\,\beta, I)$ and $\beta \sim N(0, GK^{-1}G')$. The posterior mean (see Chapters 3 and 7) of β is then

$$\hat{\beta}_K = [(P'X)'(P'X) + GKG']^{-1}(P'X)'P'y \tag{13}$$

$$= [X'\Sigma^{-1}X + GKG']^{-1}X'\Sigma^{-1}y,$$

where G is now a matrix of eigenvectors of $X'\Sigma^{-1}X$ and K is the matrix of constants. For $K = kI$ we have $\hat{\beta}_K = (X'\Sigma^{-1}X + kI)^{-1}X'\Sigma^{-1}y$ which is the ordinary ridge estimator.

In practice Σ is not known, therefore both the GLS and ridge estimators are not operational. Zellner [1962] suggested a consistent estimator of σ_{ij} as

$$\hat{\sigma}_{ij} = \frac{\hat{u}_i'\hat{u}_j}{T}, \ (i,j = 1, ..., M), \tag{14}$$

where $\hat{u}_i = y_i - X_i b_i$. Using this one can determine $\hat{\Sigma}$. Then substituting $\hat{\Sigma}$ in (11) and (13) we can write the operational versions of $\hat{\beta}$ and $\hat{\beta}_K$ as

$$\tilde{\beta} = (X'\hat{\Sigma}^{-1}X)^{-1} X'\hat{\Sigma}^{-1}y \tag{15}$$

$$\tilde{\beta}_K = (X'\hat{\Sigma}^{-1}X + \hat{G}K\hat{G}')^{-1} X'\hat{\Sigma}^{-1}y ,$$

respectively, where $\hat{G}$ denotes the matrix of eigenvectors of $X'\hat{\Sigma}^{-1}X$. For large samples these estimators will have distributions similar to $\hat{\beta}$ and $\hat{\beta}_K$. Thus their asymptotic variance covariance matrices are respectively:

$$Asy.V(\tilde{\beta}) = Asy.V(\hat{\beta}) = (X'\Sigma^{-1}X)^{-1} \tag{16}$$

and

$$Asy.V(\tilde{\beta}_K) = Asy.V(\hat{\beta}_K).$$

For the choice of ridge constants K, see Chapters 7 and 8.

In the case of small samples, Kakwani [1967] has shown that $\tilde{\beta}$ is unbiased. The technique of Kakwani's proof for unbiasedness has been used in the context of a mixed regression estimator in Chapter 3. For details about the exact efficiency of $\tilde{\beta}$ and its sampling distribution see Revankar [1974, 1976], Mehta and Swamy [1976], Phillips [1978], and Ullah and Rafiquzzaman [1977]. These results do not provide a simple formula for computing the variance covariance matrix of $\tilde{\beta}$.

Srivastava [1970] has provided, up to order $1/T^2$, the following simpler expression for the MSE matrix of the unbiased estimator $\tilde{\beta}$,

$$MtxMSE(\tilde{\beta}) = V(\tilde{\beta}) = (1 + \frac{M}{T})\Sigma_o - \frac{1}{T}\Sigma_o H \Sigma_o , \tag{17}$$

where $\Sigma_o = (X'\Sigma^{-1}X)^{-1}$ and $H = X'(D-L)X$. The matrices D and L are defined as follows. If we write $Q = \Sigma^{-1}(I - X\Sigma_o X'\Sigma^{-1})$ and partition this large $MT \times MT$ matrix into T^2 blocks as $Q = ((Q_{ij}))$, where Q_{ij} is a $T \times T$ matrix for all $i,j = 1, ..., M$, then the matrix L is defined as $L = ((Q_{ij}'))$. Similarly, if $F = X\Sigma_o X'\Sigma^{-1}$ is partitioned like Q, then $D = \Sigma^{-1}(I \otimes \sum_{i=1}^{M} F_{ii})$ where I is the $M \times M$ identity matrix. It is clear from the above expression that the $Asy.V(\hat{\beta}) = Asy.V(\tilde{\beta}) = (X'\Sigma^{-1}X)^{-1}$. The finite sample properties of ridge-type versions of $\tilde{\beta}$ denoted by $\tilde{\beta}_K$ are not yet known. In particular, when $M = 1$ we have the usual regression model, and it can be verified that (17) reduces to $\sigma_{11}(X_1'X_1)^{-1}$.

10.2.1 A Stein-rule Estimator for the SUR Model

The transformed model $P'y = P'X\beta + P'u$ of (10) is in the form of a standard single equation regression model considered in earlier chapters. Thus the Stein-rule estimators derived in Chapter 6 are directly applicable for the parameter β in the above model.

For example, the weighted MSE for $\hat{\beta} = (X'\Sigma^{-1}X)^{-1}X'\Sigma^{-1}y = [(P'X)'(P'X)]^{-1}(P'X)'P'y$ is

$$WMSE(\hat{\beta}) = E(\hat{\beta}-\beta)'W(\hat{\beta}-\beta)$$

$$= Tr[(P'X)'(P'X)]^{-1}W = Tr[(X'\Sigma^{-1}X)^{-1}W], \qquad (18)$$

where W is a positive definite weight matrix. Now, as in Chapter 6, eq. (30) if we consider a class of scalar transform estimators of the form $q\hat{\beta}$, where q is a scalar, the optimal value for q which minimizes WMSE is

$$q^* = 1 - \frac{Tr(\Sigma_o W)}{Tr(\Sigma_o W)+\beta'W\beta}, \quad \Sigma_o = (X'\Sigma^{-1}X)^{-1}. \qquad (19)$$

Then the minimum WMSE estimator is

$$\tilde{\beta}^* = [1 - \frac{g}{g+\beta'W\beta}]\,\hat{\beta}, \qquad (20)$$

where $g = Tr(\Sigma_o W)$. We could also consider a linear transform estimators of the form $q_o i + q_1\hat{\beta}$ as analyzed in Chapter 6 after equation (25).

In the Bayesian framework if $P'y \sim N(P'X\beta,I)$, and the prior $\beta \sim N(0,\tau^2\Sigma_o)$ where $\tau^2 = \frac{\beta'W\beta}{g}$, then the posterior mean of β is

$$\tilde{\beta}^* = (\Sigma_o^{-1} + \frac{1}{\tau^2}\Sigma_o^{-1})^{-1}(P'X)'P'y$$

$$= (1 + \frac{g}{\beta'W\beta})^{-1}(X'\Sigma^{-1}X)^{-1}X'\Sigma^{-1}y$$

$$= (1 - \frac{g}{g+\beta'W\beta})\hat{\beta} \qquad (21)$$

as given in (20). Note that $\tilde{\beta}^*$ can be seen as a special case of the generalized ridge estimator. If we take $K = \tau^2 G'\Sigma_o G$ and recall that $G'G = I$, then the prior of β in the ridge case, viz. the normal prior $N(0,GKG')$, coincides with the normal prior in the Stein-rule case, and the ridge estimator β_K coincides with $\tilde{\beta}^*$.

The estimator $\tilde{\beta}^*$ involves unknown Σ and β, and therefore it is not operational. Zellner and Vandaele (1974) suggest the following approximation

$$\tilde{\beta}_{zv} = [1 - \frac{\hat{g}}{\tilde{\beta}'W\tilde{\beta}}]\tilde{\beta}, \qquad (22)$$

where $\hat{g} = Tr(\hat{\Sigma}_p W) - Tr[(X'\hat{\Sigma}^{-1}X)^{-1}W]$, $\tilde{\beta} = (X'\hat{\Sigma}^{-1}X)^{-1}X'\hat{\Sigma}^{-1}y$, and the subscript zv refers to the authors. The estimator $\tilde{\beta}_{zv}$ is not necessarily a minimum WMSE estimator.

Srivastava [1973] provides the bias to order T^{-1}, and MtxMSE to order T^{-2}, of $\tilde{\beta}_{zv}$. His results, under the assumption that W is a $p \times p$ matrix of order T, are

$$E(\tilde{\beta}_{zv} - \beta) = -\frac{Tr(W\Sigma_o)\beta}{\beta' W \beta}. \tag{23}$$

$$MtxMSE(\tilde{\beta}_{zv}) = E(\tilde{\beta}_{zv} - \beta)(\tilde{\beta}_{zv} - \beta)'$$

$$= \left[1 + \frac{M}{T} - 2\frac{Tr(W\Sigma_o)}{\beta' W \beta}\right]\Sigma_o - \frac{1}{T}\Sigma_o H \Sigma_o$$

$$+ \frac{Tr(W\Sigma_o)}{(\beta' W \beta)^2}[2(\beta\beta' W\Sigma_o + \Sigma_o W \beta\beta') + (Tr W \Omega)\beta\beta'],$$

where H is as defined after eq. (17) above.

Using the MtxMSE expression we can obtain WMSE of $\tilde{\beta}_{zv}$ to order T^{-1}, as

$$WMSE(\tilde{\beta}_{zv}) = E(\tilde{\beta}_{zv} - \beta)' W(\tilde{\beta}_{zv} - \beta) = Tr[W\ MtxMSE(\tilde{\beta}_{zv})]$$

$$= \left[1 + \frac{M}{T} - \frac{Tr(W\Sigma_o)}{\beta' W \beta} + 4\frac{\beta' W \Sigma_o W \beta}{(\beta' W \beta)^2}\right] Tr(W\Sigma_o)$$

$$- \frac{1}{T} Tr[W\Sigma_o H \Sigma_o]. \tag{24}$$

From the expression of MtxMSE $(\tilde{\beta})$ in (17) we can write the

$$WMSE(\tilde{\beta}) = (1 + \frac{M}{T}) Tr(W\Sigma_o) - \frac{1}{T} Tr(W\Sigma_o H \Sigma_o). \tag{25}$$

Thus

$$WMSE(\tilde{\beta}) - WMSE(\tilde{\beta}_{zv}) = \left[Tr(W\Sigma_o) - 4\frac{\beta' W \Sigma_o W \beta}{\beta' W \beta}\right]\frac{Tr(W\Sigma_o)}{\beta' W \beta}. \tag{26}$$

This difference is positive, that is $\tilde{\beta}_{zv}$ is better than $\tilde{\beta}$, if

$$\lambda_{\min} Tr(W\Sigma_o) - 4 > 0, \text{ or } \lambda_{\min}\ Tr(W\Sigma_o) > 4, \tag{27}$$

where $\lambda_{\min}$ is the minimum eigenvalue of the matrix $(\Sigma_o W)^{-1}$. In obtaining the condition in (27) we have used the result that the minimum of $(\beta' A \beta / \beta' C \beta)$ for all β is the value of λ which satisfies the minimum of $|A - \lambda C| = 0$, C.R.Rao [1973, p74].

In a special case when $W = TI$ the above condition (27) reduces to

$$\lambda_{\min} Tr(\Sigma_o) = \lambda_{\min} Tr(X'\Sigma^{-1}X)^{-1} > 4 \tag{28}$$

where $\lambda_{\min}$ is simply the minimum eigenvalue of $\Sigma_o^{-1} = X'\Sigma^{-1}X$. Furthermore, if $M = 1$, i.e. there is only one equation, (28) becomes

$$\lambda_{\min} Tr(X_1' X_1)^{-1} > 4.$$

This inequality is obviously harder to satisfy (conservative) compared to equation (42) of Chapter 6, where the right hand side is 2.

10.3 TEMPORAL CROSS-SECTION MODEL

Consider $p_i = p$ regressors in each of the equations of the model (1), and write the i-th equation for i-th cross-sectional unit, such as a region, a business, a consumer, etc., as:

$$y_i = X_i \beta_i + u_i, \; i = 1, ..., M \; . \tag{29}$$

The disturbance vector u_i is such that $Eu_i = 0$, and $E(u_i u_i') = \sigma_{ii}$, whereas $E(u_i u_j') = 0$ for $i \neq j$. The model (29) is a single equation model, with data on M cross-sectional units and T time periods. The problem is to estimate pM regression coefficients. This model in a compact form is

$$y = X\beta + u,$$

and its transformed form, similar to (10), is

$$P'y = P'X\beta + P'u.$$

When dealing with cross-section and time series data, where each individual cross-section sample is small so that sharp inferences about the coefficients are not possible, it is a common practice in applied work to pool all data together, and estimate a common regression. The basic motivation for pooling time series and cross-section data is that if the model is properly specified, pooling provides more efficient estimation, inference, and possibly prediction.

To decide the question whether or not to pool, a test for the equality of regression coefficients across equations is performed. The null hypothesis to be tested is H_o: $\beta_1 = \cdots = \beta_M$. For convenience, we can express this null hypothesis as: $\beta_1 - \beta_2 = \beta_2 - \beta_3 = 0 = ... = \beta_{p-1} - \beta_p = 0$ and write it as a constraint: $R\beta = 0$, where

$$R = \begin{vmatrix} I & -I & 0 & \cdot & \cdot & \cdot & 0 \\ 0 & I & -I & \cdot & \cdot & \cdot & 0 \\ \vdots & & & & & & \\ 0 & \cdot & & \cdot & \cdot & I & -I \end{vmatrix} \tag{30}$$

is a $(M-1)p \times Mp$ matrix of fixed constants. To test the null hypothesis we can use the Toro-Vizcarrondo and Wallace's F statistic given in equation (20) of Chapter 3. In our context this statistic is

$$\frac{(R\hat{\beta})'[R(X'\Sigma^{-1}X)^{-1}R']^{-1}R\hat{\beta}/Mp}{(y-X\hat{\beta})'(y-X\hat{\beta})/M(T-p)} \sim F(Mp, M(T-p)), \tag{31}$$

where $\hat{\beta} = (X'\Sigma^{-1}X)^{-1}X'\Sigma^{-1}y$ and

$$\Sigma = \begin{bmatrix} \sigma_{11}I & \cdot & \cdot & \cdot & \cdot & 0 \\ \vdots & & & & & \\ 0 & \cdot & \cdot & \cdot & \cdot & \sigma_{MM}I \end{bmatrix}. \tag{32}$$

If $E(u_i u_i') = \sigma^2 I$ then the above test will be independent of the Σ matrix, and one can use F tables. In general, Σ is not known, and so one needs to use $\hat{\sigma}_{ii} = (y_i - X_i b_i)'(y_i - X_i b_i)/(T-p)$ to estimate Σ in the above test statistic. For large samples one can appropriately use the above F statistic of (31) with $\Sigma = \hat{\Sigma}$.

If the hypothesis $R\beta = 0$ is accepted, then one pools the data and estimates the common regression coefficient. Assume that: $\beta_1 = \cdots = \beta_M = \beta^+$, where β^+ is a $p \times 1$ vector. Now we can write the model (29) in a compact matrix form as

$$y = Z\beta^+ + u\ , \tag{33}$$

where y and u are $MT \times 1$ vectors as before, and

$$Z = \begin{bmatrix} X_1 \\ \vdots \\ X_M \end{bmatrix}$$

is an $MT \times p$ matrix. Then the pooled OLS estimator of β^+ is

$$b^+ = (Z'Z)^{-1}Z'y = (\sum_{i=1}^{M} X_i'X_i)^{-1} \sum_{i=1}^{M} X_i'y_i\ . \tag{34}$$

Transforming the model (33) as $P'y = P'Z\beta^+ + P'u$, we can write the GLS estimator of β^+ as

$$\hat{\beta}^{+} = (Z'\Sigma^{-1}Z)^{-1}Z'\Sigma^{-1}y$$

$$= \left[\sum_{i=1}^{M} \frac{X_i'X_i}{\sigma_{ii}}\right]^{-1} \sum_{i=1}^{M} \frac{X_i'y_i}{\sigma_{ii}} . \tag{35}$$

In the "pooling" model considered by Mundlak [1961], it is assumed that $\sigma_{ii} = \sigma^2$ and therefore he has $\hat{\beta}^{+} = b^{+}$. Thus the estimator $\hat{\beta}^{+}$ generalizes Mundlak's "fixed effect" (FE) estimator.

In the so-called "error components" (EC) model we assume that the errors have a special structure: $u_{it} = \epsilon_i + v_{it}$, where $Ev_{it} = 0$, $E(v_{it}^2) = \sigma_{ii}$, $E\ \epsilon_i = 0$, and $E\ \epsilon_i^2 = \sigma^2$. Thus $E\ u_{it} = 0$, $E\ u_{it}^2 = \sigma^2 + \sigma_{ii}$ and

$$\begin{aligned} E\ u_{it}u_{jt'} &= 0, \quad \text{if } i \neq j. \\ &= \sigma^2, \quad \text{if } i = j, \text{ and } t \neq t'. \end{aligned}$$

This implies that $E\ u_i = 0$, $E\ u_i u_j' = 0$ for $i \neq j$. Now, let us define notation Σ_{ii} as:

$$E\ u_i u_i' = \begin{bmatrix} \sigma^2+\sigma_{ii} & \cdots & \sigma^2 \\ \vdots & & \vdots \\ \sigma^2 & \cdots & \sigma^2+\sigma_{ii} \end{bmatrix} = \sigma^2\iota\iota' + \sigma_{ii}I = \Sigma_{ii} ,$$

where ι is a $T \times 1$ column vector of ones. Further, define a block diagonal matrix

$$E\ uu' = \Sigma = \begin{bmatrix} \Sigma_{11} & \cdots & 0 \\ \vdots & & \vdots \\ 0 & \cdots & \Sigma_{MM} \end{bmatrix} .$$

The GLS estimator of β^{+} can be written, in this case, as

$$\hat{\beta}^{+} = (Z'\Sigma^{-1}Z)^{-1}Z'\Sigma^{-1}y$$

$$= \left[\sum_{i=1}^{M} (X_i'\Sigma_{ii}^{-1}X_i)\right]^{-1} \sum_{i=1}^{M} (X_i'\Sigma_{ii}^{-1}y_i). \tag{36}$$

Using the following matrix inversion result

$$\Sigma_{ii}^{-1} = (\sigma^2\iota\iota' + \sigma_{ii}I)^{-1} = \frac{I}{\sigma_{ii}} - \frac{\sigma^2\iota\iota'}{\sigma_{ii}(\sigma_{ii} + \sigma^2\iota'\iota)} , \tag{37}$$

(Hint: to verify this, check that the inverse times Σ_{ii} matrix yields I) we can write

$$\hat{\beta}^{+} = \sum_{i=1}^{M} q_i b_i , \tag{38}$$

where $b_i = (X_i'X_i)^{-1}X_i'y_i$ is simply the OLS estimator obtained from only the i-th cross-sectional unit, and the weights q_i are given by

$$q_i = \frac{1/(\sigma^2+\sigma_{ii}T^{-1})}{\sum_{j=1}^{M}[1/(\sigma^2+\sigma_{jj}T^{-1})]} . \tag{39}$$

In this framework $\hat{\beta}^+$ is a weighted sum of b_i's.

10.4 RANDOM COEFFICIENT REGRESSION MODEL

This model was considered by Swamy [1970, 1971]. Assume $p_i = p$ and the vector β_i in (1) having p elements to be

$$\beta_i = \bar{\beta} + \eta_i, \; i = 1, ..., M , \tag{40}$$

where β_i is random with the mean vector $\bar{\beta}$ and the diagonal variance covariance matrix D for all i, i.e. $E\eta_i = 0$ and $E\eta_i\eta_i' = D = Diag.\ (d_{11} \cdots d_{pp})$. Also, $E\eta_i\eta_j' = 0$ for $j \neq i$, $j = 1, ..., M$. The η_i and u_i are jointly independent. Further, $E\ u_i = o$, $E\ u_iu_i' = \sigma_{ii}I$ and $E\ u_iu_j' = 0$ for $i \neq j$.

Under these assumptions the model (1) can be written as $y_i = X_i\beta_i + u_i = X_i\bar{\beta} + X_i\eta_i + u_i$. Hence, combining the errors together we write

$$y_i = X_i\bar{\beta} + w_i , \tag{41}$$

where the combined error in the i-th equation (i.e., i-th cross-section) is $w_i = u_i + X_i\eta_i$, such that $E\ w_i = 0$, and

$$\begin{aligned} E\ w_iw_j' &= \sigma_{ii}I + X_iDX_i' = \Omega_{ii}, \quad \text{for } j = i \\ &= 0 \qquad\qquad\qquad\quad , \quad \text{for } j \neq i. \end{aligned}$$

If we further write the model for all i together as

$$y = Z\bar{\beta} + w, \tag{42}$$

where y and Z are as defined before, and $w = (w_1, ..., w_M)'$ then all MT disturbances are such that $Ew = 0$, and

$$Eww' = \begin{bmatrix} \sigma_{11}I+X_1DX_1' & \cdots & 0 \\ \vdots & & \\ 0 & & \sigma_{MM}I+X_MDX_M' \end{bmatrix} = \Omega , \tag{43}$$

where the covariance matrix of the $p \times 1$ vector β_i of random coefficients is

$$D = \begin{vmatrix} d_{11} & \cdot & \cdot & \cdot & 0 \\ & \cdot & \cdot & & \\ & \cdot & & \cdot & \\ & \cdot & & \cdot & \\ 0 & & & & d_{pp} \end{vmatrix} . \tag{44}$$

The OLS and GLS estimators, respectively, are

$$\bar{b}_o = (Z'Z)^{-1}Z'y = (\sum_{i=1}^{M} X_i'X_i)^{-1}(\sum_{i=1}^{M} X_i'y_i), \tag{45}$$

$$\bar{b} = (X'\Omega^{-1}X)^{-1}X'\Omega^{-1}y = [\sum_{i=1}^{M} (X_i'\Omega_{ii}^{-1}X_i)]^{-1} \sum_{i=1}^{M} (X_i'\Omega_{ii}^{-1}y_i), \tag{46}$$

where

$$\Omega_{ii} = \sigma_{ii}I + X_iDX_i' . \tag{47}$$

Using the matrix inversion result for Ω_{ii}, we write similar to (37)

$$\begin{aligned}\Omega_{ii}^{-1} &= (\sigma_{ii}I + X_iDX_i')^{-1} \\ &= \frac{1}{\sigma_{ii}} I - \frac{1}{\sigma_{ii}} X_i \left[\frac{X_i'X_i}{\sigma_{ii}} + D^{-1}\right]^{-1} \frac{X_i'}{\sigma_{ii}} .\end{aligned} \tag{48}$$

Thus pre-multiplying by X_i' and post-multiplying by X_i we have

$$\begin{aligned}X_i'\Omega_{ii}^{-1}X_i &= \frac{X_i'X_i}{\sigma_{ii}} \left[\frac{X_i'X_i}{\sigma_{ii}} + D^{-1}\right]^{-1} \frac{X_i'X_i}{\sigma_{ii}} \\ &= \frac{X_i'X_i}{\sigma_{ii}} \left[I - \left[\frac{X_i'X_i}{\sigma_{ii}} + D^{-1}\right]^{-1} \frac{X_i'X_i}{\sigma_{ii}}\right] .\end{aligned} \tag{49}$$

Using this

$$\bar{b} = \sum_{i=1}^{M} \Lambda_i b_i \tag{50}$$

which is a matrix weighted linear combination of the OLS estimator $b_i = (X_i'X_i)^{-1} X_i'y_i$, which is distributed normally with mean β_i and covariance matrix $V(b_i) = \sigma_{ii}(X_i'X_i)^{-1}$. The matrix of weights is

$$\wedge_i = [\sum_{j=1}^{M} (D + \sigma_{jj}(X_j'X_j)^{-1})^{-1}]^{-1}(D + \sigma_{ii}(X_i'X_i)^{-1})^{-1} , \qquad (51)$$

such that $\sum_{i=1}^{M} \wedge_i = I$.

The pooled estimator $\bar{b}$ under the assumption that the vector of coefficients β_i, for each cross-section, is distributed about the mean vector $\bar{\beta}$ and covariance matrix D, is a best linear unbiased estimator of β.

Several types of shrinkage estimators for $\bar{\beta}$ can be developed along the lines discussed in Chapters 6, 7, and 8. For example, if we write the transformed model: $P'y = P'X\bar{\beta} + P'u$, and assume that the prior distribution of $\bar{\beta}$ is normal, $N(0, GK^{-1}G')$, the "generalized" ridge estimator (GRE) is

$$\bar{b}_{GRE} = [(P'X)'P'X + GKG']^{-1}(P'X)'P'y \qquad (52)$$

$$= (X'\Omega^{-1}X + GKG')^{-1} X'\Omega^{-1}y ,$$

where G is a matrix of eigenvectors of $X'\Omega^{-1}X$ and K is a matrix of ridge biasing parameters. For $K = kI$ we have the family of "ordinary" ridge estimators. Our notation G and K is slightly different from the notation in Chapter 7. Note that this G incorporates Ω, and this K incorporates σ^2 also. This modification of the notation is helpful in simplifying the following discussion.

We may use equation (97) of Chapter 3, and write the condition for the MSE matrix superiority of the shrinkage estimator as follows:

$$MtxMSE(\bar{b}) - MtxMSE(\bar{b}_{GRE}) \text{ is positive definite,}$$

if and only if

$$\bar{\beta}' [(X'\Omega^{-1}X)^{-1} + 2GK^{-1}G']^{-1}\bar{\beta} \le 1. \qquad (52a)$$

This condition is similar to equation (12b) of Chapter 7, except for the slight change of notation mentioned above.

The estimator $\bar{b}$ of (50) contains unknown σ_{ij} values and D. The unknown σ_{ii} can be estimated by

$$\hat{\sigma}_{ii} = \hat{u}_i'\hat{u}_i/(T-K) ,$$

where $\hat{u}_i = [I - X_i(X_i'X_i)^{-1}X_i']y_i$. Now $\hat{\sigma}_{ii}$ is an unbiased estimator of σ_{ii} obtained by dividing the sum of squares of OLS residuals for i-th cross-sectional regression by the corresponding degrees of freedom. The unknown covariance matrix of the errors η_i of the random coefficients model (40) can be estimated by the following difference between two matrices:

$$\hat{D} = [\sum_{i=1}^{M} b_i b_i' - \frac{1}{M}\sum_{i=1}^{M} b_i \sum_{i=1}^{M} b_i'] (M-1)^{-1} - \frac{1}{M}\sum_{i=1}^{M} \hat{\sigma}_{ii}(X_i'X_i)^{-1} . \qquad (53)$$

Note that the estimator $\hat{D}$ may contain some negative elements because the expression (53) is in the form of a difference between two matrices. The above estimator for D has been obtained from the relation

$$E(b_i - \bar{\beta})(b_i - \bar{\beta})' = D + \sigma_{ii}(X_i'X_i)^{-1} . \qquad (54)$$

An operational (OP) version of $\bar{b}$ of (50) with estimated values of σ_{ii} and D is defined by

$$\bar{b}_{OP} = \sum_{i=1}^{M} \Lambda_i b_i , \qquad (55)$$

where

$$\Lambda_i = [\sum_{j=1}^{M} \{\hat{D} + \hat{\sigma}_{jj}(X_i'X_i)^{-1}\}^{-1}]^{-1} [\hat{D} + \hat{\sigma}_{ii}(X_i'X_i)^{-1}]^{-1} . \qquad (56)$$

The estimator in (55) is asymptotically normally distributed about $\bar{\beta}$ and variance-covariance matrix $Asy.V(\bar{b}_{OP}) = [\sum_{j=1}^{M} \{D + \sigma_{jj}(X_i'X_i)^{-1}\}^{-1}]^{-1}$. Thus the asymptotic standard errors of estimates can be obtained as the square root of the diagonal elements of the first term in (56) as:

$$Asy.V(\bar{b}_{OP}) = [\sum_{j=1}^{M} \{\hat{D} + \hat{\sigma}_{jj}(X_i'X_i)^{-1}\}^{-1}]^{-1} . \qquad (57)$$

Swamy [1971] extends the model (41) by assuming that the disturbance u_{it} may be autoregressive with $\rho_i (0 < |\rho_i| < 1)$ as the coefficient of autoregression.

Swamy's [1973] paper further extends the model by assuming that the covariance between individuals may be nonzero and that the serial correlation between a pair of observations in different time periods declines geometrically with the distance involved. The model with these additional assumptions certainly has many more applications to real life situations. Unfortunately, it's much more difficult to estimate, because of the many additional unknowns involved (ρ_i and σ_{ij} $(i \neq j)$ for $i, j = 1, ..., M$). If sample size is small this would create the additional problem of inadequate degrees of freedom. Swamy [1973] proposes efficient GLS estimators for $\bar{\beta}$ along with consistent estimates of ρ_i, σ_{ij}, and D.

The model in (41) can be compared with the error component (EC) models considered in Section 10.3 above. If the intercept alone is random in (41) we consider $x_{1it} = 1$ for all $i = 1, ..., M$, $t = 1, ..., T$, and $\eta_i = (\eta_{1i}\ 0...0)'$. This reduces (41) to Nerlove's EC model. Furthermore, if we assume that the intercept is nonstochastic across equations (i.e., across i), the EC model becomes Mundlak's fixed effect (FE) model where the errors η_{1i} are constant. The FE model is also known as the "dummy variables" model, or the analysis of covariance (ANOCOVA). Recently Mundlak [1978] has provided a unified framework for considering EC, FE and related models.

10.5 LINDLEY'S HYPERPARAMETER MODEL

As mentioned earlier, the pooled estimation requires the pretesting of the constancy of regression coefficients across the equations. As noted by Maddala [1971], this raises the question of what significance level to use when deciding whether or not to pool. Also, if estimates obtained from highly aggregated data are used as descriptive measures of broad behavioral tendencies, then they may be misleading because of aggregation bias.

An alternative model that is discussed here is due to Lindley [1971], Lindley and Smith [1972], and Smith [1973], and has been discussed earlier in Chapters 3, 6 and 9. The structure of this model is hierarchical. Its application has been explored in psychology by Novick et al. [1972] and in econometrics by Ullah and Raj [1979], Raj and Ullah [1981], Trivedi [1980], and Parikh and Trivedi [1980]. The idea behind the model is to use all available prior information for the estimation of each regression parameter while at the same time allowing for differences across the equations.

The model can be written as

$$y_i = X_i \beta_i + u_i , i - 1, \dots, M, \tag{58}$$

such that given β_i, we have $y_i \sim N(X_i\beta_i, \sigma_{ii}I)$. Next, given a $p \times 1$ vector β_o, the common prior for all $\beta_i \sim N(\beta_o, \Omega_o)$, and at the second (possibly last) stage the prior is $\beta_o \sim N(\mu, \Omega_1)$. The vectors β_i, β_o, and μ are each $p \times 1$ vectors of unknown parameters. β_o and μ are called hyperparameters by Lindley [1971].

The above statements about them are essentially the statements of prior beliefs and "beliefs about beliefs." The prior beliefs about β_i are that they are the same across the cross-sections, an assumption which Lindley calls *exchangeability between equations.* Ω_o and Ω_1 are prior variance covariance matrices. If $\Omega_1^{-1} \to 0$, the second stage prior is said to be diffuse. The case $\Omega_1^{-1} \to 0$ has been considered earlier in Section 3-6.5.

Under the assumption of known variances σ_{ii} and Ω_o, and a diffuse prior about β_o, the posterior mean of β_i (Bayes' estimator) can be written (following Section 3-6.5 of Chapter 3) as

$$\beta_i^* = (\sigma_{ii}^{-1} X_i'X_i + \Omega_o^{-1})^{-1}(\sigma_{ii}^{-1} X_i'X_i b_i + \Omega_o^{-1} \beta_o^*), \tag{59}$$

where $b_i = (X_i'X_i)^{-1}X_i'y_i$ is the OLS estimator and β_o^* is the pooled estimator (posterior mean) of β_o given as

$$\beta_o^* = \sum_{i=1}^{M} \wedge_i^* b_i$$

$$\wedge_i^* = \left[\sum_{j=1}^{M} (\sigma_{jj}^{-1} X_j' X_j + \Omega_o^{-1})^{-1} \sigma_{jj}^{-1} X_j' X_j \right]^{-1} (\sigma_{ii}^{-1} X_i' X_i + \Omega_o^{-1})^{-1} X_i' X_i \sigma_{ii}^{-1} \tag{60}$$

For $M = 1$ equation (59) is similar to equation (105) in Section 3.6.5.

Note that $V(b_i) = \sigma_{ii}(X_i'X_i)^{-1}$. Thus $\sigma_{ii}^{-1}(X_i'X_i)$ is the precision matrix of the data. Similarly, Ω_o^{-1} is the precision matrix of the prior information, and therefore $\sigma_{ii}^{-1} X_i'X_i + \Omega_o^{-1}$ is the total precision of both the sample and prior information. Further, $(\sigma_{ii}^{-1} X_i'X_i + \Omega_o^{-1})^{-1} \sigma_{ii}^{-1} X_i'X_i$ is the relative precision of the sample information. The weight matrix $\wedge_i^*$ of b_i in (60) is the normalized relative precision matrix such that $\sum_{i=1}^{M} \wedge_i^* = I$.

We can also write β_i^* as

$$\beta_i^* = \beta_o^* + [I - (\sigma_{ii}^{-1} X_i'X_i + \Omega_o^{-1})^{-1} \Omega_o^{-1}](b_i - \beta_o^*) \tag{61}$$

(also see equation (105) in Chapter 3).

The expression for β_i^* is a weighted matrix combination of the OLS estimator b_i and the estimator β_o^* of (60) obtained by pooling all M equations, the pooling being made possible by the exchangeability assumption. As expected, and it can be easily verified that β_o^* is identical to the pooled estimator $\bar{b}$ obtained in the context of the random coefficient regression model in the earlier section. Thus our Bayesian point estimator β_i^* involves *shrinkage in the direction of the estimated overall mean* β_o^*, and in this sense β_i^* is based on information about all M groups. It also represents a "compromise" between the OLS estimator of β_i and the pooled estimator of the hyperparameter β_o. The estimate of micro parameters like β_i would be useful in providing a sounder basis for prediction for individual cross-sections. On the other hand, the pooled estimate β_o^* can be the basis of aggregate policy or generalization.

Note that the estimator β_i^* of (61) depends on unknown variances σ_{ii} and Ω_o. To deal with this problem, Lindley-Smith [1972] and Novick et al. [1972] specify a Wishart prior distributions on Ω_o^{-1} as follows:

$$\Omega_o^{-1} \sim W(\rho, \bar{R}), \tag{62}$$

where W denotes a Wishart prior distribution with the degrees of freedom parameter ρ, and the location matrix $\bar{R}$. They also specify the following inverse Chi-square prior for σ_{ii}:

$$\nu_i \lambda_i^* / \sigma_{ii} \sim \chi^2(\nu_i), \text{ independently for } i = 1, \ldots, M. \tag{63}$$

For a clear discussion of the Wishart distribution the reader is referred to Zellner [1971]. A Wishart distribution is a distribution of the matrix of variances and covariances. It generalizes the usual Chi-square distribution of variances. In particular, when the Wishart matrix is 1×1 it is a Chi-square variable. This is why we have Chi-squares in (63). The degrees of freedom parameters ν_i and ρ, the prior location matrix $\bar{R}$, and λ_i^* remain to be specified by the investigator. Once this is done, the joint and marginal posterior distribution of (β_i, σ_{ii}), may be obtained by numerical integration.

An alternative approach, suggested by Lindley [1971] and discussed by Lindley and Smith [1972], involves the solution of joint model equations for (β_i, σ_{ii}), and the resultant modes are treated as approximations to the posterior mean. Lindley and Smith give the modal equations and discuss an iterative solution technique. Novick et al. [1972] also provide a discussion for the application of their technique (see Trivedi [1980]). The properties of the model estimates have been analyzed by Smith [1973] and Ullah [1980a].

In this chapter we have considered inter-relationships between Zellner's seemingly unrelated (SUR) model, temporal cross-sectional model and random coefficients model. We have also applied shrinkage-type concepts to these models, and indicated situations where shrinkages are advisable. For example, equations (28) and (52a) give the appropriate conditions for dominance over the usual estimators.

EXERCISES

10.1 Consider a set of M seemingly unrelated regression equations model in the vector form as $y = X\beta + u$. Show that the McElroy's [1977] measure of goodness of fit

$$R^2 = 1 - \frac{\hat{u}'(\Sigma^{-1} \otimes I)\hat{u}}{y'(\Sigma^{-1} \otimes J)y}$$

lies between 0 and 1, where $\hat{u}$ is the generalized least squares residuals, $J = I - \frac{\iota\iota'}{T}$; ι is a column of unit elements, and each equation has an intercept term. Further $\Sigma = ((\sigma_{ij}))$; $i, j = 1, \ldots, M$ is the variance covariance matrix of disturbances across the equations. Suggest a test for the significance of the population multiple correlation coefficient.

10.2

(a) Consider an M equations hierarchical model $y_i \sim N(X_i\beta_i, \sigma_{ii} I)$, $\beta_i \sim N(\beta_o, D)$ and β_o is random with diffuse prior; $i = 1, \ldots, M$, y_i is a $T \times 1$ vector and β_i is a $p \times 1$ vector. Further suppose a conjugate Wishart distribution for D^{-1}, i.e., $D^{-1} \sim W(\rho, \bar{R})$ and independent conjugate inverse χ^2 distributions for the σ_{ii}'s, $\nu_i\lambda_i^*/\sigma_{ii} \sim \chi^2(\nu_i)$. Show that the modal equations for approximate Bayesian estimators $\hat{\beta}_i$, $\hat{\sigma}_{ii}$ and $\hat{D}$ for β_i, σ_{ii} and D respectively are

$$\hat{\beta}_i = (\hat{\sigma}_{ii}X_i'X_i + \hat{D}^{-1})^{-1}(\hat{\sigma}_{ii}X_i'X_ib_i + \hat{D}^{-1}\hat{\beta}_o)$$

$$\hat{\sigma}_{ii} = \{\nu_i\lambda_i^* + (y_i - X_i\hat{\beta}_i)'(y_i - X_i\hat{\beta}_i)\}/(T + \nu_i + 2)$$

$$\hat{D} = \{\bar{R} + \sum_{i=1}^{M} (\hat{\beta}_i - \hat{\beta}_o)(\hat{\beta}_i - \hat{\beta}_o)'\}/(M + \rho - p - 2)$$

where b_i is the OLS estimator and

$$\hat{\beta}_o = \sum_{i=1}^{M} \{\sum_{j=1}^{M} (X_j'\hat{\sigma}_{jj}^{-1}X_j + \hat{D}^{-1})^{-1}X_j'\hat{\sigma}_{jj}^{-1}X_j\}^{-1}(X_i'\hat{\sigma}_{ii}^{-1}X_i + \hat{D}^{-1})^{-1}X_i'\hat{\sigma}_{ii}^{-1}X_ib_i$$

Reference: Lindley and Smith [1972 Section 5.2].

(b) What would $\hat{\sigma}_{ii}$ be if we consider a diffuse prior for σ_{ii}?

(c) If $y \sim N(X_i\beta_i, \sigma_{ii}I)$, $\beta_i \sim N(\beta_o, D)$ and β_o is fixed, then show that the posterior mean of β_i is

$$\beta_i^* = [\sigma_{ii}^{-1}X_i'X_i + D^{-1}]^{-1}(\sigma_{ii}^{-1}X_i^1X_ib_i + D^{-1}\beta_o) .$$

Compare this with (59) and OLS. Also compare the MSE matrices [Hint: See (97) of Chapter 3].

10.3 Consider an investment function $I_{it} = \beta_{1i} + \beta_{2i}F_{it-1} \quad \beta_{3i}C_{it-1} + u_{it}$ where I is gross investment, F is the value of firm and C is the stock of plant and equipment; $i = 1, ..., M$ and $t = 1, ..., T$. Using the data in Table 10.1 for 10 major U.S. Corporations over the period 1935-54 obtain (i) Zellner's estimator (ii) Swamy's estimator of the 3×1 vector β_o under the assumption that $\beta_i \sim N(\beta_o, D)$ where $\beta_i = [\beta_{1i}\ \beta_{2i}\ \beta_{3i}]'$ and D is a diagonal matrix (iii) Lindley and Smith modal estimator of β_o under the exchangeable prior $\beta_i \sim N(\beta_o, D)$ and β_o diffuse. Further assume vague priors for D and $\sigma_{ii} = v(u_{it})$. [Hint: Approximate model equations for estimation are as in the Exercise 10.2. Substitute $\nu_i = 0$, $\rho = 1$ $\bar{R} = .001$ as in approximation for vague priors on D and σ_{ii}] (iv) Mundlak's estimator. State your assumptions in each case.

10.4 Formulate a Monto Carlo project to study the sampling properties of various estimators discussed in this chapter.

TABLE 10.1

	G.M.			U.S. Steel			G.E.			Chrysler		
	I	F_{-1}	C_{-1}	I	F_{-1}	C_{-1}	I	F_{-1}	C_{-1}	I	F_{-1}	C_{-1}
1935	317.6	3078.5	2.8	209.9	1362.4	53.8	33.1	1170.6	97.8	40.29	417.5	10.5
36	391.8	4661.7	52.6	355.3	1807.1	50.5	45.0	2015.8	104.4	72.76	837.8	10.2
37	410.6	5387.1	156.9	469.9	2676.3	118.1	77.2	2803.3	118.0	66.26	883.9	34.7
38	257.7	2792.2	209.2	262.3	1801.9	260.2	44.6	2039.7	156.2	51.60	437.9	51.8
39	330.8	4313.2	203.4	230.4	1957.3	312.7	48.1	2256.2	172.6	52.41	679.7	64.3
40	461.2	4643.9	207.2	361.6	2202.9	254.2	74.4	2132.2	186.6	69.41	727.8	67.1
41	512.0	4551.2	255.2	472.8	2380.5	261.4	113.0	1834.1	220.9	68.35	643.6	75.2
42	448.0	3244.1	303.7	445.6	2168.6	298.7	91.9	1588.0	287.8	46.80	410.9	71.4
43	499.6	4053.7	264.1	361.6	1985.1	301.8	61.3	1749.4	319.9	47.40	588.4	67.1
44	547.5	4379.3	201.6	288.2	1813.9	279.1	56.8	1687.2	321.3	59.57	698.4	60.5
45	561.2	4840.9	265.0	258.7	1850.2	213.8	93.6	2007.7	319.6	88.78	846.4	54.6
46	688.1	4900.9	402.2	420.3	2067.7	232.6	159.9	2208.3	346.0	74.12	893.8	84.8
47	568.9	3526.5	761.5	420.5	1796.7	264.8	147.2	1656.7	456.4	62.68	579.0	96.8
48	529.2	3254.7	922.4	494.5	1625.8	306.9	146.3	1604.4	543.4	89.36	694.6	110.2
49	555.1	3700.2	1020.1	405.1	1667.0	351.1	98.3	1431.8	618.3	78.98	590.3	147.4
50	642.9	3755.6	1099.0	418.8	1677.4	357.8	93.5	1610.5	647.4	100.66	693.5	163.2
51	755.9	4833.0	1207.7	588.2	2289.5	342.1	135.2	1819.4	671.3	160.62	809.0	203.5
52	891.2	4924.9	1430.5	645.5	2159.4	444.2	157.3	2079.7	726.1	145.00	727.0	290.6
53	1304.4	6241.7	1777.3	641.0	2031.3	623.6	179.5	2371.6	800.3	174.93	1001.5	346.1
54	1486.7	5593.6	2226.3	459.3	2115.5	669.7	189.6	2759.9	888.9	172.49	703.2	414.9

Notes to Table 10.1

I = gross investment* = additions to plant and equipment plus maintenance and repairs in millions of dollars deflated by P_1.

F = value of the firm† = price of common and preferred shares at Dec. 31 (or average price of Dec. 31 and Jan. 31 of the following year) times number of common and preferred shares outstanding plus total book value of debt at Dec. 31 in millions of dollars deflated by P_2.

C = stock of plant and equipment* = accumulated sum of net additions to plant and equipment deflated by P_1 minus depreciation allowance deflated by P_3 in these definitions.

TABLE 10.1 (Continued)

	Atlantic Richfield			I.B.M.			Union Oil		
	I	F_{-1}	C_{-1}	I	F_{-1}	C_{-1}	I	F_{-1}	C_{-1}
1935	39.68	157.7	183.2	20.36	197.0	6.5	24.43	138.0	100.2
36	50.73	167.9	204.0	25.98	210.3	15.8	23.21	200.1	125.0
37	74.24	192.9	236.0	25.94	223.1	27.7	32.78	210.1	142.4
38	53.51	156.7	291.7	27.53	216.7	39.2	32.54	161.2	165.1
39	42.65	191.4	323.1	24.60	286.4	48.6	26.65	161.7	194.8
40	46.48	185.5	344.0	28.54	298.0	52.5	33.71	145.1	222.9
41	61.40	199.6	367.7	43.41	276.9	61.5	43.50	110.6	252.1
42	39.67	189.5	407.2	42.81	272.6	80.5	34.46	98.1	276.3
43	62.24	151.2	426.6	27.84	287.4	94.4	44.28	108.8	300.3
44	52.32	187.7	470.0	32.60	330.3	92.6	70.80	118.2	318.2
45	63.21	214.7	499.2	39.03	324.4	92.3	44.12	126.5	336.2
46	59.37	232.9	534.6	50.17	401.9	94.2	48.98	156.7	351.2
47	58.02	249.0	566.6	51.85	407.4	111.4	48.51	119.4	373.6
48	70.34	224.5	595.3	64.03	409.2	127.4	50.00	129.1	389.4
49	67.42	237.3	631.4	68.16	482.2	149.3	50.59	134.8	406.7
50	55.74	240.1	662.3	77.34	673.8	164.4	42.53	140.8	429.5
51	80.30	327.3	683.9	95.30	676.9	177.2	64.77	179.0	450.6
52	85.40	359.4	729.3	99.49	702.0	200.0	72.68	178.1	466.9
53	81.90	398.4	774.3	127.52	793.5	211.5	73.86	186.8	486.2
54	81.43	365.7	804.9	135.72	927.3	238.7	89.51	192.7	511.3

Notes to Table 10.1 (Cont.)

P_1 implicit price deflator of producers' durable equipment (base 1947)‡

P_2 implicit price deflator of GNP (base 1947)‡

P_3 depreciation expense deflator = 10 year moving average of wholesale price index of metals and metal products (base 1947)§

* Moody's Industrial Manual and Annual Reports of Corporations.

† Bank and Quotation Record and Moody's Industrial Manual.

‡ Survey of Current business, July 1956, July 1957.

§ Historical statistics of the U.S., 1789-1945, and Economic Report of the President, January 1957, p. 161.

TABLE 10.1 (Continued)

	Westinghouse			Goodyear			Diamond Match		
	I	F_{-1}	C_{-1}	I	F_{-1}	C_{-1}	I	F_{-1}	C_{-1}
1935	12.93	191.5	1.8	26.63	290.6	162	2.54	70.91	4.50
36	25.90	516.0	.8	23.39	291.1	174	2.00	87.94	4.71
37	35.05	729.0	7.4	30.65	335.0	183	2.19	82.20	4.57
38	22.89	560.4	18.1	20.89	246.0	198	1.99	58.72	4.56
39	18.84	519.9	23.5	28.78	356.2	208	2.03	80.54	4.38
40	28.57	628.5	26.5	26.93	289.8	223	1.81	86.47	4.21
41	48.51	537.1	36.2	32.08	268.2	234	2.14	77.68	4.12
42	43.34	561.2	60.8	32.21	213.3	248	1.86	62.16	3.83
43	37.02	617.2	84.4	35.69	348.2	274	.93	62.24	3.58
44	37.81	626.7	91.2	62.47	374.2	282	1.18	61.82	3.41
45	39.27	737.2	92.4	52.32	387.2	316	1.36	65.85	3.31
46	53.46	760.5	86.0	56.95	347.4	302	2.24	69.54	3.23
47	55.56	581.4	111.1	54.32	291.9	333	3.81	64.97	3.90
48	49.56	662.3	130.6	40.53	297.2	359	5.66	68.00	5.38
49	32.04	583.8	141.8	32.54	276.9	370	4.21	71.24	7.39
50	32.24	635.2	136.7	43.48	274.6	376	3.42	69.05	8.74
51	54.38	723.8	129.7	56.49	339.9	391	4.67	83.04	9.07
52	71.78	864.1	145.5	65.98	474.8	414	6.00	74.42	9.93
53	90.08	1193.5	174.8	66.11	496.0	443	6.53	63.51	11.68
54	68.60	1188.9	213.5	49.34	474.5	468	5.12	58.12	14.33

Source: J. C. G. Boot and G. M. Dewit [1960].

11

Simultaneous Equations Model

This chapter assumes that the reader has some familiarity with this topic from Econometrics textbooks, e.g., Theil [1971]. We provide a concise introduction to the breadth of this specialized topic. Unfortunately, the space limitations do not permit a clearer discussion of motivations for all the concepts.

11.1 MODEL SPECIFICATION

This chapter is concerned with the problem of identification and estimation in a simultaneous equations model. Consider a set of M structural equations of a simultaneous equations model in K predetermined variables. We use standard notation in Econometrics literature, which should not be confused with ridge notation. We have:

$$\gamma_{11}y_{1t} + \ldots + \gamma_{M1}y_{Mt} + \beta_{11}x_{1t} + \ldots + \beta_{K1}x_{Kt} = u_{1t}$$

$$\gamma_{1M}y_{1t} + \ldots + \gamma_{MM}y_{Mt} + \beta_{1M}x_{1t} + \ldots + \beta_{KM}x_{Kt} = u_{Mt}, \tag{1}$$

where y_{mt} is the $t^{th}\,(t = 1, \ldots, T)$ observation on the $m^{th}\,(m = 1, \ldots, M)$ current endogenous or jointly dependent variable and x_{kt} is the $t^{th}\,(t = 1, \ldots, T)$ observation of the $k^{th}\,(k = 1, \ldots, K)$ predetermined variable. The γ's and β's are structural coefficients of the endogenous and predetermined variables. The endogenous variables are the variables which are to be determined from the system of equations, whereas the predetermined variables are determined from outside of the system. Usually, the predetermined variables consist of fixed exogenous, lagged exogenous and lagged endogenous variables. It is because of this that we use a new letter K instead of p to denote the set of predetermined variables.

To give the real flavor of a simultaneous equations model, take Klein's

[1950] Model I of the U.S. economy. Table 11.1 at the end of this chapter gives the complete data.

1. Consumption Equation

$$C_t = \gamma_{12}P_t + \gamma_{13}(W_t+W_t') + \beta_{11} + \beta_{12}P_{t-1} + u_{1t}, \tag{1a}$$

2. Investment Equation

$$I_t = \gamma_{22}P_t + \beta_{21} + \beta_{22}P_{t-1} + \beta_{23}K_{t-1} + u_{2t}, \tag{1b}$$

3. Labor Equation

$$W_t = \gamma_{32}X_t + \beta_{31} + \beta_{32}X_{t-1} + \beta_{33}(t-1931) + u_{3t}, \tag{1c}$$

4. National Accounts: Definition 1

$$X_t = C_t + I_t + G_t, \tag{1d}$$

5. Profit and Loss Account: Definition 2

$$P_t = X_t - W_t - T_t, \tag{1e}$$

6. Change in Capital Stock = Investment: Definition 3

$$K_t - K_{t-1} = I_t, \tag{1f}$$

where γ's and β's are the so-called structural coefficients, u's are disturbances; X is the total production of private industry; C is aggregate consumption; P is total profits, W is total wages paid by private industry, W' is the government wage bill; I is net investment, K is capital stock, T is taxes; G is government expenditure; and t is a time trend.

The model is dynamic in the sense that it contains some lagged variables. It consists of three behavioral equations (1a) to (1c) and three identities (1d) to (1f), expressed in six endogenous or jointly dependent variables and eight predetermined variables.

The six endogenous variables are: C, P, W, I, K, and X. The eight predetermined variables are comprised of five exogenous variables: G, Intercept, t, T, and W', and three lagged (one period) endogenous variables: K_{-1}, P_{-1}, and X_{-1}.

In all cases not all of the endogenous and predetermined variables in each equation of the simultaneous equation model are present. The structural coefficients (the γ's and β's) of all those variables that do not appear in an equation are assumed to be zero. The coefficients of identities (1d) to (1f) in the model are assumed to be fixed and known. Further, in each behavioral equation one of the γ's is assumed to be unity, and the endogenous variable associated with that unit coefficient is the left-hand endogenous (dependent) variable. This assumption is known as a normalization condition for the model.

The model (1) can be written in the matrix notation as

$$Y\Gamma + XB = U, \tag{2}$$

where

$$Y = \begin{bmatrix} y_{11} & \cdots & y_{M1} \\ \vdots & & \\ y_{1T} & \cdots & y_{MT} \end{bmatrix} \text{ and } X = \begin{bmatrix} x_{11} & \cdots & x_{K1} \\ \vdots & & \\ x_{1T} & \cdots & x_{KT} \end{bmatrix} \tag{3}$$

are the matrices of observations on M jointly dependent and K predetermined variables. Further, in (2) we have denoted

$$\Gamma = \begin{vmatrix} \gamma_{11} & \cdots & \gamma_{1M} \\ \vdots & & \\ \gamma_{M1} & \cdots & \gamma_{MM} \end{vmatrix} \text{ and } B = \begin{vmatrix} \beta_{11} & \cdots & \beta_{1M} \\ \vdots & & \\ \beta_{K1} & \cdots & \beta_{KM} \end{vmatrix}, \tag{4}$$

which are the matrices of structural coefficients, and

$$U = \begin{bmatrix} u_{11} & \cdots & u_{M1} \\ \vdots & & \\ u_{1T} & \cdots & u_{MT} \end{bmatrix} \tag{5}$$

denotes the matrix of structural disturbances. It should be observed that the successive columns of Γ and B provide the coefficients of different equations in (1), the first column of Γ and the first column of B give the coefficients in the first structural equation, and the last column of Γ and the last column of B give the coefficients of the Mth equation.

The so-called reduced form of the simultaneous equation system (1) is derived by solving the set of structural equations for jointly dependent variables Y in terms of the predetermined variables (X) and the structural disturbances (U). This can be done most conveniently in the matrix notation by post-multiplying both sides of (2) by Γ^{-1}. Thus

$$Y = X\Pi + V, \tag{6}$$

where

$$\Pi = \begin{vmatrix} \pi_{11} & \cdots & \pi_{1M} \\ \vdots & & \\ \pi_{K1} & \cdots & \pi_{KM} \end{vmatrix} = -B\Gamma^{-1}, \text{ and } V = \begin{vmatrix} v_{11} & \cdots & v_{M1} \\ \vdots & & \\ v_{1T} & \cdots & v_{MT} \end{vmatrix} = U\Gamma^{-1} \tag{7}$$

are the matrices of reduced form coefficients and reduced form disturbances, respectively. The successive columns of Π and V provide the coefficients and disturbances in different equations. It should be noted that the existence of the reduced form requires Γ to be a nonsingular matrix, i.e., the determinant value of Γ should not be zero. This is a rather reasonable condition, because if Γ were singular or close to singularity, any small changes in exogenous variables

X would lead to very large changes in endogenous variables Y. In fact, this kind of sensitivity of Y is not observed in reality.

The model (2) and its reduced form (6), at the time point t, can be written (respectively) as

$$y_t'\Gamma + x_t'B = u_t', \tag{8}$$

and

$$y_t' = x_t'\pi + v_t', \tag{9}$$

where y_t', x_t', u_t', and v_t' are the tth rows of Y, X, U, and V, respectively. Similarly, a particular equation of the model (say the first) can be written as

$$Y\gamma\cdot_1 + X\beta\cdot_1 = u_1, \tag{10}$$

where $\gamma\cdot_1$, $\beta\cdot_1$, and u_1 are the first columns of Γ, B and U, respectively.

11.2 FURTHER SPECIFICATIONS OF THE SYSTEM

We propose to study the stochastic simultaneous equations in (1) written in matrix notation (2) under the following assumptions.

Assumption 1 There are M jointly dependent and K predetermined variables connected by a set of M interdependent linear equations of the form (1). Γ in (2) is a nonsingular matrix. There are T row vectors of observations on all jointly dependent and predetermined variables in (3), viz,

$$y_t' = [y_{1t}, \ldots, y_{Mt}] \text{ and } x_t' = [x_{1t}, \ldots, x_{Kt}], \tag{11}$$

respectively, are given for $t = 1, \ldots, T$.

It should be observed that each endogenous variable in Y occurs without lag at least once in some equation; for otherwise the number of jointly dependent variables would be less than M.

Assumption 2 The elements of each row of the $T \times K$ matrix X are independently distributed of the structural disturbances in the corresponding row of U.

Or, alternatively

Assumption 3 The elements of X are nonstochastic and fixed in repeated samples.

It should be noted that Assumption 3 does not permit the presence of lagged endogenous variables in X. However, in most cases (especially in large sample analysis) the independence of predetermined variables and structural disturbances is all that is required.

Assumption 4 The rank of X is $K \leqq T$.

Assumption 4 implies that the columns of X are linearly independent and $X'X$ is nonsingular. For most econometric models, including Klein's we have adequate sample size: $K \leq T$. However, for some large econometric models it is possible that $K > T$.

Assumption 5 If the elements of X are stochastic we would require that each individual element of $\operatorname{plim}_{T\to\infty} (1/T)X'X$ is finite and that this is a nonsingular matrix. In case X is purely nonstochastic we require $\lim_{T\to\infty} (1/T)X'X$ to be finite and nonsingular.

The equations of the structural system (1) are characterized by disturbances $u_{1t}, \cdots, u_{Mt}$. As indicated earlier the purpose of these disturbances is to take into account the effect of those variables which are not explicitly introduced in the equations. We make the following assumption about the stochastic behavior of these disturbances.

Assumption 6 The T row vectors of the structural disturbances

$$u_t' = [u_{1t}, ..., u_{Mt}] \tag{12}$$

for $t = 1, ..., T$, of the matrix U, introduced in (5) are independent random drawings from an M-dimensional normal population with

(i) means: $Eu_{it} = 0$,
(ii) variances: $Vu_{it} = \sigma_{ii}$, and
(iii) covariances: $Cov\,[u_{it}, u_{jt'}] = 0$ if $t \neq t'$
$= \sigma_{ij}$ *if* $t = t'$,

where $i, j = 1, ..., M$ and $t, t' = 1, ..., T$.

Thus the structural disturbances have zero mean and they are homoscedastic. Further, this assumption does not permit any serial- or auto-correlation among disturbances. The contemporaneous disturbances in different structural equations are correlated, so that σ_{ij} denotes the covariance between the disturbances of the ith and the jth equations in the same time period.

In matrix notation of (2) we assume

$$EU = 0, \tag{13}$$

where 0 is a $T \times M$ zero matrix and the covariance matrix of the contemporaneous disturbances may be written as

$$\Sigma = (1/T)EU'U = \begin{bmatrix} \sigma_{11} & \cdots & \sigma_{1M} \\ \vdots & & \\ \sigma_{M1} & \cdots & \sigma_{MM} \end{bmatrix}. \tag{14}$$

We require Σ *to be a positive definite matrix* if the model contains no identity relations (otherwise nonnegative definiteness).

Since the reduced form disturbances, given by $V = U\Gamma^{-1}$ in (7), are linear functions of the structural disturbances, it follows that they are also distributed according to an M-dimensional normal law with

$$EV = 0, \tag{15}$$

0 being a $T \times M$ zero matrix, and the covariance matrix of the contemporaneous reduced form disturbances is

$$\Omega = (1/T)EV'V = \Gamma'^{-1}\Sigma\Gamma^{-1}, \tag{16}$$

which is also positive definite, if Σ is positive definite.

11.3 PROBLEM OF IDENTIFICATION

If we have ample data information available in the form of a long time series of observations on y_t and x_t, we can determine (or estimate) parameters Π and Ω of the reduced form equation (6). The problem of identification is then to see whether we can determine (or estimate) the structural parameters Γ, B, and Σ from the estimates of reduced form parameters,

$$\Pi = -B\Gamma^{-1} \text{ and } \Omega = \Gamma'^{-1}\Sigma\Gamma^{-1}.$$

As is clear, the total known values from Π and Ω are $MK + \frac{1}{2} M(M+1)$ in number, and there are $M^2 + MK + \frac{1}{2} M(M+1) - M$ unknowns to be determined. Thus, the excess of unknowns over knowns equals $M(M-1)$. Hence without some *a priori* information (restrictions) on the structure Γ, B, and Σ, it is impossible to determine the unknowns, or in other words, identify the parameters. We shall therefore analyze below the identification of parameters of each equation of (1) under different types of linear *a priori* restrictions.

11.3.1 Identification by Zero Restrictions

One approach to identification is to impose zero restrictions; that is, equating some elements of the matrix Γ and B to be zero. This approach is based on the fact that each structural equation of the complete system represents some hypothesis about the behavior of certain groups of individuals or variables. For example, in Klein's Model I of the United States economy, the first equation is a hypothesis regarding the behavior of consumers whereas the second and third are related with investment and demand for labor. Therefore, it is rather obvious that all variables of the complete system will not necessarily be represented in each equation. For example, only the variables in Klein's model which reflect consumer behavior would appear in a consumption function, and others would appear with zero coefficients.

Let us then suppose that $m_1 + 1 \leq M$ jointly dependent and $K_1 \leq K$ predetermined variables enter any one (say, the first) structural equation of the system. In addition, we wish to identify one endogenous variable as the "left-hand side" variable by letting its associated coefficient equal minus one. Further, with some rearrangements we can write

$$Y = [y_1 \; Y_1 \; Y_1^*], \; X = [X_1 \; X_1^*] \tag{17}$$

and

$$\gamma_{\cdot 1} = \begin{bmatrix} 1 \\ -\gamma_1 \\ 0 \end{bmatrix} \quad \beta_{\cdot 1} = \begin{bmatrix} -\beta_1 \\ 0 \end{bmatrix}, \tag{18}$$

where Y_1 and Y_1^* are matrices of T observations on m_1 included and $m_1^* = M - m_1 - 1$ excluded jointly dependent variables, respectively. Similarly X_1 and X_1^* are matrices of T observations on K_1 included and $K_1^* = K - K_1$ excluded predetermined variables. The vectors $\gamma_{\cdot 1}$ and $\beta_{\cdot 1}$ are the first column of Γ and B, respectively, with γ_1 and β_1 of order $m_1 \times 1$ and $K_1 \times 1$. Zero vector in $\gamma_{\cdot 1}$ is of order $m_1^* \times 1$ whereas in $\beta_{\cdot 1}$ it is of order $K_1^* \times 1$.

Now the first equation of the system, from (2), can be written as

$$y_1 = Y_1\gamma_1 + X_1\beta_1 + u_1, \tag{19}$$

where u_1 is the first column of U.

The reduced form equation $Y = X\Pi + V$ can be correspondingly partitioned as

$$[y_1 \; Y_1 \; Y_1^*] = [X_1 \; X_1^*] \begin{bmatrix} \pi_{11} & \Pi_{11} & \Pi_{11}^* \\ \pi_{21} & \Pi_{21} & \Pi_{21}^* \end{bmatrix} + [v_1 \; V_1 \; V_1^*], \tag{20}$$

where π_{11}, π_{21} are $K_1 \times 1$ and $K_1^* \times 1$ vectors, Π_{11}, Π_{21} are $K_1 \times m_1$ and $K_1^* \times m_1$ matrices, respectively, of reduced form parameters. Further, v_1, V_1 and V_1^* are, respectively, $T \times 1$, $T \times m_1$, and $T \times m_1^*$ disturbances corresponding to y_1, Y_1, and Y_1^*.

Our next aim is to obtain the relationship between the parameters of the above single equation with the reduced form parameters. For this we first write such a relationship in the complete system, viz,

$$\Pi\Gamma + B = 0 \text{ and } V\Gamma = U. \tag{21}$$

then taking only the first columns of Γ and B we get

$$\Pi\gamma_{\cdot 1} + \beta_{\cdot 1} = 0 \tag{22}$$

or

$$\begin{bmatrix} \pi_{11} & \Pi_{11} & \Pi_{11}^* \\ \pi_{21} & \Pi_{21} & \Pi_{21}^* \end{bmatrix} \begin{bmatrix} 1 \\ -\gamma_1 \\ 0 \end{bmatrix} + \begin{bmatrix} -\beta_1 \\ 0 \end{bmatrix} = 0. \tag{23}$$

This provides the following relations or identifiability relationships between the structural parameters and the reduced form parameters for the first equation, i.e.,

$$\pi_{11} = \Pi_{11}\gamma_1 + \beta_1 \text{ and } \pi_{21} = \Pi_{21}\gamma_1. \tag{24}$$

Also, from (21)

$$[v_1 \; V_1 \; V_1^*] \begin{bmatrix} 1 \\ -\gamma_1 \\ 0 \end{bmatrix} = u_1$$

or

$$v_1 = V_1\gamma_1 + u_1. \tag{25}$$

Note that these relationships have been obtained after using *a priori* zero restrictions on the coefficients of the first equation. We shall now investigate the conditions under which γ_1 and β_1 can be determined.

The equations $\pi_{21} = \Pi_{21}\gamma_1$ are a set of K_1^* nonhomogeneous equations in m_1 unknowns γ_1. Thus a necessary and sufficient condition for its unique solution is that

$$\text{Rank } \Pi_{21} = \text{Rank}[\pi_{21} \; \Pi_{21}] = m_1. \tag{26}$$

This is called the "rank condition" for identification. Since the rank of Π_{21} is $\leq \min(K_1^*, m_1)$, that is, cannot exceed the number of its rows or columns, and Π_{21} is $K_1^* \times m_1$, it follows that we must have

$$K_1^* \geq m_1. \tag{27}$$

This is a *necessary* condition for the identifiability of the equation and it has been called the *order condition.* The condition (27) implies that "the number of predetermined variables excluded from the equation must not be smaller than the number of jointly dependent variables included in the equation less one."

If $K_1^* > m_1$ we say that the equation is *over-identified* and the equation is called *just-identified* if $K_1^* = m_1$. Of course, the equation is not identifiable if K_1^* is smaller than m_1. Over-identification normally provides multiple solutions, (exact) just identification provides a unique solution and under-identification provides no consistent solution to the estimation problem.

11.3.2 Identification by General Linear Homogeneous and Non-Homogeneous Restrictions

General linear homogeneous restrictions on Γ and B include restrictions of the type $\gamma_{21} - \gamma_{31} = 0$, $\gamma_{41} = -\beta_{41}$, etc. The zero restrictions considered above are then special cases. To develop identification by general linear homogeneous restrictions we consider A to be an $(M+K) \times M$ matrix.

$$A = \begin{bmatrix} \Gamma \\ B \end{bmatrix},$$

where Γ and B are as given in (4). Further let $\delta_{\cdot 1} = (1, \gamma_{21}, \ldots, \gamma_{Ml}, \beta_{11}, \ldots, \beta_{K1})'$ be the $(M+K) \times 1$ first column vector of A related to the first equation. Express the restriction on the first equation as

$$\delta_{\cdot 1}' R = 0, \tag{28}$$

where R is an $(M+K) \times r$ matrix, and r is the number of columns, which is equal to the number of restrictions. For example if $r = 2$, with two restrictions: $\gamma_{21} - \gamma_{31} = 0$ and $\gamma_{41} = -\beta_{41}$, then the complete matrix of constraints can be written as

$$R = \begin{bmatrix} 0 & 0 \\ 1 & 0 \\ -1 & 0 \\ 0 & 1 \\ \vdots & 0 \\ & \cdot \\ 0 & -1 \\ \vdots & 0 \\ & \cdot \\ 0 & 0 \end{bmatrix}.$$

The rank condition (necessary and sufficient) of identification of the structural parameters of the first equation under (28) is then

$$\text{Rank}(A'R) = M - 1. \tag{29}$$

However, since multiplying a matrix by another cannot increase its rank, a necessary condition (also a "*general*" *order condition)* for identification is

$$\text{Rank}(R) \geq M - 1. \tag{29a}$$

This is easier to check in practice, because it does not involve the Π matrix. A further necessary condition for identification, which is necessary for the "general order condition" to be met, is

$$r \geq M - 1. \tag{29b}$$

Thus if $\text{Rank}(R) > M - 1$ the equation is over-identified, if $\text{Rank}(R) = M - 1$ it is just-identified, and it is under-identified if

Rank(R) $< M - 1$. Similar statements will be true if we replace Rank(R) by the number of restrictions, r in (29b).

In the case when the restrictions are linear nonhomogeneous we can write them as

$$\delta'_{\cdot 1} R = c,$$

where c has at least one nonzero element. These kinds of restrictions are quite common in econometric work. The restriction could be, say, $\gamma_{21} + \gamma_{31} = 1$, and so on. For such a case the rank condition becomes

$$\text{Rank}(A'R) = M,$$

which is different from (29) because the normalization is different here. For proofs and other problems in identification see Maddala [1977], Theil [1971], Schmidt [1976] and Fisher [1966].

11.4 ESTIMATION OF STRUCTURAL SYSTEMS

Now we consider the estimation of the complete system $Y\Gamma + XB = U$ given in (2). There are two ways to estimate this system: (i) by estimating each equation by taking into account the identifiability restrictions on only that equation; and (ii) by estimating all equations together by taking into account all the identifiability restrictions in the system. In the former case the estimators are called limited information (LI) estimators, whereas in the latter case they are called full information (FI) estimators. We shall first consider the LI estimators.

11.4.1 Limited Information Estimators

Let us rewrite the first equation (19) of the system (2) as

$$y_1 = Y_1\gamma_1 + X_1\beta_1 + u_1 = Z_1\delta_1 + u_1;\ Z_1 = [Y_1\ X_1] \text{ and } \delta_1 = \begin{bmatrix}\gamma_1\\ \beta_1\end{bmatrix}, \tag{30}$$

where y_1 is a $T \times 1$ vector having the (left-hand) jointly dependent variables, Z_1 is a $T \times p_1$ matrix where $p_1 = m_1 + K_1$ whose first m_1 columns contain Y_1 which is our $T \times m_1$ matrix of m_1 right-hand jointly dependent variables, and the last K_1 columns of Z_1 contain X_1 which is a $T \times K_1$ matrix of included predetermined variables. Also, γ_1 and β_1 are $m_1 \times 1$ and $K_1 \times 1$ coefficient vectors, respectively, and δ_1 is a $p_1 \times 1$ vector. Further, the reduced form corresponding to y_1 from the left side and Y_1 from the right side can be written from (20) as follows, where we continue to concentrate our attention on the first equation only.

$$y_1 = X_1\pi_{11} + X_1^*\pi_{21} + v_1$$

$$Y_1 = X_1\Pi_{11} + X_1^*\Pi_{21} + V_1, \tag{31}$$

where X_1^* is a $T \times K_1^*$, $K_1^* = K - K_1$, matrix of excluded predetermined variables; π_{11}, π_{21} are $K_1 \times 1$ and $K_1^* \times 1$ vectors; Π_{11}, Π_{21} are $K_1 \times m_1$ and $K_1^* \times m_1$ matrices; and v_1 and V_1 are a $T \times 1$ vector and a $T \times m_1$ matrix, respectively.

From the identifiability restriction (24), the relationship between the structural parameters of the first equation $y_1 = Y_1\gamma_1 + X_1\beta_1 + u_1$, and the reduced form parameters can be written as

$$\pi_1 = J_1\delta_1, \tag{32}$$

where

$$\pi_1 = \begin{bmatrix}\pi_{11}\\ \pi_{21}\end{bmatrix} \text{ and } J_1 = \begin{bmatrix}\Pi_1 & \begin{matrix} I \\ 0 \end{matrix}\end{bmatrix};\ \Pi_1 = \begin{bmatrix}\Pi_{11}\\ \Pi_{21}\end{bmatrix} \tag{33}$$

are matrices of orders $K \times 1$ and $K \times p_1$, respectively, with $p_1 = m_1 + K_1$. The matrices Π_1, I, and 0 in J_1 are of orders $K \times m_1$, $K_1 \times K_1$, and $K_1^* \times K_1$, respectively. Generally speaking, the subscript 1 refers to the first equation of a simultaneous equation model.

Now we discuss alternative limited information estimators.

11.4.2 A. Indirect Least Squares (ILS) Estimator

To obtain the ILS estimator of γ_1 and β_1 we use the relationship (32), $\pi_1 = J_1\delta_1$. In the first step of ILS we replace π_1 and J_1 by their respective OLS estimates from (31). They are

$$\hat{\pi}_1 = (X'X)^{-1}X'y_1, \text{ and } \hat{J}_1 = [\hat{\Pi}_1 \ \begin{smallmatrix} I \\ 0 \end{smallmatrix}];\ \hat{\Pi}_1 = (X'X)^{-1}X'Y_1, \tag{34}$$

where $X = [X_1\ X_1^*]$ as given in (17). Thus we can write $\hat{\pi}_1 = \hat{J}_1\delta_1$. Now in the second step we solve for δ_1. If the equation is just identified, i.e., $m_1 = K_1^*$, then $\hat{J}_1$ becomes square nonsingular and δ_1 will have a unique solution, viz,

$$\delta_1^* = \hat{J}_1^{-1}\,\hat{\pi}_1, \tag{35}$$

which gives

$$\gamma_1^* = \hat{\Pi}_{21}^{-1}\hat{\pi}_{21} \text{ and } \beta_1^* = \hat{\pi}_{11} - \hat{\Pi}_{11}\hat{\gamma}_1.$$

This is known as the ILS estimator.

If the equation is over-identified, i.e., $m_1 < K_1^*$ or $p_1 = m_1 + K_1 < K$, then $\hat{J}_1$ will not be a square matrix, and there can be more than one solution for δ_1. One way to get δ_1 in this case is to use the unique Moore-Penrose generalized inverse J_1^+ of $\hat{J}_1$. Since J_1 has the column rank $m_1 + K_1$ we can write

$$\delta_1^* = (\hat{J}_1'\hat{J}_1)^{-1}\hat{J}_1'\hat{\pi}_1 = \hat{J}_1^+\hat{\pi}_1, \tag{36}$$

This is the ILS for the over-identified case, and was obtained by Khazzoom [1975].

11.4.3 B. Two Stage Least Squares (2SLS) Estimator

The 2SLS estimator is popular among econometricians because of the relative ease with which it can be computed. For equation (30) 2SLS is another estimator for δ_1 in the over-identified case. It can be obtained from the estimated version of (32) given by $\hat{\pi}_1 = \hat{J}_1\delta_1$ by premultiplying by X and applying least squares to solve the equation $X\hat{\pi}_1 = X\hat{J}_1\delta_1$ as

$$\begin{aligned} d_1 &= (\hat{J}_1'X'X\hat{J}_1)^{-1}\hat{J}_1'X'X\hat{\pi}_1 = (\hat{J}_1'X'X\hat{J}_1)^{-1}\hat{J}_1'X'y \\ &= \begin{bmatrix} \hat{\Pi}_1'X'X\hat{\Pi}_1 & \hat{\Pi}_1'X'X_1 \\ X_1'X\hat{\Pi}_1 & X_1'X_1 \end{bmatrix}^{-1} \begin{bmatrix} \hat{\Pi}_1'X'X\hat{\pi}_1 \\ X_1'X\hat{\pi}_1 \end{bmatrix}, \end{aligned} \tag{37}$$

where we have used

$$X_1 = [X_1\ X_1^*]\binom{I}{0} = X\binom{I}{0}. \tag{38}$$

The estimator d_1 in (37) is in fact Theil's 2SLS estimator even though his two stages are not apparent there. We may now derive it in an alternative way to bring out his two stages:

Let us write the structural equation (30) as

$$y_1 = [X\Pi_1\ X_1]\delta_1 + v_1 = XJ_1\delta_1 + v_1 = \bar{Z}_1\delta_1 + v_1, \tag{39}$$

where

$$\bar{Z}_1 = XJ_1 = [X\Pi_1\ X_1] = EZ_1, \tag{40}$$

where we use $Y_1 = X\Pi_1 + V_1$, and the identifiability relationship $v_1 = V_1\gamma_1 + u_1$ given in (25). The vector v_1 is such that $Ev_1 = 0$ and $Ev_1v_1' = \omega_{11}I$. Now, given $\Pi_1 = \hat{\Pi}_1 = (X'X)^{-1}X'Y_1$ at the first stage, the OLS estimator of δ_1 in (39) is

$$\begin{aligned} d_1 &= (\hat{\bar{Z}}_1'\hat{\bar{Z}}_1)^{-1}\hat{\bar{Z}}_1'y_1 = (\hat{J}_1'X'X\hat{J}_1)^{-1}\hat{J}_1'X'X\hat{\pi}_1 \\ &= (Z_1'X(X'X)^{-1}X'Z_1)^{-1}Z_1'X(X'X)^{-1}X'y_1, \end{aligned} \tag{41}$$

where $\hat{\bar{Z}}_1 = X\hat{J}_1 = [X\hat{\Pi}_1\ X_1] = [X(X'X)^{-1}X'Y_1\ X_1] = X(X'X)^{-1}X'Z_1$ by using

(40). The reader should compare the line preceding (37) with the line preceding (41).

Thus it can be shown that both the 2SLS and ILS estimators are *minimum norm* least squares estimators (see Chapter 1 for the definition of a norm). The difference is that while in the 2SLS case the quadratic norm is defined with respect to $X'X$, in the ILS case it is with respect to I. The two estimators will be identical when $X'X = I$, and when the equation is just (exactly) identified. The main motivation for using 2SLS is that it can be proved to be consistent, whereas OLS is not.

11.4.4 C. Minimum Expected Loss Estimator (MELO)

Let us rewrite the relationship (32) between structural parameters δ_1 of the first equation and reduced form parameters π_1 and J_1 after premultiplying by X as

$$X\pi_1 = XJ_1\delta_1. \tag{42}$$

Let $\hat{\delta}_1$ be any estimate of δ_1, and $\epsilon_1 \equiv X\pi_1 - XJ_1\hat{\delta}_1$ be the measure of the extent to which the exact relations $\pi_1 = J_1\delta_1$ fail to be satisfied when $\hat{\delta}_1$ replaces δ_1. Consider our loss to be

$$L^o = \epsilon_1'\epsilon_1 = (X\pi_1 - XJ_1\hat{\delta}_1)'(X\pi_1 - XJ_1\hat{\delta}_1)$$

$$= (\delta_1 - \hat{\delta}_1)'J_1'X'XJ_1(\delta_1 - \hat{\delta}_1), \tag{43}$$

which represents a weighted quadratic loss function with the weights given by $J_1'X'XJ_1$, depending on X and Π_1.

Now, given a posterior probability density for $[\pi_1\ \Pi_1]$ that possesses finite first and second moments, the value of δ_1 which minimizes the (posterior) expected loss, viz,

$$EL^o = E(\pi_1'X'X\pi_1) - 2\hat{\delta}_1'E(J_1'X'X\pi_1) + \hat{\delta}_1'E(J_1'X'XJ_1)\hat{\delta}_1 \tag{44}$$

is

$$\hat{\delta}_1 = [E(J_1'X'XJ_1)]^{-1}E(J_1'X'X\pi_1). \tag{45}$$

To evaluate $\hat{\delta}_1$, one has to have a specific posterior probability density function (pdf) for $[\pi_1\ \Pi_1] = \Pi_o$. If we use Assumption 6 and thus consider a usual normal reduced form system; that is, assume that the rows of V are normally and independently distributed each with zero mean vector and an $M \times M$ covariance matrix Ω, and if we employ a diffuse prior pdf, $p(\Pi_o, \Omega) \alpha |\Omega|^{-(M+1)/2}$, then the marginal posterior pdf of Π_o is in the following matrix student-t form (see Chapter 3 for a special case).

$$p(\Pi_o|Y) \propto [S_o + (\Pi_o-\hat{\Pi}_o)'X'X(\Pi_o-\hat{\Pi}_o)]^{-m_1/2}, \tag{46}$$

where $\hat{\Pi}_o = (X'X)^{-1}X'Y$ and $S_o = (Y-X\hat{\Pi}_o)'(Y-X\hat{\Pi}_o)$. Zellner [1978] has shown that when the above posterior pdf in (46) is used to evaluate the posterior expectations in (45), the result is

$$\tilde{\delta}_1 = \begin{vmatrix} \tilde{\gamma}_1 \\ \tilde{\beta}_1 \end{vmatrix} = \begin{vmatrix} \hat{\Pi}_1'X'X\hat{\Pi}_1 + K\bar{S}_{22} & \hat{\Pi}_1'X'X_1 \\ X_1'X\hat{\Pi}_1 & X_1'X_1 \end{vmatrix}^{-1} \begin{vmatrix} \hat{\Pi}_1'X'X\hat{\pi}_1 + K\bar{S}_{12} \\ X_1'X\hat{\pi}_1 \end{vmatrix}, \tag{47}$$

where $\hat{\Pi}_o = (\hat{\pi}_1\ \hat{\Pi}_1)$ and

$$\bar{S}_o = (Y-X\hat{\Pi}_o)'(Y-X\hat{\Pi}_o)/(\nu-2) = \begin{bmatrix} \bar{s}_{11} & \bar{S}_{12}' \\ \bar{S}_{21} & \bar{S}_{22} \end{bmatrix} \tag{48}$$

with $\nu = T - K_1 - m_1 > 2$, is a condition required for the existence of the (finite) second moment of the posterior pdf given above.

11.4.5 D. k-Class Estimators

Using (34), note that $\hat{\Pi}_1'X'X_1 = Y_1'X_1$. Further, $\hat{\Pi}_1'X'X\hat{\Pi}_1 = Y_1'X(X'X)^{-1}X'Y_1 = Y_1'Y_1 - \hat{V}_1'\hat{V}_1$, where $\hat{V}_1 = Y_1 - X\hat{\Pi}_1$ is the estimated OLS residual matrix such that $\hat{V}_1'X = 0$. Also, $\bar{S}_{22} = \hat{V}_1'\hat{V}_1/(\nu-2)$. Thus we can write $\tilde{\delta}_1$ of (47) in a slightly general form as

$$\tilde{\delta}_{1k*} = \begin{bmatrix} Y_1'Y_1 - k^*\hat{V}_1'\hat{V}_1 & Y_1'X_1 \\ X_1'Y_1 & X_1'X_1 \end{bmatrix}^{-1} \begin{bmatrix} Y_1' - k^*\hat{V}_1' \\ X_1' \end{bmatrix} y_1, \tag{49}$$

where k^* is any arbitrary stochastic or nonstochastic scalar. The estimator $\tilde{\delta}_{1k*}$ is Theil's k-class estimator. For $k^* = 1 - \frac{K}{\nu-2} = 1 - \frac{K}{T-K_1-m_1-2}$, $\tilde{\delta}_{1k*}$ is the MELO estimator given in (47), and $k^* \to 1$ as $T \to \infty$. Further, for $k^* = 0$ and 1 we get the OLS and 2SLS estimators, respectively. The notation K represents the number of predetermined variables, and should not be confused with the K in ridge regression. The subscript 1 refers to the first structural equation. There is no loss of generality in using the formula (49) for other structural equations in the system of M equations in (2).

Theil [1971] gives a sufficient condition for the consistency of the general k-class estimator. This is $\text{plim}(k^*-1) = 0$ as $T \to \infty$. Obviously, the OLS choice of $k^* = 0$ does not satisfy this condition. The limiting distribution of the general k-class estimator $\tilde{\delta}_{1k*}$ is the same as that of the 2SLS estimator provided $\text{plim}\,\sqrt{T}\,(k^*-1) = 0$, that is, it is normal with the mean vector δ_1 and asymptotic variance covariance matrix

$$\text{Asy.}V(\tilde{\delta}_{1k^*}) = \text{Asy.}V(d_1) = \sigma_{11}\begin{bmatrix}\Pi_1'X'X\Pi_1 & \Pi_1'X'X_1 \\ X_1'X\Pi_1 & X_1'X_1\end{bmatrix}^{-1} = \sigma_{11}(\bar{Z}_1'\bar{Z}_1)^{-1}, \quad (50)$$

where $\sigma_{11}I = Eu_1u_1'$. The condition plim $\sqrt{T}\,(k^*-1) = 0$ is satisfied by the 2SLS and MELO estimators. With respect to the finite sample properties of the MELO estimator, see Zellner and Park [1979]; and for other members of the k-class see Sawa [1972], Anderson and Sawa [1979], Phillips [1980], Richardson [1968], Sargan [1975,1976], Ullah and Nagar [1974], Nagar [1959], Nagar and Ullah [1970,1973] and the references therein.

11.4.6 E. Ridge-type Estimators at the Second Stage of 2SLS

An advantage in writing the first structural equation $y_1 = Y_1\gamma_1 + X_1\beta_1 + u_1$ as $y_1 = \bar{Z}_1\delta_1 + v_1$ is that $\bar{Z}_1 = [X\Pi_1 X_1] = XJ_1 = X[\Pi_1 \; {}^{I}_{0}]$ becomes a $T \times p_1$ matrix, with $p_1 = m_1 + K_1$, is the number of nonstochastic regressors. Thus the so-called Gauss-Markov theorem is fully applicable to this equation, although it is not applicable to the first structural equation $y_1 = Y_1\gamma_1 + X_1\beta_1 + u_1$ having stochastic regressors.

It can be shown that δ_1 is identifiable if $\bar{Z}_1$ has full column rank. This is because the rank of $\bar{Z}_1$ = rank of $J_1 = K_1 + m_1$ if rank of $\Pi_{21} = m_1$, which is the so-called rank condition (26) for identification. Thus, a new set of estimators of δ_1 can easily be developed corresponding to the results discussed in earlier chapters for the regression case.

A generalized ridge estimator (GRE) of δ_1 in $y_1 = \bar{Z}_1\delta_1 + v_1$, where $Ev_1 = 0$ and $Ev_1v_1' = \omega_{11}I$, can be written as

$$d_1^{\text{GRE}} = (\bar{Z}_1'\bar{Z}_1 + H_1C_1H_1')^{-1}\bar{Z}_1'y_1$$

$$= H_1\Delta H_1'd_1, \quad (51)$$

where H_1 is the matrix of the eigenvectors of $\bar{Z}_1'\bar{Z}_1$ such that $H_1'H_1 = I$ and $\Delta =$ diag (q_ℓ), a diagonal matrix of "shrinkage fractions" $q_\ell = \gamma_\ell(\gamma_\ell + c_{1\ell})^{-1}$, where the γ_ℓ's with $\ell = 1, \ldots, p_1$, are the eigenvalues of $\bar{Z}_1'\bar{Z}_1$, and the $c_{1\ell}$'s are the diagonal elements of the matrix of ridge constants C_1. An ordinary ridge estimator (ORE) is a special case of the GRE, where $C_1 = c_1I$. This is

$$d_1^{\text{ORE}} = (\bar{Z}_1'\bar{Z}_1 + c_1I)^{-1}\bar{Z}_1'y_1. \quad (52)$$

Unfortunately, the notation K,k for ridge constants cannot be retained in this chapter because K is a standard notation for total number of predetermined variables, whereas k is standard for k-class estimation in econometrics.

Various interpretations and derivations of the ridge estimators have been given in Chapter 7 and they can all be easily modified for the equation $y_1 = \bar{Z}_1\delta_1 + u_1$. We reconsider here only the Bayesian interpretation.

For a given Π_1, consider $y_1 \sim N(\bar{Z}_1\delta_1, \omega_{11}I)$. Further given $\bar{\delta}_1$, assume the prior distribution of $\delta_1 \sim N(\bar{\delta}_1, \tau^2 A_1)$. Then as shown in Chapters 3 and 7, the mean of the posterior density of δ_1 (Bayesian estimator), given four things: A_1, ω_{11}, τ^2, and the data, is

$$d_1^* = \left[\frac{\bar{Z}_1'\bar{Z}_1}{\omega_{11}} + \frac{A_1^{-1}}{\tau^2}\right]^{-1}\left[\frac{\bar{Z}_1'\bar{Z}_1}{\omega_{11}} d_1 + \frac{A_1^{-1}\bar{\delta}_1}{\tau^2}\right]$$

$$= \bar{\delta}_1 + \left[I - (A_1^{-1} + \frac{\tau^2}{\omega_{11}}\bar{Z}_1'\bar{Z}_1)^{-1}A_1^{-1}\right](d_1 - \bar{\delta}_1). \tag{53}$$

If $\bar{\delta}_1 = 0$ and $A_1 = \dfrac{\omega_{11}}{\tau^2} H_1 C_1^{-1} H_1'$, then d_1^* reduces to the GRE of (51).

Given Π_1, the choices for the constant matrix C_1 in the GRE, or choices for the scalar constant c_1 in the ORE of (52) follow directly from the results in Chapters 7 and 8. For given $\bar{Z}_1$ the ridge estimators are operational. Unfortunately, Π_1, and hence $\bar{Z}_1$ are not known in practice. Thus, to make the estimators (51) to (53) operational we can use $\hat{\Pi}_1 = (X'X)^{-1}X'Y_1$ in $\bar{Z}_1$, which would give $\hat{\bar{Z}}_1 = [X\hat{\Pi}_1\ X_1] = X\hat{J}_1$. Then the operational ORE is denoted by an additional hat in (52) as

$$\hat{d}_1^{\text{ORE}} = (\hat{\bar{Z}}_1'\hat{\bar{Z}}_1 + c_1 I)^{-1}\hat{\bar{Z}}_1' y_1$$

$$= (\hat{J}_1'X'X\hat{J}_1 + c_1 I)^{-1}\hat{J}_1'X'y_1$$

$$= [Z_1'X(X'X)^{-1}X'Z_1 + c_1 I]^{-1}Z_1'X(X'X)^{-1}X'y_1. \tag{54}$$

The operational GRE for the parameters of the first structural equation is

$$\hat{d}_1^{\text{GRE}} = [Z_1'X(X'X)^{-1}X'Z_1 + H_1 C_1 H_1']^{-1}Z_1'X(X'X)^{-1}X'y_1.$$

These estimators may be called second stage ridge estimators because we use the usual OLS estimator of reduced form parameters Π_1 at the first stage. For the special case where the ridge constants c_1 or $C_1 = 0$, the ridge estimators reduce to the 2SLS estimator

$$d_1 = (\hat{J}_1'X'X\hat{J}_1)^{-1}\hat{J}_1'X'y_1 = (Z_1'X(X'X)^{-1}X'Z_1)^{-1}Z_1'X(X'X)^{-1}X'y_1.$$

Note that MSE properties associated with choices of c_1, for given Π_1, will not hold for $\hat{\Pi}_1$. Further discussion about it is in the last section of this chapter.

It has been shown by Mehta and Swamy [1978] that the operational estimator $\hat{d}_1^{ORE}$ possesses finite moments of all positive integer orders, *whatever the degree of over-identification* of the equation (30) or degrees of freedom may be. This is an improvement relative to the estimators that possess no finite moments or possess finite moments depending on the degree of over-identification or degrees of freedom. For example, the 2SLS estimator possesses no moments (i.e., it can have infinite MSE) when the equation is exactly identified, i.e. when $K_1^* = m_1$. And 2SLS possesses the second moment only if the degree of over-identification is at least two, i.e., $K_1^* - m_1 \geq 2$, Mariano [1972]. Zellner [1978] has pointed out that a sufficient condition for the existence of the second-order moment of his MELO estimator is that the degrees of freedom $T - m_1 - K_1$ should be at least two.

From a result in Chapter 7, we also note that, with respect to the matrix mean squared error (MtxMSE) criterion, the estimator d_1^{ORE} for the first equation is preferred to the corresponding OLS estimator d_o under certain conditions. We have $MtxMSE(d_o) - MtxMSE(d_1^{ORE})$ = a positive definite matrix, if and only if

$$\frac{\delta_1'[(\bar{Z}_1'\bar{Z}_1)^{-1} + 2c_1^{-1}]^{-1}\delta_1}{2\omega_{11}} \leq \frac{1}{2}, \tag{55}$$

where $d_o = (\bar{Z}_1'\bar{Z}_1)^{-1}\bar{Z}_1'y_1$ and ω_{11} is the variance of y_1 used in (53) above.

11.4.7 E.1 A Second Stage Stein-Type Estimator for 2SLS

A Stein-type estimator can be obtained as a special case of (53). Consider $\delta_1 \sim N(0, \tau^2(\bar{Z}_1'\bar{Z}_1)^{-1})$. Then, substituting $A_1 = (\bar{Z}_1'\bar{Z}_1)^{-1}$ and $\bar{\delta}_1 = 0$ in (53) we get

$$d_1^* = \left[1 - \frac{\omega_{11}}{\tau^2+\omega_{11}}\right]d_1 . \tag{53a}$$

This estimator shrinks the estimator d_1 towards zero, and it is analogous to the Bayesian estimator given in equation (20) of Chapter 6 in the regression context.

It also follows from equation (21) of Chapter 6 that one can replace $(\tau^2+\omega_{11})$ in (53a) by its unbiased estimator: $\delta_1'\bar{Z}_1'\bar{Z}_1\delta_1/(m_1+K_1)$. Then the estimator d_1^* becomes

$$d_1^* = \left[1 - \frac{(m_1+k_1)\omega_{11}}{\delta_1'\bar{Z}_1'\bar{Z}_1\delta_1}\right] d_o$$

$$= \left[1 - \frac{(m_1+k_1)\omega_{11}}{\pi_1'X'X\pi_1}\right] d_o, \tag{55a}$$

where $d_o = (\bar{Z}_1'\bar{Z}_1)^{-1}\bar{Z}_1'y_1$ is the OLS estimator for the first equation, and where we note from (32) and (34) that $\delta_1'\bar{Z}_1'\bar{Z}_1\delta_1 = \delta_1'J_1'X'XJ_1\delta_1 = \pi_1'X'X\pi_1$.

Note that the estimator d_1^* is not yet operational. To make it operational, we can substitute the least squares estimator of ω_{11} and π_1 from the "reduced form" equation $y_1 = x\pi_1 + v_1$. Then, as operational version of d_1^* can be formulated as

$$\hat{d}_1^S = \left[1 - \frac{g_1\hat{v}_1'\hat{v}_1}{\hat{\pi}_1'X'X\hat{\pi}_1}\right] d_1,$$

where $\hat{\pi}_1 = (X'X)^{-1}X'y_1$, $\hat{v}_1 = y - X\hat{\pi}_1$, d_1 is the 2SLS estimate, g_1 is an arbitrary scalar and the superscript S refers to Stein-type.

The estimator $\hat{d}_1^S$ is also derived in Ullah and Srivastava [1980] as an operational version of an optimal estimate (qd_o) where q is obtained by minimizing the mean squared error of (qd_o). An alternative estimator has been obtained by Zellner and Vandaele [1975] by minimizing the MSE of (qd_1) instead of (qd_o). Srivastava, Tiwari and Pandey [1978] have worked out a small sigma approximation for the bias and MSE of Zellner and Vandaele's estimator. However, the expressions are so involved that it is difficult to deduce any clear inference.

Ullah and Srivastava [1980] have obtained the small-σ approximate $\text{WMSE} = E(\hat{\delta}_1^S-\delta_1)'\bar{Z}_1'\bar{Z}_1(\hat{\delta}_1^S-\delta_1)$. They have shown that

$$\text{WMSE}(\hat{\delta}_1^S) - \text{WMSE}(d_1) \le 0,$$

so long as

$$0 \le g_1 \le \frac{2r_1}{T-K+2}\,[(L-3)r_2 + (2p_1-K+1)],$$

and the upper bound g_1 is positive; where $r_1 = \sigma_{11}/\omega_{11}$ is the ratio of structural variance to reduced form variance, $p_1 = m_1 + K_1$ and $L = K - (m_1+K_1)$ is the degree of over-identification in the equation under consideration. Furthermore, in the above expression we have $-r_1^{1/2} \le r_2 = \dfrac{\sigma_{(1)}'h}{\sigma_{11}} \le r_1^{1/2}$, where the vectors $\sigma_{(1)}$ and h are the first columns of Σ and Γ^{-1}, respectively. Thus the condition of dominance of the Stein-type estimator $\hat{\delta}_1^S$ over the 2SLS estimator d_1 depends on the unknown ratios r_1 and r_2. This implies that the 2SLS estimator cannot be uniformly dominated. In a special case when we have the "recursive model," we have $r_1 = r_2 = 1$, and the condition for dominance reduces to $0 \leq g_1 \leq \frac{2}{T-K+2} [p_1-2]; p_1 > 2$. When $m_1 = 0$, i.e., there is no endogenous variable in the equation, $p_1 = K_1$, then this condition compares directly with the dominance condition for the Stein-type estimator in the regression context discussed in Chapters 6 and 13. Thus, our generalization of the Stein rule methods to simultaneous equation problems is appealing and consistent with its special cases.

11.4.8 F. Two Stage Ridge Estimators

We expect that if the data matrix $X'X$ is ill-conditioned due to multicollinearity then the value of $\bar{Z}_1'\bar{Z}_1 = J_1'X'XJ_1$ will decrease. This means that when the predetermined variables are collinear then the estimator $(\bar{Z}_1'\bar{Z}_1)^{-1}\bar{Z}_1'y_1$ is not reliable. In such a situation the ridge-type and Bayesian estimators might be more useful.

However, note that ill conditioning of $X'X$ affects the first stage OLS estimator itself. Thus one can use ridge-type, Stein-type or Bayesian estimators for the reduced form parameter matrix Π_1, denoted by $\tilde{\Pi}_1$, instead of the OLS estimator $\hat{\Pi}_1$. These estimators can be obtained as a direct outcome of the results in the regression case discussed in earlier chapters. For example, the GRE of π_1 in the first reduced form equation $y_1 = X\pi_1 + v_1$ can be written as $\tilde{\pi}_{1C} = (X'X+GCG')^{-1}X'y_1$, where G is the matrix of eigen vectors of $X'X$, and C is the matrix of ridge constants. Using $\tilde{\pi}_{1C}$ we can get a new estimator of $\bar{Z}_1$ which is $\tilde{\bar{Z}}_1 = [X\tilde{\pi}_{1C}\ X_1]$.

With this $\tilde{\bar{Z}}_1$ based on GRE, we can get a new set of estimators of δ_1 in $y_1 = \bar{Z}_1\delta_1 + v_1$. For example, instead of the 2SLS estimator we can get $\tilde{d}_1 = (\tilde{\bar{Z}}_1'\tilde{\bar{Z}}_1)^{-1}\tilde{\bar{Z}}_1'y_1$ which we may call a "first stage ridge and second stage least squares" estimator. If we use $\tilde{\bar{Z}}_1$ in d_1^{GRE} given in (51), we will get $\tilde{d}_1^{\text{GRE}} = (\tilde{\bar{Z}}_1'\tilde{\bar{Z}}_1 + H_1C_1H_1')^{-1}\tilde{\bar{Z}}_1'y_1$, which may be called a "two stage generalized ridge" estimator. All this implies is that the various kinds of improved estimators can be developed for 2SLS estimator. Some work related to this is reported in Lee and Trivedi [1978].

11.4.9 G. Instrumental Variable Estimators

Let us write the structural equation (19) as

$$y_1 = Y_1\gamma_1 + X_1\beta_1 + u_1 = Z_1\delta_1 + u_1, \tag{56}$$

where $Z_1 = [Y_1\ X_1]$ and $\delta_1 = (\gamma_1,\beta_1)'$. One obvious approach for estimating this equation would then be to use OLS. This gives

$$\hat{\delta}_1 = (Z_1'Z_1)^{-1}Z_1'y_1, \tag{57}$$

where the inverse exists if Z_1 has rank $m_1 + K_1$. However, as indicated earlier this estimator is inconsistent. The reason is that Y_1 in Z_1 on the right hand side is correlated with u_1, because $Y_1 = X\Pi_1 + V_1$ and $v_1 = V_1\gamma_1 + u_1$. Thus one of the assumptions of the Gauss-Markov theorem is violated. In addition we note that the OLS approach ignores the distinction between explanatory endogenous and included exogenous variables in the Z_1 matrix. It also ignores all information available concerning variables not included in the equation being estimated, i.e., it ignores identifiability restrictions.

In view of the above problems in the context of the structural equation, several methods of estimation have been suggested which take into account the identifiability restriction on the equation under consideration, and eliminates the problem of dependence of Y_1 on u_1. One general estimation method, which contains many known estimators as special cases, is known as the Generalized Instrumental variable (GIVE) estimator. According to this method one looks for those variables as a replacement for Y_1 which are highly correlated with Y_1 and at the same time are uncorrelated with u_1 in the limit. That is, choose $\hat{Z}_1 = [\hat{Y}_1\ X_1]$ for $Z_1 = [Y_1\ X_1]$ such that in the population,

$$\operatorname*{plim}_{T\to\infty} \frac{\hat{Z}_1 u_1}{T} = 0. \tag{58}$$

Such a matrix $\hat{Z}_1$ is called a matrix of instrumental variables in which $\hat{Y}_1$ is the matrix of instrumental variables corresponding to Y_1, and X_1 is the matrix of nonstochastic variables as before. Thus a choice is made only with respect to Y_1.

Given the data, our aim is to choose δ_1 such that in the sample we have

$$\frac{\hat{Z}_1'\hat{u}_1}{T} = 0 \text{ or } \hat{Z}_1'(y_1 - Z_1\hat{\delta}_1) = 0$$

or

$$\hat{Z}_1'y_1 = \hat{Z}_1'Z_1\hat{\delta}_1, \tag{59}$$

where the second Z_1 on the right hand side does not have a hat, and where we denote $\hat{u}_1 = y_1 - Z_1\delta_1$ corresponding to an estimator of δ_1. The above expression is similar to the usual normal equations in OLS, viz. $Z_1'y_1 = Z_1'Z_1\hat{\delta}_1$. The solution of $\hat{\delta}_1$ from the normal equations (59) is

$$d_1^{\text{GIVE}} = (\hat{Z}_1'Z_1)^{-1}\hat{Z}_1'y_1. \tag{60}$$

This estimator is consistent. (Note: no hat on the second Z_1 in parentheses.)

It is obvious that different choices of $\hat{Y}_1$ would give different $\hat{Z}_1$ and hence different consistent instrumental variable (IV) estimators. In the "just identified" case, however, all the consistent estimators including 2SLS, ILS, IV, etc. would be identical.

Recall that Z_1 denotes the set of all regressors $[Y_1\ X_1]$ in the first structural equation. Looking at the choices of $\hat{Z}_1$ where Y_1 in Z_1 are replaced by corresponding instruments, we note that one of the general class of instrumental variables can be chosen as

$$\begin{aligned}\hat{Z}_1 &= [\hat{Y}_1\ X_1] \\ &= [NY_1\ X_1] \\ &= NZ_1 + (I-N)Z_1D, \end{aligned} \tag{61}$$

where

$$D = \begin{bmatrix} 0 & 0 \\ 0 & I \end{bmatrix}$$

such that $Z_1D = [0\ X_1]$. Thus in this case Y_1 is replaced by its linear combination NY_1, where N is an arbitrary (chosen) matrix.

Alternatively, $\hat{Z}_1$ in (61) can be thought to be a weighted matrix combination of $Z_1 = [Y_1\ X_1]$ and $Z_1D = [0\ X_1]$. The problem of choosing $\hat{Z}_1$ is then one of considering various possibilities for N such that NZ_1 is uncorrelated with u_1.

Some special cases are of interest. For example if $N = X(X'X)^{-1}X'$ then the GIVE becomes the 2SLS estimator. If $N = (1-\hat{k})I + \hat{k}X(X'X)^{-1}X'$ then we get Theil's k-class estimator. Since $\hat{k} = 0$ implies $N = I$, the GIVE becomes the OLS estimator in this case. If $\hat{k}$ is the minimum characteristic root of the determinantal equation $|Y_o'M_1Y_o - \lambda Y_o'MY_o| = 0$; $Y_o = [y_1\ Y_1]$, $M_1 = I - X_1(X_1'X_1)^{-1}X_1$, $M = I - X(X'X)^{-1}X'$, and we consider it in $N = (1-\hat{k})I + \hat{k}X(X'X)^{-1}X'$ then the GIVE becomes the limited information maximum likelihood (LIML) estimator.

Now if we consider generalized ridge-type modification of N denoted as N_C and defined by $X(X'X+GCG')^{-1}X' = N_C$, then the GIVE formula (60) becomes

$$d_1^{\text{GIVE}} = [Z_1'N_CZ_1 + D'Z_1'(I-N_C)Z_1]^{-1}[Z_1'N_C + D'Z_1'(I-N_C)]y_1, \tag{62}$$

which we have called earlier to be the "first stage ridge and second stage least squares" estimator. Finally, if we consider $N = S(S'S)^{-1}S'$, where $S = [XG_r, X_1]$, and G_r denotes a $K \times r$ matrix consisting of the first r eigenvectors of $X'X$, then the GIVE becomes Kloek-Mennes' [1960] two-stage "principal components" estimator. Instead of omitting the remaining $K - r$

principal components they are down-weighted in (62).

Note that the full matrix of all "predetermined" regressors from all M simultaneous equations is denoted by X. We are concerned here with the eigenvalues of $X'X$, not just $X_1'X_1$ involved in the first equation being estimated by (62). There is no loss of generality in applying this method to all M equations, one at a time. The eigenvectors G need to be computed only once as in Kloek and Mennes.

11.4.10 Problem of Undersized Sample and Multicollinearity

A sample is said to be "undersized" if the number of exogenous (or predetermined) variables in a system of simultaneous equations exceeds the number of observations. The issue of estimating the structural coefficients when the sample is undersized has been an important concern of econometricians in the past two decades. This issue is particularly important for very large econometric models which have been implemented in practice. The essence of the difficulty caused by the undersized samples lies in the inability to estimate the reduced form parameters, because the relevant matrix cannot be inverted.

As $X'X$, appearing in the 2SLS estimator, is a $K \times K$ square matrix, it will be singular unless the rank of X is K. This requires, among other things, $T > K$. Also, the 2SLS estimator exists only if the matrix $Z_1'X(X'XZ)^{-1}X'Z_1$ has rank $p_1 = m_1 + K_1$, which requires $K \geq p_1$, provided that X has full column rank (i.e., rank K). In econometrics, many large models (e.g., the Brookings Model) have the number of predetermined variables K exceeding the number of observations T, and so the rank of X is less than K. This is a situation commonly known as "undersized sample" in the literature. Thus in addition to the usual rank condition ($K \geq p_1$) for the 2SLS to exist, we now impose the extra condition ($K \leq T$). If this additional condition is not fulfilled, some modification of the conventional 2SLS is obviously needed, because the usual rank identifiability condition (29) will be violated as well. Note that this problem is not only confined to 2SLS estimation. Limited-information maximum likelihood, three stage least squares, full-information maximum likelihood and linearized maximum likelihood (discussed in the next section) all directly or indirectly use the condition $K \geq p_1$ where p_1 = Rank of $Z_1'X(X'X)^{-1}X'Z_1$. In other words, they rely on the existence of 2SLS estimates, and hence are inapplicable whenever 2SLS estimation is impossible.

For the 2SLS case various modified estimation methods have been suggested of which two are particularly noteworthy. The first involves a *modified* 2SLS (or an instrumental variable procedure) in which the list of predetermined variables is truncated at the initial stage; a second method involves using the principal components of the predetermined variables in 2SLS estimation. Truncation methods have been discussed by Fisher [1965a, 1965b] and Mitchell and Fisher [1970]. Principal components methods have been discussed by

Kloek and Mennes [1960], Amemiya [1966], Dhrymes [1974], Kadiyala and Nunns [1976], and Theil [1971]. Both of these types of procedures may be thought to be special cases of a class of GIVE given in an earlier section. However, we note that the solution of the undersized sample problem by these authors has been successful only with respect to tackling the problem of estimation at the first stage, while ignoring it at the second stage. Even while dealing with first stage estimation, these authors impose additional restrictions on the reduced form parameters which may conflict with the identifying restrictions imposed on the structural coefficients. Furthermore, the estimators of structural coefficients based on these restricted reduced form parameters generally will not even possess finite means. In finite samples the mean square error measure does not provide a basis for comparing the estimators when some of them have infinite second-order moments. If the undersized sample problem occurs among the predetermined variables, only the first stage estimation needs to be adjusted, and their implications have been analyzed in the preceding section. If all the variables exhibit a certain degree of collinearity, then a remedy at only the first stage is not adequate. Neeleman [1973] was perhaps the first econometrician to tackle the problem directly. He proposed a so-called Generalized Two Stage Least Squares procedure (G2SLS) which entails repeated use of the generalized inverse technique for the inversion of singular matrices in both stages. However, in practice, the matrices to be inverted may not be exactly singular but only nearly so. Therefore, G2SLS may be identical to 2SLS. In order to avoid this, Neeleman proposed replacing the near-singular matrix by an approximate matrix of less than full rank. The approximation is done by choosing a linear combination of a subset of the columns of the original matrix such that the norm of the difference between the new matrix and the original one is a minimum. This turns out to be an expensive way of getting truncated principal components estimates at both stages.

These deficiencies are corrected to some extent by the Swamy and Holmes [1971] estimator which possesses finite second-order moments, is not based on conflicting restrictions, and tackles the problem at both stages of estimation. Their estimator is

$$\hat{d}_{SH} = [Z_1'X(X'X)^- X'Z_1 + c_1 I]^{-1} Z_1'X(X'X)^- X'y_1 \qquad (63)$$

where $(X'X)^-$ is any generalized inverse of $X'X$ instead of the Moore Penrose generalized inverse $(X'X)^+$, and with SH representing Swamy-Holmes. This estimator is in the form of $\hat{d}_1^{ORE}$ given in (54) except for $(X'X)^-$. However, it can be verified that the SH estimator is not formally different from the 2SPC if the rank of X is $r < K$, or from OLS if $K \geq T$.

The method of Generalized Ridge Regression has Restricted Least Squares and Truncated Principal Components as special cases, and can be conveniently thought of as another possible estimator for the first stage of reduced form estimation. Yet, the extension of the ridge regression technique at both stages of the two stage least squares procedure is desirable in order to tackle both the problem of undersized samples and of multicollinearity. Such a two stage ridge procedure has been mentioned earlier and a detailed study of its properties remains to be carried out.

Recently, Theil and Laitinen [1980] have proposed a maximum entropy (ME) technique to estimate singular $X'X$ matrices arising from undersized samples. They fitted a continuous ME distribution (except for two discontinuous points) to the values of K variables arranged in an ascending order of magnitude, i.e., to the order statistics in the sample. The resulting $X'X$ matrix can be proved to be nonsingular, primarily because the ME technique involves adding a ridge type correction to the diagonal. Vinod [1981b] modifies the continuity assumption and uses measurement errors to determine the correction along the diagonal as follows. If the ith variable is measured correct to h_i places to the right of the decimal point, define $d_i = 0.5(10)^{-h_i}$. The diagonal correction is proved to be simply $d_i^2/3$. Professor Theil and his students (notably, J. Meisner) have produced many new results regarding the implications of the ME approach to simultaneous equation estimation.

√11.5 FULL INFORMATION ESTIMATORS

The estimators discussed earlier were "limited information," which do not take into account all of the restrictions on the structural parameters in the remaining equations of the system. To generate a "joint" or "full information" estimator we write the analogue associated with each of the M structural equations of the model:

$$\begin{aligned} y_1 &= Z_1\delta_1 + u_1 \\ &\vdots \\ y_M &= Z_M\delta_M + u_M \end{aligned} \tag{64}$$

or in a compact notation:

$$y = Z\delta + u,$$

where y is an $MT \times 1$ vector,

$$Z = \begin{vmatrix} Z_1 & \dots & 0 \\ \cdot & \cdot & \\ \cdot & \cdot & \\ \cdot & \cdot & \\ 0 & & Z_M \end{vmatrix}, \quad Z_i = [Y_i \; X_i],$$

so Z is an $MT \times p$ matrix of "included endogenous" and "predetermined" variables, $p = \sum_{i=1}^{M} p_i = \sum_{i=1}^{M} (m_i+K_i)$, and δ is a $p \times 1$ vector. Also, u is an $MT \times 1$ vector such that

$$Eu u' = \Sigma \otimes I_T, \tag{65}$$

where Σ is as defined before and I_T is a $T \times T$ identity matrix.

Let $\hat{Z}$ be the matrix of instrumental variables (see eq. (61) above)

$$\hat{Z} = \begin{vmatrix} \hat{Z}_1 & \dots & 0 \\ & \ddots & \\ 0 & & \hat{Z}_M \end{vmatrix} ; \hat{Z}_i = [\hat{Y}_i \ X_i] = NZ_i + (I_T - N)Z_i D, \tag{66}$$

where $Z_i D = [0 \ X_i]$ and N is an arbitrary $T \times T$ matrix from (61) which is a function of X. Alternatively, we can write a compact expression in terms of Z instead of Z_i as:

$$\hat{Z} = (I_M \otimes N)Z + [I_M \otimes (I_T - N)]Z \otimes D. \tag{67}$$

Unfortunately, this $\hat{Z}$ matrix of instruments is not appropriate for the full information model (64) because it ignores the joint dependence of errors u and their dispersion matrix Σ. For this purpose, let us define a weight matrix

$$W = (\Sigma^{-1} \otimes I_T)\hat{Z} = \begin{bmatrix} W_1 & \dots & 0 \\ & \ddots & \\ 0 & & W_M \end{bmatrix}, \tag{68}$$

which is an $MT \times p$ matrix of weighted instruments. Note that in the single equation case $W_1 = \frac{1}{\sigma_{11}} \hat{Z}_1$, but considering either $\hat{Z}_1$ or $\hat{Z}_1/\sigma_{11}$ in that case did not make any difference because the right hand side of(58) is zero.

The normal equations which provide the solution for FGIVE (full information GIVE) of δ can be written as a generalization of (59):

$$W'Z\delta_{\text{FGIVE}} = W'y, \tag{69}$$

or substituting W of (68) into (69) we have

$$(\hat{Z}'(\Sigma^{-1} \otimes I_T)Z)\delta_{\text{FGIVE}} = \hat{Z}'(\Sigma^{-1} \otimes I_T)y.$$

It should be noted that if $\sigma_{ij} = 0$, $i \neq j$, and/or $X_1 = \dots = X_M$, then the solution for $\delta_1, \dots, \delta_M$ will be the same as in the single equation case.

Now various estimators, viz., the full information maximum likelihood (FIML), and the three-stage least squares (3SLS) are special cases of solutions of the above normal equations. For example, we can get 3SLS as follows. In (65) and (66) we choose

$$N = X(X'X)^{-1}X', \text{ and } \Sigma = \hat{\Sigma}, \tag{70}$$

where $\hat{\sigma}_{ij} = (y_i - Z_i d_i)'(y_j - Z_j d_j)/T$ is the (i,j)th element of $\hat{\Sigma}$ and d_i

represents the 2SLS estimator applied to the ith equation. Then an operational version of W of (68) for 3SLS becomes:

$$W = (\hat{\Sigma}^{-1} \otimes I_T)(I_M \otimes X(X'X)^{-1}X')Z$$
$$= (\hat{\Sigma}^{-1} \otimes X(X'X)^{-1}X')Z. \qquad (71)$$

To verify this use two things. First, $(A \otimes B)(C \otimes D) = AC \otimes BD$ provided the matrices are of proper orders. Second, that when $N = X(X'X)^{-1}X'$, we have

$$(\hat{\Sigma}^{-1} \otimes I_T)[I_M \otimes (I_T - N)]Z \otimes D = [\hat{\Sigma}^{-1} \otimes (I_T - N)]Z \otimes D = 0$$

by using (38). (Note: Does this remind you of the familiar result $MX = 0$?) When W is from (71) FGIVE from (69) becomes the 3SLS estimator given by the solution of the normal equations

$$[Z'(\hat{\Sigma}^{-1} \otimes X(X'X)^{-1}X')Z]\delta_{3SLS} = Z'(\hat{\Sigma}^{-1} \otimes X(X'X)^{-1}X')y. \qquad (72)$$

This is

$$\delta_{3SLS} = (Z'\hat{R}Z)^{-1}Z'\hat{R}y, \qquad (73)$$

where $\hat{R} = \hat{\Sigma}^{-1} \otimes X(X'X)^{-1}X'$. The name three stage refers to the fact that $\hat{\Sigma}$ is estimated from the second stage residuals.

Our next task is to find the normal equations for FIML by appropriate choice of W in (69), which in turn needs Σ and Z. We continue to use $\hat{\Sigma}$ from second stage residuals. To begin with, the estimate $\tilde{Z}$ may be based on the parameter estimates from 2SLS.

Next, if $\tilde{Z}_i = [X\tilde{\Pi}_i\ X_i]$ where $\tilde{\Pi}_i$ is obtained from $\tilde{\Pi} = -\hat{B}\hat{\Gamma}^{-1}$ ($\Pi = -B\Gamma^{-1}$, by definition), $\hat{B}$ and $\hat{\Gamma}$ having been obtained from any appropriate limited information estimation method, then $W = (\hat{\Sigma}^{-1} \otimes I)\tilde{Z}$ will give rise to the following normal equations:

$$[\tilde{Z}'(\hat{\Sigma}^{-1} \otimes I)Z]\delta_{FIML} = \tilde{Z}'(\hat{\Sigma}^{-1} \otimes I)y. \qquad (74)$$

The reader may wonder: why this is called a maximum likelihood (ML) estimator? It has been verified in the literature that if we write the likelihood function and maximize it we get (74). To make this plausible, the reader may note that the familiar GLS estimator $(X'\Omega^{-1}X)^{-1}X'\Omega^{-1}y$ is similar in appearance, and is also ML.

The solution of this is the first step of FIML estimation. If we iterate on this equation, with respect to $\tilde{Z}$ and $\hat{\Sigma}$, then the converging solution is called the FIML estimator. One could also get an iterative 3SLS by iterating with respect to Σ only in (72). Note that the iterative 3SLS will not be identical with the FIML estimator.

There are several ways in which the above estimators can be "improved," in an analogous way to the single equation case. Specifically, the techniques are:

(i) Improving the solution for δ from (69)

(ii) Improving instruments

(iii) Improving both the solution and instruments.

Considering (i), we note that $\hat{Z}'(\Sigma^{-1}\otimes I)Z$ in $[\hat{Z}'(\Sigma^{-1}\otimes I)Z]\delta_{FGIVE} = \hat{Z}'(\Sigma^{-1}\otimes I)y$, for any $\hat{Z}$, can become singular or ill-conditioned. In these cases, 3SLS and FIML estimators would then be unidentified or would perform poorly. In practice, it is not uncommon for one or more eigenvalues of $\hat{Z}'(\Sigma^{-1}\otimes I)Z$ to be excessively small without any detectable degree of perfect singularity in either $X'X$ or Σ. This has been dealt with earlier when discussing undersized sample problems. Also, note that the singularity points of the random matrix $\hat{Z}'(\Sigma^{-1}\otimes I)Z$ cause trouble for the integrals defining the moments of the estimators of δ in the sense that they may diverge. In other words, $\hat{\delta}_{\mathrm{FGIVE}}$ may be a poor estimator, because it is under-identified and/or have infinite MSE.

A solution for the above problems is to use the full information ridge-type GIVE (FRGIVE) estimator, which is the solution of

$$[\hat{Z}'(\Sigma^{-1}\otimes I)Z+gI_p]\delta_{\mathrm{FRGIVE}} = \hat{Z}'(\Sigma^{-1}\otimes I)y, \tag{75}$$

where g is the scalar biasing parameter similar to k of ordinary ridge estimator (ORE). One can use a constant matrix to write the GRE. In the special case corresponding to the 3SLS choice of $\hat{Z}$ and Σ we will get the ridge type 3SLS (R3SLS) estimator from the solution of δ from normal equations:

$$[Z'(\hat{\Sigma}^{-1}\otimes X(X'X)^{-1}X')Z+gI_p]\delta_{\mathrm{R3SLS}} = Z'(\hat{\Sigma}^{-1}\otimes X(X'X)^{-1}X')y. \tag{76}$$

Note that the asymptotic distribution of $\sqrt{T}\ (\delta_{\mathrm{R3SLS}} - \delta)$ is the same as that of $\sqrt{T}\ (\delta_{\mathrm{3SLS}} - \delta)$, that is, it is normal with mean δ and the asymptotic variance covariance matrix

$$\underset{T\to\infty}{plim}\ T.\ [Z'(\Sigma^{-1}\otimes X(X'X)^{-1}X')Z]^{-1}. \tag{77}$$

This result is true, however, when $g = c/T = 0(1/T)$, where c is any nonnegative constant. Maasoumi [1980] provides the order $1/\sqrt{T}$ approximations for MSE (δ_{R3SLS}). On the basis of this approximation he determines a conservative range for g for which $\mathrm{MSE}(\delta_{\mathrm{R3SLS}}) - \mathrm{MSE}(\delta_{\mathrm{SLS}}) \leq 0$. This is $0 \leq g \leq \dfrac{2}{Th^2}$ where h^2 is given by the restriction $\delta'\delta \leq h^2$. Note that $g \to 0$, i.e., R3SLS $\to$ 3SLS, if the sample information is dominant ($T \to \infty$) or if the prior information about δ is weak ($h^2 \to \infty$). It follows intuitively from this result that on the basis of approximate MSE, the ridge 2SLS estimator $\hat{d}_1^{\mathrm{ORE}}$ in (54) will dominate the 2SLS estimator d_1 if $0 \leq c_1 \leq \dfrac{2}{Th_1^2}$ where h_1^2 satisfies

$\delta_1'\delta_1 \leq h_1^2$, and c_1 is the ridge biasing parameter. More extensive analysis of the behavior of MSE of R3SLS for both stochastic and nonstochastic g remains a subject for future research.

Now we turn to the technique mentioned in (iii) before considering (ii). This improvement can be obtained by using $\tilde{Z}_{iR} = [\tilde{Y}_{iR}\ X_i]$, where $\tilde{Y}_{iR} = X\tilde{\Pi}_{iR}$ and $\tilde{\Pi}_{iR}$ is obtained by applying ridge-type estimators discussed in earlier sections. The full information two-stage ridge type GIVE will then be the solution of the following normal equations:

$$[\tilde{Z}_R'(\hat{\Sigma}^{-1}\otimes I)Z+g^*I_p]\delta = \tilde{Z}_R'(\hat{\Sigma}^{-1}\otimes I)y, \tag{78}$$

where R is for ridge. The technique (ii) for improvement mentioned after (74) above is a special case of this where $g^* = 0$.

In this chapter we considered ridge and Stein-type improved estimators in the simultaneous equations context. As was mentioned at various places in this chapter, the analytical properties of these improved estimators need further investigations. Some work has been attempted by Mehta and Swamy [1978], Maasoumi [1977, 1978, 1980], Srivastava, Tiwari and Pandey [1978], Ullah and Srivastava [1980], and Lee [1979], and these have been discussed in this chapter. We have shown that some of the difficulties (e.g., infinite variance) associated with the traditional estimators like 2SLS can be eliminated by the newer methods.

EXERCISES

11.1 Consider the stochastic model

$$C_t = \alpha + \beta Y_t + u_t \quad , \quad (t = 1, \ldots, T)$$

$$Y_t = C_t + I_t \quad ,$$

where Y_t is real net national income in period t, C_t real consumption in period t, and I_t real investment in period t. The disturbances $u_t \sim N(0,\sigma^2)$ and they are independent for $t = 1, \ldots, T$.

(a) Obtain the two stage least squares estimator of the marginal propensity to consume β in the above model.

(b) Show that the sampling distribution of the two stage least squares estimator has no integral moments of any order.

(c) Develop ridge-type estimators of β and show that all the moments of this estimator exist.

11.2 In the model

$$y_{1t} = \gamma_{12}y_{2t} + \gamma_{13}y_{3t} + \beta_{11}x_{1t} + \beta_{12}x_{2t} + u_{1t}$$

$$y_{2t} = \gamma_{21}y_{1t} + \gamma_{23}y_{3t} + u_{2t}$$

$$y_{3t} = \gamma_{31}y_{1t} + \beta_{32}y_{2t} + u_{3t}$$

where y_{it} are endogenous variables, the x_{it} are exogenous variables and the u_{it} are serially independent random disturbances with zero means and nonsingular covariance matrix.

(a) Verify the identifiability of each equation.

(b) Suppose that $\beta_{11} + \beta_{12} = 1$. What would be its effect on the identifiability of the first equation?

11.3 Formulate a Monte Carlo project to study the sampling properties of various ridge-type, Stein-type and usual estimators (limited and full information). References: Raj [1980] and Hendry [1980].

11.4 For the Klein's six equation model 1 of U.S. economy use the data in Table 11.1 and calculate the usual limited information estimators, Stein-type estimators and ridge-type estimators. Determine your ridge constant by ridge trace method. Also suggest the data dependent choices of k, the ridge constant. Source: Klein [1950].

The six simultaneous equations are given in the text following equation (1) at the beginning of this chapter. It is advisable to rewrite them in a general notation with the six endogenous variables denoted by y_1, y_2, etc.

TABLE 11.1

Year	C	P	W	I	K_{-1}	X	W'	G	T
1920	39.8	12.7	28.8	2.7	180.1	44.9	2.2	2.4	3.4
21	41.9	12.4	25.5	—.2	182.8	45.6	2.7	3.9	7.7
22	45.0	16.9	29.3	1.9	182.6	50.1	2.9	3.2	3.9
23	49.2	18.4	34.1	5.2	184.5	57.2	2.9	2.8	4.7
24	50.6	19.4	33.9	3.0	189.7	57.1	3.1	3.5	3.8
25	52.6	20.1	35.4	5.1	192.7	61.0	3.2	3.3	5.5
26	55.1	19.6	37.4	5.6	197.8	64.0	3.3	3.3	7.0
27	56.2	19.8	37.9	4.2	203.4	64.4	3.6	4.0	6.7
28	57.3	21.1	39.2	3.0	207.6	64.5	3.7	4.2	4.2
29	57.8	21.7	41.3	5.1	210.6	67.0	4.0	4.1	4.0
1930	55.0	15.6	37.9	1.0	215.7	61.2	4.2	5.2	7.7
31	50.9	11.4	34.5	—3.4	216.7	53.4	4.8	5.9	7.5
32	45.6	7.0	29.0	—6.2	213.3	44.3	5.3	4.9	8.3
33	46.5	11.2	28.5	—5.1	207.1	45.1	5.6	3.7	5.4
34	48.7	12.3	30.6	—3.0	202.0	49.7	6.0	4.0	6.8
35	51.3	14.0	33.2	—1.3	199.0	54.4	6.1	4.4	7.2
36	57.7	17.6	36.8	2.1	197.7	62.7	7.4	2.9	8.3
37	58.7	17.3	41.0	2.0	199.8	65.0	6.7	4.3	6.7
38	57.5	15.3	38.2	—1.9	201.8	60.9	7.7	5.3	7.4
39	61.6	19.0	41.6	1.3	199.9	69.5	7.8	6.6	8.9
1940	65.0	21.1	45.0	3.3	201.2	75.7	8.0	7.4	9.6
41	69.7	23.5	53.3	4.9	204.5	88.4	8.5	13.8	11.6

12

Canonical Correlations and Discriminant Analysis with Ridge-Type Modification

Statistical Multivariate analysis is often regarded as consisting primarily of the topics called canonical correlations, discriminant functions, principal components and factor analysis. The principal components regression was already noted to be special case of generalized ridge regression. Factor analysis is an extension of principal components analysis and will not be discussed here. These multivariate techniques often require inversion of ill-conditioned matrices. A ridge-type modification to improve the conditioning of these matrices has been suggested and studied in the literature. In this chapter we outline some of these developments and their extensions. We find that ridge-type modifications improve the stability and reliability of the estimated relations.

12.1 CANONICAL CORRELATIONS

Hotelling's [1936] canonical correlations analysis is a generalization of the least squares problem, where there are two or more "dependent" variables. In general, there are two sets of variables

$$X_1' = (x_1,\ldots,x_r), \tag{1}$$

$$X_2' = (x_{r+1},\ldots,x_p)\ , \tag{2}$$

where the variables in the first set may be thought to be the "dependent" variables. However, it is better to abandon the classification of variables into "dependent" and "regressor" variables because of the basic symmetry between the two sets considered by canonical correlations analysis.

Define two "index numbers" or weighted linear combinations involving the unknown parameters α_i and β_j as follows

$$w_1 = \sum_{i=1}^{r} \alpha_i x_i \ , \tag{3}$$

$$w_2 = \sum_{j=1}^{p-r} \beta_j x_{j+r} \ , \tag{4}$$

where the number of variables in the second set is $p - r$.

We assume a p-variate normal structure, $X \sim N(\chi,\Sigma)$, and write conformable subvectors and submatrices as:

$$\begin{pmatrix} X_1 \\ X_2 \end{pmatrix} \sim N\left(\begin{bmatrix} \chi_1 \\ \chi_2 \end{bmatrix} , \begin{bmatrix} \Sigma_{11} & \Sigma_{12} \\ \Sigma_{21} & \Sigma_{22} \end{bmatrix} \right) , \tag{5}$$

where we assume that Σ_{11} and Σ_{22} are nonsingular.

The parameters α_i and β_j are to be chosen in such a way that w_1 and w_2 are "best fitting" linear combinations in some appropriate sense. Hotelling [1936] defined a "canonical correlation coefficient" between the weighted sums $w_1 = \alpha' X_1$ and $w_2 = \beta' X_2$

$$\rho = \frac{cov\,(\alpha' X_1, \beta' X_2)}{[Var\,(\alpha' X_1)\,Var\,(\beta' X_2)]^{1/2}} \ , \tag{6}$$

which is defined for $\min(r,p-r)$ matching pairs of variables w_1 and w_2, and which satisfies $|\rho| \leq 1$ similar to any correlation coefficient. When $|\rho|$ is large, the strength of the relationship between w_1 and w_2 is large. Hotelling suggested that we should determine α and β by maximizing $|\rho|$ subject to the following constraints,

$$\alpha' \Sigma_{11} \alpha = 1 = \beta' \Sigma_{22} \beta \ , \tag{7}$$

to make sure that the α and β do not depend on the scalings of X_1 or X_2, and that the solution for α and β is unique. The Lagrangian for a constrained maximization is given by

$$Lagr = \alpha' \Sigma_{12} \beta + \mu_1 \left(\alpha' \Sigma_{11} \alpha - 1 \right) + \mu_2 \left(\beta' \Sigma_{22} \beta - 1 \right) . \tag{8}$$

The first order conditions by matrix differentiation are

$$0 = \frac{\partial Lagr}{\partial \alpha} = \Sigma_{12}\,\beta - 2\mu_1\,\Sigma_{11}\,\alpha\,, \tag{9}$$

$$0 = \frac{\partial Lagr}{\partial \beta} = \Sigma_{21}\,\alpha - 2\mu_2\,\Sigma_{22}\,\beta\,, \tag{10}$$

$$0 = \frac{\partial Lagr}{\partial \mu_1} = \alpha'\,\Sigma_{11}\,\alpha - 1\,, \tag{11}$$

$$0 = \frac{\partial Lagr}{\partial \mu_2} = \beta'\,\Sigma_{22}\,\beta - 1\,. \tag{12}$$

In (9) and (10) premultiplying by α' and β' respectively we have

$$\alpha'\,\Sigma_{12}\,\beta = 2\mu_1\alpha'\,\Sigma_{11}\,\alpha = 2\mu_1\,, \tag{13}$$

$$\beta'\,\Sigma_{21}\,\alpha = 2\mu_2\beta'\,\Sigma_{22}\,\beta = 2\mu_2\,, \tag{14}$$

where the second equalities of (13) and (14) use (11) and (12) respectively.

Since Σ is a symmetric covariance matrix $\Sigma_{12} = \Sigma_{21}'$. Hence the equation (14) can be written after taking a transpose of both sides as:

$$\alpha'\,\Sigma_{21}'\,\beta = \alpha'\,\Sigma_{12}\,\beta = 2\mu_2\,. \tag{15}$$

Substituting the right side of (15) on the left side of (13) we have $\mu_1 = \mu_2$. We may replace $2\mu_1$ and $2\mu_2$ in (9) and (10) by ρ and note that we have the following two matrix equations which will help us determine α and β.

$$\Sigma_{12}\,\beta - \rho\,\Sigma_{11}\,\alpha = 0 \tag{16}$$

$$\Sigma_{21}\,\alpha - \rho\,\Sigma_{22}\,\beta = 0\,. \tag{17}$$

The trivial solution is found when $\alpha = \beta = 0$. For a nontrivial solution premultiply (16) by $\Sigma_{21}\,\Sigma_{11}^{-1}$ to yield

$$\Sigma_{21}\,\Sigma_{11}^{-1}\,\Sigma_{12}\,\beta - \rho\,\Sigma_{21}\,\alpha = 0\,.$$

From (17) we have $\Sigma_{21}\,\alpha = \rho\,\Sigma_{22}\,\beta$. Substituting this, we have

$$\Sigma_{21}\,\Sigma_{11}^{-1}\,\Sigma_{12}\,\beta - \rho^2\,\Sigma_{22}\,\beta = 0\,. \tag{18}$$

Premultiplying by a nonsingular Σ_{22}^{-1} we have

$$\Sigma_{22}^{-1}\,\Sigma_{21}\,\Sigma_{11}^{-1}\,\Sigma_{12}\,\beta - \rho^2\beta = 0\,, \tag{19}$$

which is in the form $A\beta = \rho^2\beta$ where ρ^2 is some positive constant and A is a real symmetric matrix of dimension $(p-r)\times(p-r)$. In matrix algebra, the nontrivial solutions $\beta_{(j)}$, $j = 1,\ldots,p-r$, to $A\beta = \rho^2\beta$ are obtained by choosing $\beta_{(j)}$ to be the eigenvectors of A which correspond with the eigenvalues $\rho_{(j)}^2$, $j = 1,\ldots,(p-r)$.

For the first solution associated with the largest eigenvalue $\rho_{(1)}^2$ we have $A\beta_{(1)} = \rho_{(1)}^2\beta_{(1)}$ where both $\rho_{(1)}^2$ and $\beta_{(1)}$ are known from the eigenvalue eigenvector decomposition of A. From (16) we have $\rho\alpha = \Sigma_{11}^{-1}\,\Sigma_{21}\,\beta$ which can be used to obtain the solution $\alpha_{(i)}$ for $i = 1,...,r$ which corresponds with $\beta_{(i)}$ and $\rho_{(i)}^2$ to be

$$\alpha_{(i)} = \Sigma_{11}^{-1}\,\Sigma_{21}\,\beta_{(i)}[\rho_{(i)}]^{-1}\,, \tag{20}$$

where all terms on the right hand side are known from the above mentioned eigenvalue problem, provided $r \leq (p-r)$.

Note that we have used the suggestive notation of ρ for the Lagrangian coefficients $2\mu_1$ and $2\mu_2$ which were shown to be equal to each other above. Now we verify that the ρ is indeed the canonical correlation coefficient defined in equation (6) by the same symbol ρ. For the ith solution, we have

$$var\,(\alpha_{(i)}'X_1) = \alpha_{(i)}'\,\Sigma_{11}\,\alpha_{(i)} = \delta_{ij} \tag{21}$$

$$var\,(\beta_{(i)}'X_2) = \beta_{(i)}'\,\Sigma_{22}\,\beta_{(i)} = \delta_{ij}\,, \tag{22}$$

where δ_{ij} is the so-called Kronecker's delta, which equals unity when $i = j$, and is zero otherwise. This follows from $\alpha'\,\Sigma_{11}\,\alpha = 1 = \beta'\,\Sigma_{22}\,\beta$ in (7) above. For the covariance between the weighted sums $\alpha_{(i)}'X_1$ with $i = 1,...,r$ and $\beta_{(j)}'X_2$ with $j - 1,...,(p-r)$, it is obvious that there are exactly $\min(r,p-r)$ number of ordered pairs of weighted sums $\alpha_{(i)}'X_1$ and $\beta_{(j)}'X_2$. In the context of equation (20) we have noted that we can solve it, provided $r \leq p-r$, which means that $r = \min(r,p-r)$. Now consider the covariances for the first r pairs of variables:

$$cov\,(\alpha_{(i)}'X_1,\beta_{(j)}'X_2) = \alpha_{(i)}'\,\Sigma_{12}\,\beta_{(j)} = \rho_{(i)}\,, \tag{23}$$

for $i = j = 1,...,r$; where we use (13) or (14) to write $\alpha'\,\Sigma_{12}\,\beta = \rho$, which in turn yields the last equality. Substituting (21), (22) and (23) in the definition of ρ from (6) verify that ρ is indeed a correlation coefficient, except for the fact that its sign is arbitrary. The eigenvalue of A from (19) gives ρ^2 rather than ρ, and the square root can be either positive or negative. It is customary to adopt the convention of defining $\rho_{(i)}$ to be the positive square roots.

The word "canonical" in canonical correlation analysis refers to thc fact that the variance covariance matrix of the $\alpha'X_1$ and $\beta'X_2$ from (21) to (23) has the simpler from (akin to the simplification obtained by canonical reductions in matrix algebra)

$$cov\,(\alpha'X_1,\,\beta'X_2) = \begin{bmatrix} I_r & R_{ho} \\ R_{ho}' & I_{p-r} \end{bmatrix}, \tag{24}$$

where $r = \min(r,p-r)$ by assumption, I_r and I_{p-r} are r and $(p-r)$ dimensional identity matrices respectively, and where R_{ho} is $r{\times}(p-r)$ matrix having the first r diagonal elements $\rho_{(i)}$ and all remaining elements zero.

From a theoretical viewpoint, Hotelling's canonical correlations analysis is an elegant generalization of the ordinary least squares (OLS) regression problem for the situation when there are more than one dependent variables.

When $r = 1$ it can be verified that we have OLS as a special case (Hint: Σ_{11} is a scalar, $\rho^2 = \Sigma_{12} \Sigma_{22}^{-1} \Sigma_{21}/\Sigma_{11}$ is the squared multiple correlation coefficient).

Like any correlation coefficient, the canonical correlation coefficients are "pure numbers" in the sense that a (nonsingular) linear transformation of the variables leaves the $\rho_{(i)}$ values unchanged.

Since the canonical correlations $\rho_{(i)}$ are defined in terms of a Lagrangian problem, they have certain optimality properties. Note that the coefficients $\alpha_{(1)}$ and $\beta_{(1)}$ represent the "best fitting" coefficients associated with the largest canonical correlation coefficient $\rho_{(1)}$. The second pair of coefficients $\alpha_{(2)}$ and $\beta_{(2)}$ satisfy the property that their elements are "uncorrelated" with the corresponding elements of the first pair in the sense that $\alpha_{(1)}' \Sigma_{11} \alpha_{(2)} = 0 = \beta_{(1)}' \Sigma_{22} \beta_{(2)}$ etc. The optimality of $\alpha_{(2)}$ and $\beta_{(2)}$ associated with $\rho_{(2)}$ is that they are also best fitting linear weights which are uncorrelated with the first set of weights $\alpha_{(1)}$ and $\beta_{(1)}$. This can be verified by writing another Lagrangian with the additional "uncorrelatedness" constraints.

12.1.1 Sample Canonical Correlations

If we have a sample of T observed data points on each of the p variables $x_1,...,x_r$, $x_{r+1},...,x_p$ it is possible to use the above theory to estimate $\rho_{(i)}$, $\alpha_{(i)}$ and $\beta_{(i)}$ by merely incorporating hats to indicate sample values in most of the above formulas.

For empirical estimations it is sufficient to compute the eigenvalues and eigenvectors of $A = \Sigma_{21} \Sigma_{11}^{-1} \Sigma_{12}$ from (19). For this we need to estimate the $p \times p$ matrix Σ by maximum likelihood methods, define the four submatrices, and use matrix multiplications to estimate A.

For numerical stability and reliability, it is customary to work with a $\hat{\Sigma}$ matrix defined in terms of standardized variables so that $\hat{\Sigma}$ is a matrix of observed correlation coefficients.

The reader is referred to Kshirsagar [1972] for a rigorous derivation of various aspects of the statistical distributions of sample canonical correlations and corresponding sample weights, and for additional references. The distribution theory for sample weights (from eigenvectors) involving Wishart-type distributions is complicated, especially in the nonnull case when the canonical correlations are large. Wilks' lambda defined by a product of r terms:

$$\Lambda = \prod_{i=1}^{r} (1 - \hat{\rho}_{(i)}^2) \,, \tag{25}$$

is popular for measuring the strength of relationship.

12.1.2 Applications of Hotelling's Canonical Correlation Analysis

Despite the theoretical appeal of Hotelling's canonical correlations analysis, the practical applications of the tool have been few. We mention Hotelling's [1936] own study which estimates a canonical correlation to *measure the strength of the relationship* between "reading skill" measured by two variables: (speed of reading and power of understanding from reading) and "arithmetic skill" measured by two variables (speed and accuracy) based on educational tests administered to 140 school children.

Waugh [1942] uses canonical correlation model to estimate the strength of the (negative) relationship between meat consumption measured by two variables (pork and beef) and meat prices for the corresponding quantities. The aim of this study is to *define best fitting index numbers* for consumption and prices. Waugh also discusses another application where the aim is to *predict* the four dimensions of the quality of wheat flour from five characteristics of wheat kernels.

More recently Glahn [1968] discusses a weather forecasting application, whereas Vinod [1969] estimates a *joint production* function for the joint production of wool and mutton from the usual capital and labor inputs. Adelman, et al. [1969] uses the canonical correlation analysis to estimate "*goals*" *of economic development* in underdeveloped countries with economic policy "instruments" such as industrialization. The canonical weights are used to suggest best instruments. E. H. Mantell [1974] uses the canonical correlation analysis to measure the labor productivity within a post office in sorting and routing of the mail by fitting a joint production function.

Considering that the canonical correlation techniques has been known for over 40 years, the number of practical applications seem to be very few. One of the reasons is the technical difficulty in understanding and explaining the method. Another reason might be multicollinearity which can give rise to unbelievable signs, such as in a study of U.S. hard winter wheat by Waugh [1942].

Psychologists and sociologists have used canonical correlations with some success, Timm [1975]. B. R. Rao's "partial" canonical correlation involves the correlation between two sets of variables after the effect of a third set of variables is removed. These are discussed in Timm and Carson [1976] with illustrations.

The number of nonzero population canonical correlation coefficients is sometimes called the "dimensionality." Mallows' [1973] C_p statistic has been extended to solve this problem by Fujikoshi and Veitch [1979] and compared with the use of Akaike's information criterion. They conclude that the C_p criterion performs best.

When the variables $x_1,...,x_r$ in X_1 or $x_{r+1},...,x_p$ in X_2 are highly correlated with each other, the correlation matrices Σ_{11}, Σ_{22} are often ill-conditioned, and their inverses are particularly sensitive to small changes in the available data. Therefore the estimated eigenvalues $\hat{\rho}^2_{(i)}$ and eigenvectors $\hat{\alpha}_{(i)}$, $\hat{\beta}_{(j)}$, for

$i = 1,...,r$ and $j = 1,...,p-r$ of matrix $A = \Sigma_{22}^{-1} \Sigma_{21} \Sigma_{11}^{-1} \Sigma_{12}$ from (19) which involves Σ_{11}^{-1} and Σ_{22}^{-1} are also sensitive to small changes in the available data. This is where the ridge-type modifications have been suggested in Vinod [1976e].

12.1.3 Canonical Ridge Model

In Vinod [1976e] the variables $x_1,...,x_r$ in X_1 and $x_{r+1},...,x_p$ in X_2 are assumed to be subject to measurement errors ϵ_i, $i = 1,...,p$, respectively. These are intrinsic errors in the observable data. For example, we may know that a change in the observed values for x_i is so small that it and a zero change cannot be distinguished by our measurements. Since Hotelling's canonical correlation has certain optimality properties which do not distinguish between small changes due to measurement errors and actual changes, Vinod incorporates them by assuming that

$$\begin{aligned}
E(\epsilon_i) &= 0\,, \quad \text{for} \quad i = 1,...,p\,. \\
E(\epsilon_i \epsilon_j) &= 0\,, \quad \text{for} \quad i \neq j = 1,...,p\,. \\
E(\epsilon_i x_i) &= 0\,, \quad \text{for} \quad \textit{all} \ \ i = 1,...,p\,. \\
Var(\epsilon_i) &= k_1\,, \quad \text{for} \quad i = 1,...,r\,. \\
Var(\epsilon_i) &= k_2\,, \quad \text{for} \quad i = r+1,...,p\,.
\end{aligned} \tag{26}$$

Under these assumptions it can be verified that the variance covariance matrix of the variables in X_1 and X_2 is given by

$$cov(X_1, X_2) = \begin{bmatrix} \Sigma_{11} + k_1 I & \Sigma_{21} \\ \Sigma_{21} & \Sigma_{22} + k_2 I \end{bmatrix}. \tag{27}$$

For this model, Vinod [1976e] proposed the following revision of the canonical correlation coefficient

$$\rho^*(k_1, k_2) = \frac{\alpha' \Sigma_{12} \beta}{\left[\alpha' \left(\Sigma_{11} + k_1 I\right) \alpha\right]^{1/2} \left[\beta' \left(\Sigma_{22} + k_2 I\right) \beta\right]^{1/2}}\,. \tag{28}$$

This involves a ridge-type modification of Σ_{11} and Σ_{22} matrices by adding k_1 and k_2 respectively to the diagonals. Note that $\rho^*(k_1,k_2) < \rho$ when $k_1 > 0$ and/or $k_2 > 0$ because the denominator is larger for ρ^*. The empirical estimates of α and β are based on the eigenvalues and eigenvectors of

$$A^* = \left(\Sigma_{22} + k_2 I\right)^{-1} \Sigma_{21} \left(\Sigma_{11} + k_1 I\right)^{-1} \Sigma_{12} \tag{29}$$

with the sample estimates of the submatrices of Σ. Clearly, Hotelling's method is a special case of this when $k_1 = k_2 = 0$. When α_i and β_i are estimated from the eigenvectors of A^*, "the fit" is poorer when k_1 and/or k_2 are strictly positive.

12.1.3.1 Choice of k_1 and k_2

The choice of k_1 and k_2 is made as follows. We assume that there are $h = 1,...,H$ $(H>4)$ sets of subsamples or comparable data replications available to the researcher. We also assume that the data are standardized.

The *canonical ridge* estimates of α_i and β_i for each k_1, k_2 may be denoted by $\alpha_i(k_1,k_2,h)$ and $\beta_i(k_1,k_2,h)$ respectively where the (subsample) identifier h is explicitly included. We define A^* and estimate $\alpha_i(k_1,k_2,h)$ and $\beta_i(k_1,k_2,h)$ for a range of k_1 and k_2 values, and compare the results to aid in choosing the appropriate k_1 and k_2.

Step 1. Our first step eliminates certain estimates of α_i and β_i from further consideration by using informal prior knowledge about the signs of $\partial x_i/\partial x_j$; $(i \neq j=1,...,p)$, which is often available. We assume that this knowledge is not precise enough or detailed enough to describe any prior distributions of the parameters. The knowledge about these partial derivatives can imply some knowledge about the signs of α_i and β_i. We reject all choices of k_1 and k_2 that violate our informal prior knowledge about the signs of α_i and β_i, whenever such knowledge is available.

Step 2. Let $b_i(k_1,k_2,h)$ and $a_i(k_1,k_2,h)$ denote the sample estimates of β_i and α_i respectively for a given choice of k_1, k_2 and h. We fix k_1, and k_2 and find the averages and variances over the H subsamples as follows:

$$\bar{b_i}(k_1,k_2) = H^{-1} \sum_{h=1}^{H} b_i(k_1,k_2,h) \tag{30}$$

$$V_i(k_1,k_2) = (H-1)^{-1} \sum_{h=1}^{H} [b_i(k_1,k_2,h)-\bar{b_i}(k_1,k_2)]^2 \,. \tag{31}$$

The variability of $b_i(k_1,k_2,h)$ over the set of H subsamples is measured by $V_i(k_1,k_2)$. A similar variability indicator for $a_i(k_1,k_2,h)$ yields $V_i(k_1,k_2)$ for $i = r+1,...,p$. The overall variability over the p coefficients α_i and β_i is measured by the overall standard deviation defined by

$$SD(k_1,k_2) = \left[\sum_{i=1}^{p} V_i(k_1,k_2)\right]^{1/2} \,. \tag{32}$$

Step 3. A "goodness of fit" for each choice of k_1 and k_2 is now measured by $\rho^*(k_1,k_2)$ from (28).

The choice of k_1 and k_2 which maximizes the differences

$$\rho^*(k_1,k_2) - SD(k_1,k_2) \tag{33}$$

will give as good a fit as possible, without sacrificing the stability of estimated coefficients α_i and β_i over the H subsamples (or replications). An unpublished simulation study by Vinod suggests that this procedure yields coefficients that are closer to the true parameters than the conventional choice $k_1 = k_2 = 0$, when the matrices Σ_{11} and Σ_{22} are ill-conditioned.

12.1.3.2 Applications of Canonical Ridge

Vinod [1976e] [1976f] gives applications of canonical ridge model to estimate joint production functions. The variability associated with $k_1 = 0 = k_2$ is extremely large in both examples. In Vinod [1976e] the $SD(0,0) = 6.0$ is reduced to $SD(0.05,0.20) = 0.04$, and the problem of unrealistic signs of estimated α_i and β_i is eliminated. The reduction in the canonical correlation coefficient from 0.9998 to $\rho(0.05,0.20) = 0.9707$ is not substantial, when judged by a plot of output side translog function against the input side translog function. Thus the sacrifice in goodness of fit is worth the added stability of the estimated coefficients.

12.2 DISCRIMINANT ANALYSIS

Linear discriminant analysis is due to R. A. Fisher [1936]. It has been extended by several authors over the years. A more recent textbook, such as Kshirsagar [1972], should be consulted for recent references and further details about the techniques and its applications. Graphically, the basic problem may be illustrated by Figure 12.1. There are two populations π_1 and π_2 and we have two random samples from these two populations. Discriminant analysis yields a description of the "optimal" straight line which is useful for classifying further observations from π_1 and π_2. The optimality criterion used by Fisher is to equate the probability of misclassification. In the above figure this amounts to equating the probability of misclassifying a member x from π_2 as a member of π_1 to the probability of the misclassification of a member 0 from π_1 as a member of π_2.

An extension to more populations and nonlinear boundaries found in the textbooks need not be repeated here. Further bibliographies are found in Lachenbruch [1975] and Cacoullos [1973].

We have the observed values $x_1,...,x_p$ from two multivariate normal populations π_1 and π_2 having means μ_1 and μ_2 and covariance matrices Σ_1 and Σ_2. It is usually assumed that the covariance matrices are equal $\Sigma = \Sigma_1 = \Sigma_2$, and the costs of misclassification and prior probabilities of membership are also

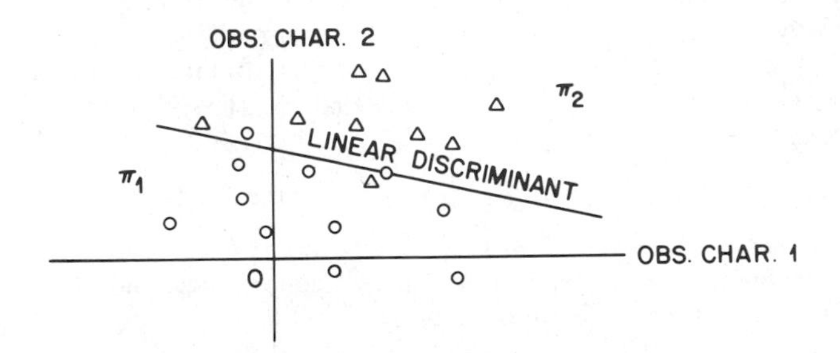

Figure 12.1 Linear Discriminant

equal for the two populations. This leads to major simplifications. Define the squared "Mahalanobis distance" between π_1 and π_2 by the quadratic form

$$D^2(\mu_1,\mu_2) = (\mu_2-\mu_1)' \Sigma^{-1} (\mu_2-\mu_1) . \tag{34}$$

Similarly the distance of observed point x of the p-dimensional space from the population π_i may be measured by the following "Mahalanobis distance":

$$D^2(x,\mu_i) = (x-\mu_i)' \Sigma^{-1}(x-\mu_i) , \tag{35}$$

for $i = 1,2$, where the means μ_i and Σ are assumed to be known. If

$$D^2(x,\mu_1) \leq D^2(x,\mu_2) \tag{36}$$

x is "closer" to π_1 than π_2, and therefore we classify x into population π_1. Otherwise we classify x into π_2. The presence of the equality in (36) is included for convenience. One may decide to randomize for the equality case by the tossing of a coin (say) to determine the classification into π_1 or π_2 as explained in Kshirsagar [1972, p. 191].

The above rule (36) can be written as

$$x' \Sigma^{-1} x - \mu_1' \Sigma^{-1} x - x' \Sigma^{-1} \mu_1 + \mu_1' \Sigma^{-1} \mu_1$$

$$\leq x' \Sigma^{-1} x - \mu_2' \Sigma^{-1} x - x' \Sigma^{-1} \mu_2 + \mu_2' \Sigma^{-1} \mu_2$$

We may cancel $x' \Sigma^{-1} x$ from both sides, write $\mu_i' \Sigma^{-1} x = x' \Sigma^{-1} \mu_i$ $(i=1,2)$, and note that (36) is reduced to

$$\mu_1' \Sigma^{-1} \mu_1 - \mu_2' \Sigma^{-1} \mu_2 \leq 2x' \Sigma^{-1}\mu_1 - 2x' \Sigma^{-1}\mu_2$$

Adding $\mu_1' \Sigma^{-1} \mu_2 - \mu_2' \Sigma^{-1} \mu_1 = 0$ to the left hand side, and rearranging we have

$$1/2(\mu_1+\mu_2)'\,\Sigma^{-1}\,(\mu_1-\mu_2) \le x'\,\Sigma^{-1}\,(\mu_1-\mu_2)\ , \tag{37}$$

where the linear expression on the right hand side is sometimes called Fisher's linear discriminant.

The rule based on (36) is sometimes called the "Minimum x^2 rule" which follows from the likelihood ratio principle. If the parameters μ_i and Σ are unknown, they have to be replaced by their sample estimates $\hat{\mu}_i$ and $\hat{\Sigma}$. The sample based minimum chi-square rule states that we classify x into π_i when it is closer to π_i in the sense that

$$D^2(x,\hat{\mu}_i) \le D^2(x,\hat{\mu}_j) \quad , \quad where \quad j \ne i\ . \tag{38}$$

For a recent summary of work in discriminant analysis and a discussion of research problems the reader is referred to Lachenbruch and Goldstein [1979] which also gives an extensive bibliography.

12.2.1 Ridge-Type Modification

Di Pillo [1976] proposes a "biased minimum chi-square rule" where the Mahalanobis distance is replaced by a ridge-type metric

$$D^{*2}(x,\hat{\mu}_i) = (x-\hat{\mu}_i)'(\hat{\Sigma} + kI)^{-1}(x-\hat{\mu}_i)\ , \tag{39}$$

of which Mahalanobis D^2 is a special case when the "biasing parameter" k is zero.

The introduction of k will reduce the variance of the estimated Mahalanobis distances. A proof of this based on the assumption of known μ_i and Σ is given by Di Pillo [1976]. The variance reduction should be obvious to those who are familiar with ridge regression by the analogy between the discriminant and linear regression, explained in Kshirsagar [1972, p. 206] and Dhyrmes [1974; p. 75]. Di Pillo also considers the derivative of the probability of misclassification (PCM*) based on ridge-type k parameter, with respect to k. He evaluates the derivative near $k = 0$ (from the positive side) and shows that this derivative is negative, suggesting that the probability of misclassification is likely to decrease with a positive k.

Monte Carlo simulations supporting the use of ridge type modification of the discriminant analysis have been reported by Di Pillo [1976] and Smidt and McDonald [1976]. In the Di Pillo simulation multivariate random normal variables are generated for sample sizes 5(5)25 for $p = 9$ or 5 based on realistic data on shellfish collected by the Department of Marine Studies of the University of Delaware. The "conditioning" of the matrix $\hat{\Sigma}$ is also varied. The value of k is always chosen to be 1 without much discussion. The results show "substantially improved performance for all combinations considered", i.e., the probability of misclassification is reduced. Di Pillo recommends using the biased chi-square rule whenever the $\hat{\Sigma}$ matrix is ill-conditioned in the sense that its trailing eigenvalues are small, and make only a "small contribution" to the

Mahalanobis distance.

Smidt and McDonald's [1976] simulation studies the question of selecting an appropriate biasing parameter k in (39). It is tempting to use the available sample to choose k to minimize error of misclassification in the sample. However, this strategy will generally suggest $k = 0$ and does not consider the performance outside the observed sample. After an estimate Σ is obtained from the entire sample, k may be selected by omitting one observation at a time and minimizing the error rate in classifying the omitted observation with $\hat{\mu}_i$ based on the remaining observations and choosing various values of k in (39). Unfortunately this may mean excessive computational burden.

EXERCISES

12.1 Consider Vinod [1968] example of joint production of wool and mutton in U.S. Compute the canonical ridge estimates for the model using the sources given at the bottom of the attached Table 12.1. Collect the data for the additional years to 1980, recompute the earlier results, and compare the forecasting performance of ordinary canonical correlations and canonical ridge.

12.2 In the derivation of equation (34) it is assumed that the covariance matrices are equal. What criterion would you use if they are unequal? How would you introduce different costs for misclassification.

12.3 Solve the variable selection problem in discriminant analysis by using Mallows C_p method and the results by Fujikoshi and Veitch [1979].

TABLE 12.1

Data for the construction of joint production function for U.S. agriculture with Capital and Labor as inputs and wool and mutton as outputs for the period 1951-1962.

Year	*Inputs*		*Outputs*			
	Labor	*Capital*	*Wool*		*Mutton*	
	(1)	(2)	(3)	(4)	(5)	(6)
1951	17703	193649	228091	253991	11416	11073.5
1952	18552	227548	233309	266909	14304	13874.9
1953	18814	206165	232258	274458	16321	15504.9
1954	13490	261888	235807	279307	16255	15442.2
1955	13390	260120	241284	282884	16553	15890.9
1956	12710	158990	242177	282677	16328	15511.6
1957	13100	262540	239101	272701	15292	14680.3
1958	13350	286690	243713	174113	14495	14205.1
1959	14280	307320	259939	294439	15528	15372.7
1960	14710	309180	166563	300163	16239	16076.6
1961	13290	302420	261249	295749	17536	17185.2
1962	12560	304810	248538	278438	17171	16655.9

Units: col. 1 = thousands of man hours, col. 2 = thousands of dollars, col. 3 and 4 = thousands of pounds, col. 5 = thousands slaughtered, col. 6 = 100,000 pounds of weight

Note: Data in col. 4 represents in addition to column 3, the pulled wool. The data in column 5 the number of sheep slaughtered and in column 6 the mutton production allowing for changing weight of sheep slaughtered over time.

Source: U.S. Agricultural Statistics, Annual for years 1951-64, U.S. Govt. Printing Office, Washington, U.S. Dept. of Agriculture.

13

Improved Estimators Under Non-Normal Errors and Robust Regression

13.1 INTRODUCTION

The objective of this chapter is to generalize the results in the earlier chapters in three respects. First, we consider a general class of estimators which contain Stein-rule estimators, ridge-type estimators, Lindley-type Bayesian estimators and various other estimators in the literature as special cases. Secondly, we analyze the properties of these estimators when the errors (disturbances) in the regression model are not normally distributed. The results for normal distribution are presented as a special case. The analysis of estimators in the nonnormal case is important because the assumption of normality is often questionable and it may have varying effects in a variety of situations (Gnanadesikan [1977]). The third objective of this chapter is to briefly introduce the reader to the growing literature on robust regression. We do not intend to provide complete coverage of this topic.

Consider the linear regression model

$$y = X\beta + u \ , \tag{1}$$

where y is a $T\times 1$ vector of observations on the dependent variable, X is a $T\times p$ matrix, β is a $p\times 1$ vector of regression coefficients and u is a $T\times 1$ disturbance vector.

We make the following assumptions

Assumption 1 The matrix X of explanatory variables is nonstochastic and of rank p.

Assumption 2 The elements of u are independently and identically distributed with first four finite moments as

$$E(u_t) = 0, \quad E(u_t^2) = \sigma^2$$

$$E(u_t^3) = \sigma^3\nu_1, \quad E(u_t^4) = \sigma^4(\nu_2+3) \; , \; t = 1...T \; , \tag{2}$$

where ν_1 and ν_2 are Pearson's measures of skewness and kurtosis (peakedness) of the distribution of disturbances.

Assumption 3 The sample size T is greater than the total number of explanatory variables p.

Notice that $\nu_1 = 0$ in (2) implies symmetry of the distribution while $\nu_2 = 0$ means that the distribution is mesokurtic (no more or less peaked than the normal).

The ordinary least squares (OLS) estimator of β in (1) is given by

$$b = (X'X)^{-1}X'y \; , \tag{3}$$

which is unbiased. Its matrix mean squared error (MtxMSE) and the weighted MSE (WMSE) are given, respectively, as

$$MtxMSE(b) = E(b-\beta)(b-\beta)' = \sigma^2(X'X)^{-1} \; ,$$

$$WMSE(b) = E(b-\beta)'Q(b-\beta) = \sigma^2 Tr(X'X)^{-1}Q \; , \tag{4}$$

where Q is a known positive definite matrix, and "Tr" represents the trace of the matrix [details about (3) and (4) are in Chapters 1 and 2].

Throughout this chapter, we consider the (weighted) loss function $(b-\beta)'Q(b-\beta)$ for evaluating the estimators of β in $y = X\beta + u$.

13.2 A GENERAL CLASS OF IMPROVED ESTIMATORS

Consider a general class of biased estimators as

$$\hat{\beta} = [I+aD]^{-1}b \; , \tag{5}$$

where D is any known $p \times p$ matrix, and the scalar a is

$$a = \frac{cs}{b'Bb} \; ; \quad s = (y-Xb)'(y-Xb) \; , \tag{6}$$

which depends upon the arbitrary constant c and a given positive definite matrix B. The term s is, in fact, the estimated residual sum of squares from (1), and b is the OLS estimator.

A more general form of $\hat{\beta}$ can be written by regarding "a" as an arbitrary but well defined function of $b'Bb/s$, i.e., $a = f(b'Bb/s)$ (see Brown [1971] and Casella [1980]). This will, however, introduce the problem of the choice of a particular function, f, and unless f is chosen, the estimator $\hat{\beta}$ based on $a = f(b'Bb/s)$ will not be operational. So we exclude this from our study. Even when we take $a = cs/b'Bb$, a form more general than (6) could be considered as $a = cs/(b'Bb+c_1s+c_2)$ (Strawderman [1978]) where c_1 and c_2 are two arbitrary positive constants. The results regarding the dominance condition on c will not be affected by this choice of $\hat{a}$ even though we will need to replace the $\beta' B \beta$ in the WMSE expression of $\hat{\beta}$, given in (14), by $\beta' B \beta + c_2$. In view of this, and the fact that choosing $a = cs/(b'Bb + c_1 s + c_2)$ involves the additional problem of determining positive c_1 and c_2, let us take these constants c_1 and c_2 to be zero (also see Chapter 8 for these points). Thus, $\hat{\beta}$ in (5) is the operational set of estimators for known D and B, and the c which is given later on in this chapter. The choices of D, B and c which are useful for the practitioner are also discussed later.

The estimator $\hat{\beta}$ unifies various biased estimators available in the literature. For example, if "a" in (5) is sufficiently small, then the first order linear approximation of $\hat{\beta}$ can be written as

$$\hat{\beta} \cong (I - aD)b = [I - \frac{cs}{b'Bb} D]b = \hat{\beta}_{JB} . \tag{7}$$

This belongs to the family of biased estimators given in Judge and Bock [1978] and the subscript JB refers to them. These estimators can be called first order approximate estimators.

Next, if $D = I$ in (5), we get

$$\hat{\beta} = \left(1 - \frac{cs}{b'Bb + cs}\right) b = \hat{\beta}_S , \tag{8}$$

which is the Stein-type estimator. For $B = X'X$, $b'Bb + cs = y'y - (1-c)s$, and $\hat{\beta}_S$ becomes Ullah and Ullah type double k-class estimator. These estimators have been derived in Chapter 6 by Bayesian and non-Bayesian methods. While Bayesian derivations and the likelihood approach needed the assumption of normality of disturbances, it was not needed for some of the non-Bayesian methods. The sampling properties of $\hat{\beta}_S$ were, however, obtained only under the assumption of normality.

Further, if $D = (X'X)^{-1}$, the estimators $\hat{\beta}$ become the operational ordinary ridge estimators. These are

$$\hat{\beta} = [I + \frac{cs}{b'Bb}(X'X)^{-1}]^{-1} b = \hat{\beta}_{OR} , \tag{8a}$$

where OR refers to "ordinary ridge." These estimators are discussed in Chapter 8, and their Bayesian and non-Bayesian interpretations are given in Chapter 7. Note that in Chapters 7 and 8 the matrix X was standardized such that $G'X'XG = \Lambda = \text{Diag.}(\lambda_1 \ldots \lambda_p)$ was the diagonal matrix of eigenvalues. If X is so standardized, and $B = \text{Diag.}(0 \ldots \lambda_i \ldots 0)$ is a $p \times p$ diagonal matrix with the only nonzero diagonal element λ_i at the i-th place, then $\hat{\beta}$ in (5) with $D = \Lambda^{-1}$ will become the operational generalized ridge-type estimator discussed in Chapter 8. As in the case of Stein-type estimators, we note from Chapter 7 that the ridge-type estimators can be obtained by non-Bayesian methods which do not need the assumption of normality of the errors.

Finally, consider $B = J'\Omega_0^{-1}J$ and $D = (\Omega_0 X'X)^{-1}J$ in (5), where Ω_0 is a known $p \times p$ symmetric positive definite matrix and for a known $p \times r$ matrix R of rank r, $J = I - R(R'\Omega_0^{-1}R)^{-1}R'\Omega_0^{-1}$. Then we get

$$\hat{\beta} = \hat{\beta}_L = [I + \frac{cs}{b'J'\Omega_0^{-1}Jb}(\Omega_0 X'X)^{-1}J]^{-1}b$$

$$= Rb_0 + [I + \frac{cs}{b'J'\Omega_0^{-1}Jb}(\Omega_0 X'X)^{-1}]^{-1}(b - Rb_0)] , \qquad (9)$$

where $b_0 = (R'\Omega_0^{-1}R)^{-1}R'\Omega_0^{-1}b$. This is the Lindley-type Bayes estimators in Chapter 9. Notice again that this set of estimators shrink the OLS estimator b towards Rb_0 instead of towards zero as in the cases of Stein and ridge-type estimators. In fact, for $\Omega_0 = (X'X)^{-1}$ and $\Omega_0 = I$ the estimator $\hat{\beta}_L$ provides a Lindley-type mean correction to the Stein-type estimators (for $R = \iota$, see equation (33), Chapter 6) and ridge-type estimators, respectively.

13.3 PROPERTIES OF $\hat{\beta}$ UNDER NONNORMAL DISTURBANCES

The properties of the class of estimators $\hat{\beta}$ in (5) will now be analyzed under Assumptions 1 to 3 of Section 13.1. Before looking into the results, it will be convenient to introduce the following notations

$$M = I - X(X'X)^{-1}X' , \quad \bar{M} = I - M$$

$$N = (X'X)^{-1}X'(I * \bar{M})X , \quad n = T - p \qquad (10)$$

$$q = n(n+2) + \nu_2 Tr(M * M) , \quad \phi = \frac{TrN(X'X)^{-1}QD}{Tr(X'X)^{-1}QD} \text{ and } \theta = \frac{\nu_2}{n+\nu_2} ,$$

where "*" denotes the Hadamard product of matrices, which for $I * \bar{M}$ involves multiplying each element of the I matrix with the corresponding element of the $\bar{M}$ matrix (also see Rao [1973, p. 30]). It has been shown in the explanatory note 13.2 that

$$0 \le \eta_p \le \phi \le \eta_1 \le 1 , \qquad (11)$$

where $\eta_1 \ge \eta_2 \dots \ge \eta_p$, are the eigenvalues of the matrix N in (10).

Regarding the term $\theta = \nu_2/(n+\nu_2)$ in (10), we observe that

$$0 < \theta < 1 , \quad \text{when } \nu_2 > 0 ,$$

$$\theta < 0 , \quad \text{when } \nu_2 < 0 \text{, and } n \ge 2 \qquad (12)$$

$$\theta = 0 , \quad \text{when } \nu_2 = 0 .$$

The inequality $0 < \theta < 1$ is obvious. When $\nu_2 < 0$, we note from a result in Rao ([1973], (1e.4.2), p. 57) that $2 + \nu_2 \ge 0$. Hence, $\theta < 0$ holds for all

$n \geq 2$. When $n = 1$, $\theta \leq 0$ provided $1 + \nu_2 > 0$. When $n = 0$, $s = (y-Xb)'(y-Xb) = 0$ and $\hat{\beta}$ becomes b. Thus, $n = 0$ is not an interesting case.

We can now present Kadane's small-disturbance approximations for the bias and WMSE of $\hat{\beta}$. These are

$$E(\hat{\beta}-\beta) = \text{Bias}(\hat{\beta}) = -\frac{cn\,\sigma^2}{\beta' B \beta} D\beta \tag{13}$$

up to order σ^2, and

$$E(\hat{\beta}-\beta)'Q(\hat{\beta}-\beta) = \text{WMSE}(\hat{\beta}) = \sigma^2 r_2 - \sigma^3 r_3 - \sigma^4 r_4 \tag{14}$$

up to order σ^4, where

$$r_2 = Tr(X'X)^{-1}Q\ , \quad r_3 = 2c\nu_1 \frac{\beta' D' Q (X'X)^{-1} X'(I*M)\iota}{\beta' B \beta} \tag{15}$$

and

$$r_4 = -cq\frac{\beta'D'QD\beta}{(\beta'B\beta)^2}\left[c - \frac{2(n+\nu_2)}{q}\;\frac{\beta'B\beta}{\beta'D'QD\beta}\left\{(1-\theta\phi)Tr(X'X)^{-1}QD\right.\right.$$

$$\left.\left. -2\frac{\beta'B(X'X)^{-1}QD\beta}{\beta'B\beta}\left(1-\theta\frac{\beta'BN(X'X)^{-1}QD\beta}{\beta'B(X'X)^{-1}QD\beta}\right)\right\}\right]. \tag{16}$$

Here ι denotes a column unitary vector. The results (13) and (16) are derived in the explanatory note 13.1. The MtxMSE of $\hat{\beta}$ is also derived there. These results are based on Ullah, Srivastava and Chandar [1980].

Observing from (4) that the WMSE$(b) = \sigma^2 r_2$, it is seen from (14) that

$$\text{WMSE}(\hat{\beta}) - \text{WMSE}(b) = -\sigma^3 r_3 - \sigma^4 r_4. \tag{17}$$

We shall analyze (17) for the cases $\nu_1 = 0$ and $\nu_1 \neq 0$, respectively.

Results for Symmetrical Distribution (ν_1=0)

Let us write the expression (17) for symmetrical distributions ($\nu_1 = 0$) of disturbances as

$$\text{WMSE}(\hat{\beta}) - \text{WMSE}(b) = -\sigma^4 r_4. \tag{18}$$

Now using (11), (12) and (16) we can easily show that for symmetrical leptokurtic disturbances ($\nu_2 > 0$)

$$\text{WMSE}(\hat{\beta}) - \text{WMSE}(b) \leq 0 \tag{19}$$

when

$$0 < c \leq \frac{2(n+\nu_2)}{q}\delta_p \mu_1[(1-\theta\phi)d - 2(1-\theta\eta_p)] \; ; \tag{20}$$

$$d = \frac{1}{\mu_1} Tr\,(X'X)^{-1}QD > 2 + \Delta_1$$

where δ_p and η_p are the minimum eigenvalues of the matrices $B(D'QD)^{-1}$ and N, respectively, and μ_1 is the maximum eigenvalue of $(X'X)^{-1}QD$ along with

$$\Delta_1 = \frac{2\theta}{1-\theta\phi}(\phi-\eta_p) > 0 \, . \tag{21}$$

Similarly, for symmetrical platykurtic distributions ($\nu_2<0$) (19) holds for all $n \geq 2$, and for $n = 1$ provided $1 + \nu_2 > 0$, so long as

$$0 < c \leq \frac{2(n+\nu_2)}{q}\delta_p \mu_1[(1-\theta\phi)d - 2(1-\theta\eta_1)] \; ; \quad d > 2 + \Delta_2 \tag{22}$$

where

$$\Delta_2 = \frac{2\theta}{1-\theta\phi}(\phi-\eta_1) > 0 \, . \tag{23}$$

In obtaining the conditions in (20) and (22), we also use the result that if E is any symmetric matrix and F is positive definite then $\min_{\beta}(\beta'E\beta/\beta'F\beta)$ is the minimum root of $|E - \lambda F| = 0$.

The conditions for the dominance of Stein-type and ridge-type estimators, in (8) and (8a), respectively, can be written from (20) and (22) for the non-normal symmetric distributions. Some other remarks about (20) and (22) are given below.

Remark 1 For the case of normal distribution ($\nu_2=0$), the estimators

$$\hat{\beta} = [I + \frac{cs}{b'Bb}D]^{-1}b$$

dominate the OLS estimator b, for all β and small σ, under the condition

$$0 < c \leq \frac{2\delta_p \mu_1}{n+2}(d-2) \; ; \quad d = \frac{1}{\mu_1} Tr\,(X'X)^{-1}QD > 2 \, , \tag{24}$$

where δ_p is the minimum eigenvalue of $B(D'QD)^{-1}$ and μ_1 is the maximum eigenvalue of $(X'X)^{-1}QD$. The condition (24) follows by substituting $\nu_2 = 0$ in (20) and (22). This condition, for $Q = I$ and $D = (X'X)^{-1}$, is identical with the condition based on exact MSE in Alam and Hawkes [1978]. Further, when $D = Q^{-1}X'X$ and $B = X'X$, it reduces to the exact condition given in Strawderman [1978] and Casella [1980]. Also, the condition (24) is identical with the condition (based on the exact WMSE) of Judge and Bock [1978] for their estimators in (7). This latter finding implies that the dominance condition of the estimators $\hat{\beta}$ over b are in general the same as those of approximate estimators $\hat{\beta}_{JB}$ over b. Further, since the condition (24) is identical with the condition based on the exact WMSE, the estimators $\hat{\beta}$, in fact, dominate b for all β and σ (not only for small σ). Thus, $\hat{\beta}$ is a class of minimax estimators. [See Chapter 2 for the definition of minimaxity.]

The above findings also suggest that Kadane's [1971] small-σ expansion is a useful and very simple method of deriving the conditions for dominance of the estimators $\hat{\beta}$. This technique could prove to be specially useful when one is analyzing ridge and Stein-type estimators in the situations where their exact MSE are extremely difficult, as in the case of the models in Chapters 10 and 11. The fact that Kandane's expansion would provide the condition for dominance which is identical with that based on the exact WMSE has been established in (24) for a general class of estimators in the regression case (also see Ullah [1981]). A set of rigorous mathematical conditions for this equality can perhaps be developed following the work of Berger [1976a] and Casella [1980].

The results for Stein, Lindley and ridge-type estimators $\hat{\beta}_S$ and $\hat{\beta}_{OR}$ in (8) and (8a), respectively, follow from (24) by appropriate substitutions. In general, the dominance condition (24) would depend upon the eigenvalues and therefore in many applications the data set may not satisfy the condition $d > 2$ for which $p \geq 3$ is necessary but not sufficient. The condition $d > 2$, therefore, restricts the application of the estimators $\hat{\beta}$ in practice. However, if we choose B, Q and D such that

$$B(D'QD)^{-1} = I, \text{ and } (X'X)^{-1}QD = I \ , \tag{25}$$

then we can get estimators from (5) whose dominance condition will become

$$0 < c \leq \frac{2}{n+2}(p-2) \ , \quad p > 2 \tag{26}$$

which is simple and independent of the eigenvalues of $X'X$.

To see explicitly the estimators which require (26), we first assume a given Q in the loss function $(\hat{\beta}-\beta)'Q(\hat{\beta}-\beta)$. Then (25) would give $B = X'XQ^{-1}X'X$ and $D = Q^{-1}X'X$. Substituting these values of B and D in (5) we get the following estimators

$$\hat{\beta}_1 = [I + \frac{cs}{b'X'XQ^{-1}X'Xb}Q^{-1}X'X]^{-1}b \ , \tag{27}$$

which dominate the OLS estimator b for $0 < c \leq \frac{2}{n+2}(p-2)$, and $p > 2$. For $Q = I$, e.g., $\hat{\beta}_1$ is Berger's [1976b] estimators.

Similarly, for a given choice of D, (25) gives $Q = X'XD^{-1}$ and $B = D'X'X$. Using this value of B in (5) we get the estimators

$$\hat{\beta}_2 = [I + \frac{cs}{b'D'X'Xb}D]^{-1}b \ , \tag{28}$$

which dominate over b for $0 < c \leq \frac{2}{n+2}(p-2)$ under the loss function $(\hat{\beta}_2-\beta)'X'XD^{-1}(\hat{\beta}_2-\beta)$. For example, if we consider a Stein-type estimator $(D=I)$ then it dominates OLS for $0 < c \leq \frac{2}{n+2}(p-2)$ under the loss function $(\hat{\beta}_2-\beta)'X'X(\hat{\beta}_2-\beta)$. Similarly, for a ridge-type estimator $(D=(X'X)^{-1})$ the loss function would be $(\hat{\beta}_2-\beta)'(X'X)^2(\hat{\beta}_2-\beta)$ for $0 < c \leq \frac{2}{n+2}(p-2)$. In Chapter 8, $(\hat{\beta}_2-\beta)'X'X(\hat{\beta}_2-\beta)$ and $(\hat{\beta}_2-\beta)'(X'X)^2(\hat{\beta}_2-\beta)$ are named as Mahalanobis distance and Strawderman distance, respectively.

In a special case when $Q = (X'X)^m$, where m is any arbitrary number, the estimators $\hat{\beta}_1$ in (27) become

$$\hat{\beta}_3 = [I + \frac{cs}{b'(X'X)^{2-m}b}(X'X)^{1-m}]^{-1}b \tag{29}$$

which dominate over OLS for $0 < c \leq \frac{2}{n+2}(p-2)$ and $p > 2$. This set of estimators could be useful in practice. For $m = 1$ and $m = 2$ it becomes James and Stein [1961] and Hoerl, Kennard and Baldwin's [1975] estimators, respectively, which are discussed in Chapters 7 and 8.

Remark 2 Comparing (20) and (22) with (24), it is clear that the conditions for dominance of $\hat{\beta}$ over b for symmetrical leptokurtic and platykurtic distributions are different from those for a normal distribution of errors. The conditions $d > 2 + \Delta_1$ and $d > 2 + \Delta_2$ for the nonnormal symmetric distributions appear to be stronger (harder to satisfy) than the condition $d > 2$ for the normal case.

When we consider the estimators $\hat{\beta}_1$, $\hat{\beta}_2$ and $\hat{\beta}_3$ in (27), (28) and (29), respectively, the dominance conditions (20) and (22) reduce to

$$0 < c \leq \frac{2(n+\nu_2)}{q}[(1-\theta\bar{\eta})p - 2(1-\theta\eta_p)] \; ; \quad p > 2 + \Delta_1 \tag{30}$$

for $\nu_2 > 0$, and for $\nu_2 < 0$

$$0 < c \leq \frac{2(n+\nu_2)}{q}[(1-\theta\bar{\eta})p - 2(1-\theta\eta_1)] \; ; \quad p > 2 + \Delta_2 \, , \tag{31}$$

where $\bar{\eta} = \frac{TrN}{p} = \sum_1^p \eta_i/p = \phi$, and $\eta_1 \geq \ldots \geq \eta_p$ are the eigenvalues of N. Notice that even for these estimators, the dominance conditions are quite different compared to the normal case $(\nu_2=0)$ in which we require $0 < c \leq \frac{2}{n+2}(p-2)$ and $p > 2$.

Remark 3 If we use MtxMSE expression in (79) and compare it with the MtxMSE of the OLS, then it will not be possible to get conditions on c under which the estimator $\hat{\beta}$ will be better than the OLS for all β (under normal, as well as, nonnormal errors).

Results for Skewed Distributions ($\nu_1 \neq 0$)

Let us write the expression (17) as

$$\text{WMSE}(\hat{\beta}) - \text{WMSE}(b) = -\sigma^3 r_3 - \sigma^4 r_4 ,$$

where r_4 is as given in (16) and r_3 from (15) is

$$r_3 = 2c\nu_1 \frac{\beta' D' Q (X'X)^{-1} (I * M)\iota}{\beta' B \beta} = 2c\nu_1 \frac{\beta' D' Q \alpha}{\beta' B \beta} ; \quad \alpha = cov(b, \bar{s})$$

where $\bar{s} = s/n = (y - Xb)'(y - Xb)/n$ is the error variance estimator and $n\alpha = n\ cov(b, \bar{s}) = (X'X)^{-1} X' E[u'Mu, u] = (X'X)^{-1} X'(I * M)\iota$ is a $p \times 1$ vector of covariances between b and $\bar{s}$. We note that if all the elements of $\alpha = cov(b, \bar{s})$ and β are of the same (opposite) sign, then r_3 is positive (negative), otherwise the sign of r_3 is not clear.

Recall that r_4 is positive under the conditions (20) when $\nu_2 < 0$, (22) when $\nu_2 > 0$, and (24) when $\nu_2 = 0$. Thus, under the ranges of c implied by these conditions we know that $\hat{\beta}$ dominates b for the skewed distribution of errors, in the sense that

$$\text{WMSE}(\hat{\beta}) - \text{WMSE}(b) \leq 0 ,$$

when β and $\alpha = cov(b, \bar{s})$ are of the same signs.

13.4 A REVIEW OF MINIMAXITY RESULTS UNDER NORMAL AND NONNORMAL ERRORS

In Chapter 6, it was noted that Stein [1955] was the first researcher to show that, in higher dimensional problems ($p \geq 3$), the sample mean of a multivariate normal distribution is inadmissible against MSE (expected squared error loss). This result was further extended and analyzed in the regression context by James and Stein [1961] and Brown [1966], and they showed the inadmissibility of the OLS estimator b for $p \geq 3$ regressors. Because of this deficiency several authors developed the estimators, called Stein-type improved estimators [see (8)] discussed earlier in Chapter 6, whose WMSE dominates that of the OLS estimator b. Since the OLS estimator is minimax with constant $\sigma^{-2}\,\text{WMSE}(b) = TrQ(X'X)^{-1}$, the Stein-type estimators in (8), which dominate OLS in the entire parameter space of β, were also called a minimax set of estimators. The conditions for dominance are given in (20) and (22) for the nonnormal case, and in (24) for the normal case (for $D = I$).

A second disadvantage in the OLS estimator was noted by Hoerl and Kennard. They pointed out that if the design (data) matrix $X'X$ is "nearly singular," the OLS estimator b will be "unstable" in the sense that small changes in the observations might produce large changes in b. To correct the ill-conditioning of $X'X$ due to near singularity, they proposed the ridge estimator $(X'X+kI)^{-1}X'y$ where k is a positive (nonstochastic) constant. To see that the ridge estimator is more stable than the OLS, recall the definition of "condition number" $K^{\#} = \lambda_1^{1/2}\lambda_p^{-1/2}$ or $(K^{\#})^2 = \lambda_1/\lambda_p$ of a matrix from Chapter 5, equation (4); λ_1 and λ_p are the maximum and minimum, eigenvalues of $X'X$ matrix. Large values of $K^{\#}$ imply that the matrix is ill-conditioned. Since $(\lambda_1+k)/(\lambda_p+k) < \lambda_1/\lambda_p$ for $k > 0$, the ridge estimator reduces the ill-conditioning of the design matrix $X'X$ (for further details see Chapter 7). Because of this advantage ridge estimator has become quite popular with the applied researchers. However, due to the problem of the unknown k, some research has gone in the direction of considering ridge estimators where $k = a(y)$ is data dependent (stochastic) similar to (6). This has produced a large number of operational ridge-type estimators given in (8a), some of which have been discussed earlier in the Chapter 8. These estimators also dominate the OLS estimator under certain conditions on "a" which are called minimaxity conditions. These conditions are given in (20) and (22) for the nonnormal case and in (24) for the normal case (for $D=(X'X)^{-1}$).

Both Stein-type and operational ridge estimators shrink the OLS estimator toward zero. The arbitrariness of shrinking toward zero was first criticized by Lindley [1962], who suggested shrinking toward the overall mean while dealing with Stein's estimator for the multivariate normal mean. Zellner and Vandaele [1975] introduced Lindley-type mean correction in a Stein-type estimator, and Srivastava-Ullah [1980] have studied its sampling properties. A general form of Lindley-type estimators is given in (9). The application of Lindley-type estimator in the polynomial distributed lag model has been discussed in Chapter 9.

With respect to these improved estimators, we now look into two questions: (i) what are the situations in which one would expect to gain by using Stein-type, ridge-type and Lindley-type improved estimators? (ii) are these improved estimators more stable compared to the OLS estimator?

13.4.1 Minimaxity of Stein-type Estimators (D=I)

Consider a special case $B = X'X$ and write from (8) $\hat{\beta}_S = \left[1 - \dfrac{cs}{b'X'Xb+cs}\right]b$.
Then for $Q=I$ and under the normality of errors, this estimator dominates OLS when $0 < c \le \dfrac{2}{n+2}(d-2)$; $d = \lambda_p \sum_1^p \lambda_i^{-1} > 2$ from (24) and $\lambda_1 \ge \lambda_2 \ldots \ge \lambda_p$ are the eigenvalues. For $Q = X'X$ we will get $0 < c \le \dfrac{2}{n+2}(p-2)$; $p > 2$. Thus, for the loss function with $Q = I$ the range of c depends crucially on the eigenvalue spectrum. If there is severe multicollinearity λ_p will be close to zero, and $d > 2$ may not be satisfied. This implies that for $Q=I$, the above estimator will be useful only when there is a moderate multicollinearity such that $d > 2$ is satisfied. When $Q = X'X$, the estimator will be better than the

OLS so long as there are at least three regressors. The same results hold for the James and Stein estimator $\hat{\beta}_{JS} = \left(1 - \frac{cs}{b'X'Xb}\right)b$. Further, we note from (20) and (22) that in the nonnormal case, the improvement over OLS depends not only on the eigenvalues but also on the shape of the distribution of errors.

To see the stability of Stein-type estimators, let us write from (8), $\hat{\beta}_S = (\delta X'X)^{-1}X'y$ where $\delta = (b'Bb+cs)/b'Bb$. Then it is clear that the "condition number" $K^{\#2} = \delta\lambda_1/\delta\lambda_p = \lambda_1/\lambda_p$ which is identical with the condition number of the design matrix $X'X$ in the OLS case. Thus, though a Stein-type estimator dominates the OLS estimator under a WMSE criterion, it does not improve the stability of the OLS estimates.

13.4.2 Minimaxity of Ridge-type Estimators ($D=(X'X)^{-1}$) with Stochastic k

Consider $B = X'X$ in (8a) and $Q = I$ in the WMSE. Then the operational ordinary ridge estimator is

$$\hat{\beta}_{OR} = [I + \frac{cs}{b'X'Xb}(X'X)^{-1}]^{-1}b = [X'X + \frac{cs}{b'X'Xb}I]^{-1}X'y , \qquad (32)$$

which is Lawless and Wang's [1976] estimator discussed in Chapter 8. This estimator dominates the OLS for

$$0 < c \leq \frac{2\lambda_p}{n+2}(d-2);\ d = \lambda_p^2 \sum_1^p \lambda_i^{-2} > 2, \qquad (33)$$

and when the disturbances are normal. Again, the condition $\lambda_p^2 \sum_1^p \lambda_i^{-1} > 2$ restricts the application of the above estimator under severe multicollinearity. Notice that this condition is even stronger than $\lambda_p \sum_1^p \lambda_i^{-1} > 2$ required in Stein's case. When $Q = (X'X)^2$, we get the condition $0 < c \leq \frac{2\lambda_p}{n+2}(p-2)$; $p > 2$. This case is more interesting than $Q = I$, since $p > 2$ simply means that we should have at least three regressors. However, under severe multicollinearity, the range of c would still be too small to yield a substantial gain over the OLS estimator because of the presence of λ_p. Similar observations can be made about the Hoerl, Kennard and Baldwin's [1975] estimator ($B=I$) when $Q = I$. This estimator, when $Q = (X'X)^2$ dominates OLS for $0 < c \leq \frac{2}{n+2}(p-2)$. Note that this range is not affected by multicollinearity. Notice again that, under nonnormality, the dominance of these estimators will depend upon the specific distribution of errors, as well as, the eigenvalues of $X'X$ for any choice of Q.

Regarding the stability of Lawless and Wang's [1976] estimator, we may check the condition number of its design matrix $X'X + a(y)I$ where $a(y) = cs/b'X'Xb$. It is $K^{\#2} = [\lambda_1 + a(y)]/ [\lambda_p + a(y)]$, which is less than the condition number λ_1/λ_p of the OLS design matrix $X'X$. Thus, Lawless and Wang's estimator is more stable than the OLS estimator. The same is true for Hoerl, Kennard and Baldwin's estimator.

13.4.3 Minimaxity of Lindley-type Estimators $D=(X'X)^{-1}[I - R(R'R)^{-1}R']$

Consider $\Omega_0 = I$ in (9) and $Q = I$. Then we get a Lindley-type estimator

$$\hat{\beta}_L = (X'X + \frac{cs}{b'Jb}J)^{-1}X'y \ ; \quad J = I - R(R'R)^{-1}R' , \tag{34}$$

whose dominance condition over the OLS has been analyzed in Chapter 9. As in the case of ridge-type estimators, the dominance condition of $\hat{\beta}_L$ would depend upon the eigenvalues of $X'X$ matrix when $Q = I$, but not when $Q = (X'X)^2$. In the latter case the condition is

$$0 < c \leq \frac{2}{n+2}(p-r-2); \ p - r > 2. \tag{35}$$

When the columns in the matrix R are just one, i.e., $r = 1$, then it becomes $0 < c \leq \frac{2}{n+2}(p-3); \ p > 3$. In this case it is quite similar to the Hoerl, Kennard and Baldwin's ridge estimator's dominance condition $0 < c \leq \frac{2}{(n+2)}(p-2); \ p > 2$. In general, it appears that a Lindley-type estimator will have a (narrower) conservative range of dominance; $p - 3 < p - 2$.

13.4.4 Minimaxity of Other Improved Estimators

Let us consider the estimators

$$\hat{\beta}_1 = [I + a(y)Q^{-1}X'X]^{-1}b = [X'X + a(y)X'XQ^{-1}X'X]^{-1}X'y \ ; \tag{36}$$

$$a(y) = \frac{cs}{b'X'XQ^{-1}X'Xb} \tag{37}$$

as given in (27). As shown there, this set of estimators dominates the OLS estimator for any given Q when

$$0 < c \leq \frac{2}{n+2}(p-2) \text{ where } p > 2. \tag{38}$$

This condition for dominance (minimaxity) does not depend on the eigenvalues or multicollinearity. For nonnormal errors, the appropriate conditions on c are given in (30) and (31).

Let us now look into the stability properties of $\hat{\beta}_1$ for a few choices of Q.

First, when $Q = I$, $K^{\#2} = (\lambda_1 + a(y)\lambda_1^2)/(\lambda_p + a(y)\lambda_p^2) > \lambda_1/\lambda_p$ which implies that $\hat{\beta}_1$ is less stable than the OLS for $Q = I$. When $Q = X'X$, $K^{\#2} = (\lambda_1 + a(y)\lambda_1)/(\lambda_p + a(y)\lambda_p) = \lambda_1/\lambda_p$, which shows that $\hat{\beta}_1$ for $Q = X'X$ is as stable as b. Finally, when $Q = (X'X)^2$, $K^{\#2} = (\lambda_1 + a(y))/(\lambda_p + a(y)) < \lambda_1/\lambda_p$, which implies an improvement in stability over the OLS. One can verify that the choices of $Q = (X'X)^3$, $(X'X)^4$, etc. will make $\hat{\beta}_1$ less stable because the condition numbers may increase. Thus, the choice of $Q = (X'X)^2$ improves the OLS estimator both under the WMSE criterion and the stability criterion. Notice that $Q = (X'X)^2$ in (36) may indicate a preference for the use of Hoerl, Kennard and Baldwin-type (HKB) estimators (among those having stochastic k) in practice. In general, the estimator $\hat{\beta}$ in (5) with $D = (X'X)^{-1}$, $B = X'X$ and $Q = (X'X)^2$ will dominate the OLS, and will also be stable. But in this case the dominance condition will involve the eigenvalues of $X'X$ even in the normal case.

13.4.5 Limitations of the Minimaxity Criterion

We have already mentioned the limitations of the minimaxity criterion in Section 2.8.2. It is clearly always safe to recommend the use of a minimax alternative to OLS for anyone who may not be experienced in biased regression methods. Judicious use of biased estimators in multicollinear cases where minimaxity is absent can be justified on a case by case basis. In many cases, the use of nonstochastic biasing parameters (shrinkage factors) giving admissible estimators is also justified. One has to be careful in avoiding the regions of parameter space where ridge-type methods are known to perform poorly. This is why we considered "advisability of declining deltas" and "avoiding excess shrinkage" in Tables 7.1 and 7.2 respectively.

13.5 ROBUST LOCATION ESTIMATORS FOR NONNORMAL MEAN

For nonnormal populations, the sample mean as an estimate of (the location of) the population mean is known to be sometimes disastrous. For example, the average income in a small neighborhood containing one multimillionaire may be too large to be representative of the rest of the group. This problem particularly happens when there are outliers (see Section 2.6.6).

13.5.1 Trimmed Mean

The median or the α-trimmed mean are familiar "robust" estimators of the population mean. The α-trimmed mean (for $0 < \alpha < ½$) is computed as follows for $T = 100$ and $\alpha = .05$. First step is to rank-order all observations $y_{(1)} \leq y_{(2)} \dots \leq y_{(T)}$. Second step is to delete $\alpha T (=5)$ smallest and largest observations, and finally compute the mean of the remaining (90) observations in the center

$$(y_{(6)} + \ldots + y_{(95)})/90. \tag{39}$$

Note that when $\alpha \rightarrow ½$, we have the median as a limiting case of the trimmed mean.

13.5.2 Winsorized Mean

The term *Winsorized Mean* refers to the case where the outlying observations are explicitly "brought in". In the above example where $T = 100$, $\alpha = 0.05$, we ignore the actual values of $y_{(1)}$ to $y_{(5)}$, and pretend that they are all equal to $y_{(6)}$, and further that $y_{(96)}$ to $y_{(100)}$ are made simply equal to $y_{(95)}$ The Winsorized Mean is computed in this special case as

$$(5y_{(6)} + y_{(7)} + \ldots + y_{(94)} + 5y_{(95)})/100 \tag{40}$$

Clearly, the trimmed mean may be thought to be defined by a weighted sum

$$\sum_{t=1}^{T} \omega_t y_{(t)} / \sum \omega_t \tag{41}$$

with appropriately chosen weights. For our example, $\omega_1 = \omega_2 \ldots \omega_5 = 0$, and $\omega_{96} = \omega_{97} \ldots = \omega_{100} = 0$ and all other $\omega_t = 1$. For winsorized mean some weights are stochastic. For example, $\omega_{98} = y_{(95)}/y_{(98)}$.

13.5.3 M-Estimates (Huber's Maximum Likelihood)

Huber [1964] considers a maximum likelihood estimator of the location parameter for a parent distribution that is similar to the normal in the middle but like a "double exponential" in the tails. The density of a double exponential is given by $f(y) = exp[-|y| + \text{constant}]$. The log of this (combined) likelihood function in the middle portion is proportional to $-y^2/2$ from the standard normal case. Now Huber's [1964] (minus log likelihood) "rho function" is defined as

$$\rho(y) = y^2/2\ , \quad \text{if } |y| \le K_H\ ,$$

$$= K_H |y| - (K_H^2\, 2^{-1})\ , \quad \text{if } |y| > K_H\ , \tag{42}$$

where K_H denotes a constant (upper bound), not to be confused with similar notation from ridge regression.

From this function, Huber derives his maximum likelihood estimate (*M*-Estimate) by differentiating $\rho(y)$ with respect to y. This yields the derivative function

$$\psi(y) = \rho'(y) = y \ , \quad \text{if } |y| \leq K_H \ ,$$

$$= K_H \ sign(y) \ , \quad \text{if } |y| > K_H \ , \tag{43}$$

where *sign* (y) denotes the sign of y, and $K_H = 1.345$ (say) for reasons described below after equation (50). Note that if f is the density function $\psi = -f'/f$ by definition. Now the M-estimator $\hat{\mu}_M$ is defined by the derivative of minus log likelihood function

$$\sum_{t=1}^{T} \psi \left(\frac{y_t - \hat{\mu}_M}{\hat{\sigma}} \right) = 0 \tag{44}$$

where $\hat{\sigma}$ is introduced to obtain scale equivariance with $\hat{\sigma}$ calculated from the data. A robust estimate of the scale parameter is defined in terms of MADM (median of absolute deviations from the median):

$$\hat{\sigma}(Robust) = median \, |y_t - median\,(y_t)|/0.6745. \tag{45}$$

The choice 0.6745 ensures that $\hat{\sigma}(Robust) = \sigma$, (the standard deviation) asymptotically.

Other members of the family of M-Estimators include Andrews' SINE estimate which uses

$$\psi(y) = sin\,(y/K_A) \ , \quad \text{for } |y| < K_A$$

$$= 0 \ , \quad \text{for } |y| \geq K_A \ , \tag{46}$$

where K_A is a constant. One choice of $K_A = 1.339$ has desirable properties explained after equation (50) below. This amounts to a "re-descending" ψ curve which comes down to zero for extreme observations. In other words, extreme "outliers" are thrown out.

Tukey's Bisquare is defined by

$$\psi(y) = \frac{y}{K_B} \left[1 - \left(\frac{y}{K_B}\right)^2 \right]^2 \tag{47}$$

which involves squaring twice. It too has a "re-descending" ψ curve. Graphically, it looks similar to the SINE curve in the relevant region. The choice $K_B = 4.65$ for the constant in (47) can be shown to have desirable properties noted later after equation (50). Both SINE and Bisquare are discussed in Princeton robustness study, Andrews et al. [1972].

13.5.4 Linear Combination, L-Estimate

This involves a linear combination of so-called order statistics based on an ordered sample of $y_{(1)}, y_{(2)}, \ldots y_{(T)}$. Trimmed mean is an example of this. Median is another example. Tukey's *Trimean* is a weighted average of the quantiles at 25%, 50% and 75% with weights .25, .50 and .25:

$$Trimean = .25y_{(25)} + .5\, y_{(50)} + .25y_{(75)}\,, \tag{48}$$

for the special case when $T = 100$.

13.5.5 Distribution-free Rank, R-Estimate

This involves ranking the observations and then working with the rank numbers 1, 2 ,..., 100 themselves rather than actual values of y associated with them. These are sometimes called Wilcoxon scores. There are other scoring schemes, Jaeckel [1972]. Jureckova [1977] has proved that there are scoring schemes which, in combination with specific ψ functions can yield R-estimates which are equivalent to the simpler M-estimates.

In the Princeton robustness study of Andrews et al. [1972, pp. 97-98] certain "stylized" sensitivity curves (theoretical influence functions, ψ) are reported. Our Figures 13.1 to 13.6 give these curves for the Mean, Median, 10% trimmed mean, Tukey's Trimean, Huber's M-estimate with $K_H = 1.5$, and Andrews' SINE estimate, respectively. To plot these curves, we can start with 19 average normal order statistics, and a 20th number y added. The effect of the value of y on the overall mean for all 20 observations is shown by plotting y on the horizontal axis and the mean on the vertical axis in Figure 13.1. Similar computations are given in the other figures. Essentially, they show that the mean is sensitive to extreme observations and others are not. A distinguishing feature of Andrews' SINE estimate is the "*redescending*" weight given to the extreme observations, which amounts to rejecting them from the data. The plot for Tukey's Bisquare (not drawn here) is similar to Andrews' SINE estimate.

13.5.6 Efficiency Comparisons

Let $\bar{y}*$ denote any one of the location estimators from Sections 13.5.1 to 13.5.5. Recall that the population parameter is μ. We assume that observations $y_1, \ldots y_T$ are independent, and let their density function be denoted by f. If f is normal, the likelihood function is proportional to $\exp-\sum_{t=1}^{T}(y_t-\mu)^2/2\sigma^2$. The first derivative of log likelihood with respect to μ is $\sum_{t=1}^{T}(y_t-\mu)/\sigma^2$. Its square is $\sigma^{-4}\sum_{t=1}^{T}(y_t-\mu)^2 + \sigma^{-4}\sum_{t\neq j}(y_t-\mu)(y_j-\mu)$. Now a measure of "information" regarding μ contained in the random sample $y_1,\ldots,y_T$ is defined by Sir R. A. Fisher as the expected value of the square of the first derivative of the log likelihood function, Rao [1973, p. 325]. In (42) above we have denoted the

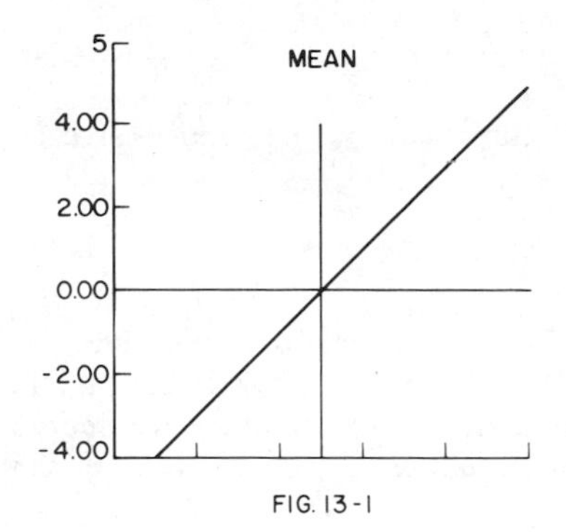

FIG. 13-1

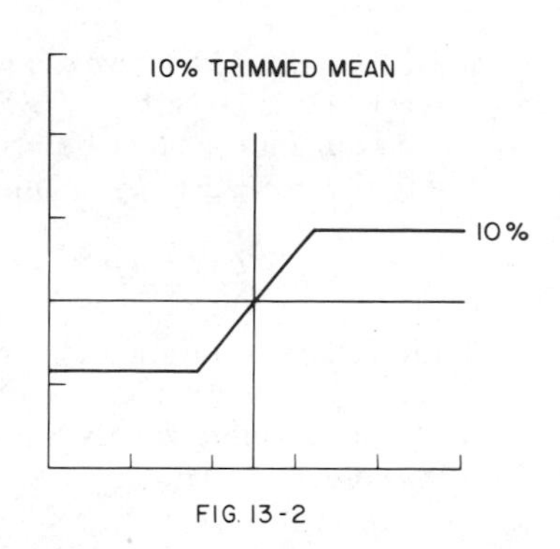

FIG. 13-2

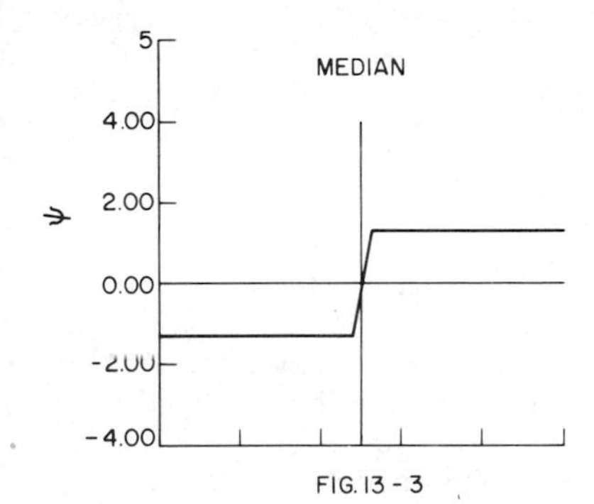

FIG. 13-3

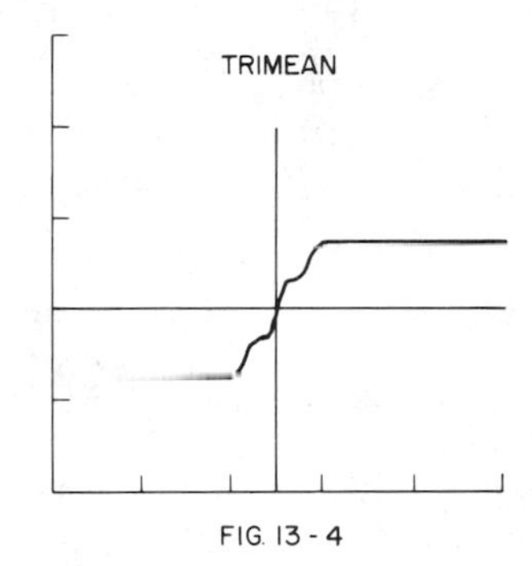

FIG. 13-4

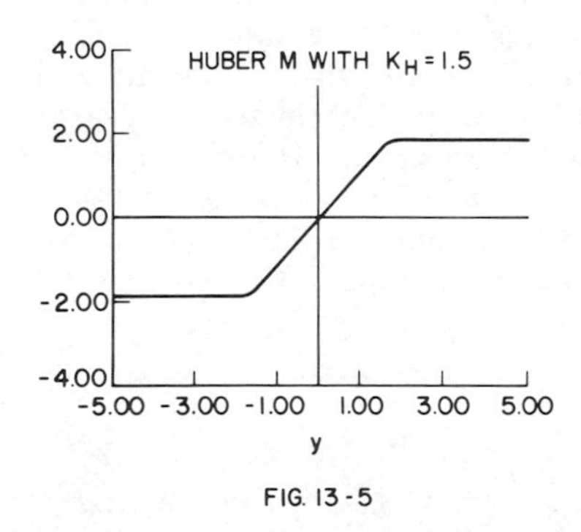

FIG. 13-5

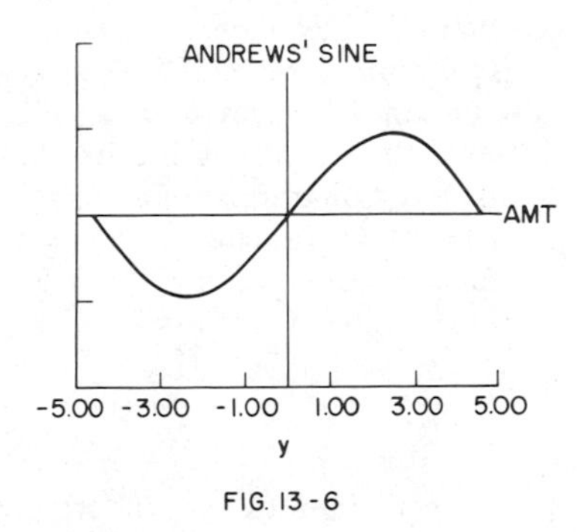

FIG. 13-6

Influence Functions

minus log likelihood by ρ and its derivative by ψ in (43). Thus, Fisherian information is given by $E\psi^2$.

For T independent observations the expectation of $(y_t-\mu)(y_j-\mu)$ for each $t \neq j$ is zero. Also, by definition

$$E(y_t-\mu)^2 = \sigma^2 \text{ implies that } E\sum_{t=1}^{T}(y_t-\mu)^2 = T\sigma^2 .$$

Thus Fisherian information in this sample from a normal parent is given by $T\sigma^2/\sigma^4 = T/\sigma^2$, which is simply reciprocal of the variance. If $\bar{y}*$ is the sample mean, its asymptotic variance is σ^2/T. In general, for lth percentile we denote $p = l/100$, $q = 1-p$, $y_{(l)} =$ the estimated percentile, $f_l =$ ordinate of the parent distribution at $y_{(l)}$. Now

$$var(y_{(l)}) = pq/(nf_{(l)}^2) . \tag{49}$$

In particular, $f_{(l)}$ for $l = 50$ (the median) is

$$\sigma/(2\pi)^{1/2} = 0.398942\sigma , \quad p = q = 1/2 \tag{49a}$$

$$Var(y_{(50)}) = (\pi/2)(\sigma^2/T)$$

$$= (1.5707)(\sigma^2/T).$$

For the normal parent distribution, the variance of median is over 57% larger than that of the mean. Thus, median is inefficient in terms of relative variance (1.5707) or relative efficiency (63.7%). For nonnormal populations, the situation changes. For a Cauchy parent, the mean is unstable (fails to satisfy the central limit theorem) and the variance is infinite, but the median is known to be a good estimate of the location. For other forms of nonnormality one may encounter in practice, (e.g. contaminated normals) the efficiency of the median seems to be too low. Thus, there is a need to consider other "robust" alternatives which are about 95% efficient in the normal case, and better than the mean in nonnormal cases.

Huber [1964] considered estimators between mean and median that are asymptotically *most robust* (minimax) in the sense that the supremum of the asymptotic variance ($T\to\infty$) is minimized when the underlying cumulative distribution F ranges over a suitable set of contaminated normal (say) distributions. If the minus log likelihood function (ρ) is convex, Huber proved that his M-estimators are "most robust" if F belongs to a convex set of cumulative distribution functions. Since the derivative of minus log likelihood function (for a robust estimator $\bar{y}*$) is denoted by ψ, the asympotic variance given by Huber in terms of ψ is

$$V(\psi,F) = \frac{E\psi^2}{(E\psi')} \geq \frac{1}{\int (f'/f)fdy} , \tag{50}$$

where f denotes the underlying density, and $E\psi^2$ is Fisherian information for the M-estimator. It is interesting to note that this variance is bounded below by σ^2 in case of the normal distribution. The denominator integral is Fisherian information associated with the maximum likelihood estimator.

We conclude this section by noting that estimates in the family of M-estimates are in some sense "most robust" asympotically normal, under suitable conditions, and have explicit expressions for asymptotic variance. These expressions can be used to determine the constants $K_H = 1.345$ for the Huber M-estimate $K_A = 1.339$ for Andrews' SINE estimate and $K_B = 4.685$ for Tukey's Bisquare estimate, such that the asymptotic efficiency of these estimators compared to the arithmetic mean is 95% in the normal case. If the departure from normality involves bimodal distributions, M-estimates with redescending weights are not recommended.

13.6 ROBUST REGRESSION FOR NONNORMAL ERRORS

If the problem of outliers is important in the location problem, it is much more serious in the multiple regression problem. Detection of outliers is much more difficult in a regression setup because they tend to be masked by the presence of many regressors. Graphical exploratory methods are recommended here. A detailed discussion of this is given by Mosteller and Tukey [1977].

The basic idea is to reduce the "influence" of extreme observations on OLS by using a regression adaptation of the various nonnormal "location" estimators discussed in the previous section. One sacrifices 5% (say) of the efficiency in the normal case in return for considerably improved efficiency in the presence of fat-tailed nonnormal errors.

13.6.1 Diagonally Weighted Least Squares

In the regression model $y = X\beta + u$, we have $E(u) = 0$. If each observation is weighted by ω_t, t=1, ...,T the weighted least squares estimator is given by applying OLS to $W^{1/2}y = W^{1/2}X\beta + W^{1/2}u$ to be

$$b_\omega = (X'WX)^{-1}X'Wy \ , \tag{51}$$

where $W = \text{diag}\ (\omega_t)$. For OLS, we have $W = I$, $b_\omega = b$.

Let the residuals for a weighted least squares estimator be denoted by

$$\hat{u}_\omega = y - Xb_\omega \ . \tag{52}$$

When the "outliers" are present, these residuals can be "large" for the outliers. The purpose of robust regression in simple terms is to trim these large residuals, which amounts to down-weighting the corresponding observations with a smaller ω_t.

M-Estimation based on Huber's family in the regression context amounts to finding estimates of β which will minimize ρ, the minus log likelihood function for the contaminated (nonnormal) parent defined as a function of the residual in (52). For example, we minimize

$$\sum_{t=1}^{T} \rho[y_t-(Xb_{\omega})_t] , \tag{53}$$

with ρ defined in (42) in the context of the location problem, and $(Xb_{\omega})_t$ denoting the t-th element of Xb_{ω}. Let x_{ti} denote (t,i)th element of X, with $t = 1,...,T$, and $i = 1,...,p$. The M-estimates of regression may be defined as a solution to a system of p equations, where the ith equation is given by

$$\sum_{t=1}^{T} x_{ti}\psi(y_t-(Xb_{\omega})_t/S_{\omega}) = 0 , \tag{54}$$

where ψ is the derivative of the ρ function, and S_{ω} is a robust estimate of the scale parameter (eg. of the standard deviation, σ). A popular estimate of S_{ω} is $\hat{\sigma}$(Robust), median of absolute deviations from the median divided by 0.6745 give above in (45).

The above set of equations can be rewritten as

$$\sum_{t=1}^{T} x_{ti}\ \omega_t^2\ \hat{u}_{\omega t} = 0 , \tag{55}$$

where

$$\omega_t^2 = \psi(\hat{u}_{\omega t}/S_{\omega})(\hat{u}_{\omega t})^{-1} . \tag{56}$$

These ω_t values can be used to transform y and X data by multiplying each variable by $\sqrt{\omega_t}$ and then applying OLS as in (51).

13.6.2 One-Step M-Estimation

For three members of the family of M-estimators (Huber-M, SINE and Bisquare), we have given the ψ functions in (43), (46) and (47), respectively. These can be used in the regression context by applying them to scaled residuals $(\hat{u}_{\omega t}/S_{\omega})$. The corresponding weights ω_t are found from (56). Thus, we find robust regression estimates for Huber-M, SINE and Bisquare.

13.6.3 LAE or L1 Estimation

The so-called least absolute error (LAE) or ($L1$ norm) estimation is equivalent to the following choice of minus log likelihood function

$$\rho(\hat{u}_{\omega}) = \sum_{t=1}^{T} (½)\ |\hat{u}_{\omega t}| , \tag{57}$$

based on a two-tailed exponential density, Blattberg anc Sargent [1971].

Estimation for $L1$ available in most computer centers is implemented by linear programming methods. An efficient procedure is given by Abdelmalek [1975]. The LAE estimation also yields robust estimates of regression coefficients, which are shown to behave well in Meyer and Glauber [1964], Ruppert and Carroll [1980], etc.

13.6.4 Regression Quantiles

A generalization of the quantiles (e.g. median) to the regression problem is suggested by Koenker and Bassett [1978]. For example, if $T = 500$ and $p = 5$, each percentile for the regression problem will be based on 5 observations. Using linear programming methods, one can generate (one hundred percentile regression lines each of which is based on five OLS estimators) the quantile regressions. The median regression (the middle of these 100 lines defined in terms of an ordering of the percentile lines) can be shown to have an efficiency given by a generalization of (49a). For examples having severe nonnormality, these regression quantiles offer an interesting robust regression method. Rather than the median regression, it is often more efficient to consider linear combinations of quantile regressions similar to the trimmed mean or the Trimean of equation (48).

13.7 ITERATIVELY REWEIGHTED LEAST SQUARES

Recently, Coleman et al. [1980] have provided a convenient computer algorithm for robust regression based on a suitable starting solution (say, OLS or $L1$) and $\hat{\sigma}$(Robust) of (45) based on the median of absolute deviations from the median. One-step M-estimation in Section 13.6.1 can be implemented from these starting residuals. It can be iterated further by re-computing the residuals and $\hat{\sigma}$(Robust) each time. Convergence properties are discussed by Coleman, et. al. [1980]. Although consensus is difficult in this field, it appears that a good research strategy is to compute converging Huber-M estimates by iteratively reweighted least squares with $K_H = 1.345$ in (43). After this solution is found, (two) iterations of Bisquare based on (47) with $K_B = 4.65$ are desirable.

Experience indicates that "redescending" weights as in SINE or Bisquare at an early stage of the iterative process may end-up with throwing out useful information. This is why our strategy calls for only one or two iterations of Bisquare after converging Huber-M regression is estimated. A combination of ridge and robust methods can be done by applying ridge regression to the transformed data, Holland [1973].

13.8 NONNORMAL ERRORS IN MULTIVARIATE MODELS

In the simultaneous equations models used mainly in econometrics literature, the assumption of normality of errors is common and convenient. Fair [1974] has considered "robust" estimation of econometric models with certain success. The conditions under which LAE estimators for econometric applications are unbiased and consistent are given by L. D. Taylor [1974].

The multivariate methods discussed in Chapter 12 including canonical correlations, discriminant analyses, etc. often involve reduction of dimensionality of multivariate data and a study of their interdependencies. An excellent discussion of data analytic, graphical and other techniques for multivariate problems without assuming normality of the variables is given by Gnanadesikan [1977]. A most interesting new tool for graphical methods involves scatter plots with influence function contours shown for Fisher's Iris setosa data in Devlin et al. [1975]. In this paper it is also shown that the usual correlation coefficient has unbounded "influence function," and hence, it is nonrobust and highly sensitive to nonnormality. Various robust estimates of correlation coefficient include Kendall's rank correlation coefficient, bivariate trimming, bivariate winsorizing, trimming by principal components etc. In their simulation, Devlin et al. suggest that none of these robust techniques dominates others. Using robust estimates of the correlation coefficients leads to a serious problem with ensuring that the resulting matrices are positive definite as they should be. Robust discriminant analysis is discussed by Lachenbruch and Goldstein [1979].

When nonnormality is a serious problem, the current statistical literature provides interesting improvements over OLS. Although theoretical purists may find the research strategy described earlier to contain some apparently arbitrary choices at various stages, an interesting question is whether or not they work well in practice. In the context of ridge-type improvements over OLS, we have described many minimax type results summarized in Section 13.4. The asymptotic minimaxity of Huber's M-estimates is noted in Section 13.5.6 based on equation (50). It appears that more research is needed. We doubt whether this research will produce mechanical tools. We will have to continue to rely on skilled use of available techniques, with adequate awareness of the properties of various estimators.

EXPLANATORY NOTES

Explanatory Note 13.1

Derivation of Results for $\hat{\beta}$

Let us rewrite the model (1) as

$$y = X\beta + \sigma v, \quad (u=\sigma v) \tag{58}$$

so that as σ approaches 0, the disturbance term tends to be small.

Thus, from (2) we have for t and t^* $(t,t^*=1,2,...,T)$

$$E(v_t)=0\,,\quad E(v_t v_{t^*})=1,\quad \text{if } t=t^*$$
$$=0,\quad \text{if } t\neq t^*$$

$$E(v_t^3)=\nu_1\,,\quad E(v_t^4)=\nu_2+3. \tag{59}$$

Now we have the following lemma:

Lemma If A is any nonstochastic matrix of order $T\times T$, then

$$E(v'Av)=TrA, \tag{60}$$

$$E(v'Av.v)=\nu_1(I*A)\iota, \tag{61}$$

$$E(v'Av.\,v')=\nu_2(I*A)+(TrA)I+A+A', \tag{62}$$

where ι is a $T\times 1$ vector with all elements unity and "$*$" represents the Hadamard product as defined earlier after equation (10).

Proof. Owing to independent and identical distribution of the elements of v, we observe that

$$E(v'Av)=E[Tr(Av')]=Tr[AEv']=TrA \tag{63}$$

which gives (60).

Next, the t^*-th element of $E(v'Av.v)$ is given by

$$\sum_{t_1,t_2}^{T} a_{t_1,t_2}E(v_{t_1}v_{t_2}v_{t^*})=a_{t^*t^*}\nu_1\,, \tag{64}$$

because the expectation term is nonzero only when $t_1=t_2=t^*$. This leads us to (61).

Similarly, the $t^*,t^{**}-th$ element of $E(v'Av.vv')$ is

$$\sum_{t_1,t_2}^{T} a_{t_1t_2}E(v_{t_1}v_{t_2}v_{t^*}v_{t^{**}})\,. \tag{65}$$

When $t^*=t^{**}$, it is easy to verify that (65) is equal to

$$TrA+(\nu_2+2)a_{t^*t^*}\,, \tag{66}$$

and when $t^*\neq t^{**}$, it is equal to

$$a_{t^*t^{**}}+a_{t^{**}t^*}\,. \tag{67}$$

Combining (66) and (67), we obtain (62) from (65).

Now for the derivation of results in (13) and (14) we first substitute (58) and (6) and note that $a=cs/b'Bb=c\sigma^2v'Mv/b'Bb$ is at least of order σ^2. Thus from (5), for sufficiently small σ in Kadane's sense, we can write

$$\hat{\beta} = (I+aD)^{-1}b = (I-aD)b + \ldots \tag{68}$$

or

$\hat{\beta} - \beta = \sigma(X'X)^{-1}X'v - T^*$, where T^* is given by

$$\frac{\sigma^2 cv'Mv \cdot D}{\beta'B\beta + 2\sigma\beta'B(X'X)^{-1}X'v + \sigma^2 v'X(X'X)^{-1}B(X'X)^{-1}X'v}\{\beta+\sigma(X'X)^{-1}X'v\} + \ldots$$

$$= \sigma^2 \frac{cv'Mv}{\beta'B\beta}\left[1 + 2\sigma \frac{\beta'B(X'X)^{-1}X_2'v}{\beta'B\beta}\right.$$

$$\left. + \sigma^2 \frac{v'X(X'X)^{-1}B(X'X)^{-1}X'v}{\beta'B\beta}\right]^{-1} D\{\beta+\sigma(X'X)^{-1}X'v\} + \ldots$$

where we have used $b - \beta = \sigma(X'X)^{-1}X'v$, and $M = I - X(X'X)^{-1}X'$.

Expanding the expression in square brackets and retaining terms to order σ^3, we find

$$\hat{\beta} - \beta = \sigma\zeta_1 + \sigma^2\zeta_2 + \sigma^3\zeta_3 , \tag{69}$$

where

$$\zeta_1 = (X'X)^{-1}X'v , \quad \zeta_2 = -\frac{cv'Mv}{\beta'B\beta}D\beta \tag{70}$$

and

$$\zeta_3 = -\frac{cv'Mv}{\beta'B\beta}\left[D(X'X)^{-1}X'v - 2\frac{\beta'B(X'X)^{-1}X'v}{\beta'B\beta}D\beta\right]. \tag{71}$$

Thus we have, to order σ^2

$$E(\hat{\beta}-\beta) = -\frac{c\sigma^2 n}{\beta'B\beta}D\beta , \tag{72}$$

where we use $Ev = 0$ and $Ev'Mv = TrM = n = T-p$. Further, upto order σ^4 we have

$$E(\hat{\beta}-\beta)'Q(\hat{\beta}-\beta) = \sigma^2 E(\zeta_1'Q\zeta_1) + 2\sigma^3 E(\zeta_1'Q\zeta_2) + \sigma^4 E(\zeta_2'Q\zeta_2 + 2\zeta_1'Q\zeta_3) \tag{73}$$

Using (60) to (62), it is easy to verify that

$$E(\zeta_1'Q\zeta_1) = E[v'X(X'X)^{-1}Q(X'X)^{-1}v] = Tr(X'X)^{-1}Q \tag{74}$$

$$\begin{aligned} E(\zeta_1'Q\zeta_2) &= -\frac{c}{\beta'B\beta}\beta'D'Q(X'X)^{-1}X'\cdot E[v'Mv.v] \\ &= -c\nu_1\frac{\beta'D'Q(X'X)^{-1}X'(I*M)\iota}{\beta'B\beta} \end{aligned} \tag{75}$$

$$\begin{aligned} E(\zeta_2'Q\zeta_2) &= \frac{c^2\beta'D'QD\beta}{(\beta'B\beta)^2}\,TrM\cdot E[v'Mv\cdot vv'] \\ &= c^2[n(n+2) + \nu_2 TrM(I*M)]\,\frac{\beta'D'QD\beta}{(\beta'B\beta)^2} \\ &= c^2q\frac{\beta'D'QD\beta}{(\beta'B\beta)^2}\,. \end{aligned} \tag{76}$$

$$\begin{aligned} E(\zeta_1'Q\zeta_3) &= \frac{-c}{\beta'B\beta}TrX(X'X)^{-1}QD\left\{I - \frac{2}{\beta'B\beta}\beta\beta'B\right\}(X'X)^{-1}X'\cdot E[v'Mv\cdot v'] \\ &= -\frac{c}{\beta'B\beta}[nTr(X'X)^{-1}QD + \nu_2 Tr(X'X)^{-1}QD\bar{N} \\ &\quad - \frac{2}{\beta'B\beta}\{n\beta'B(X'X)^{-1}QD\beta + \nu_2\beta'B\bar{N}(X'X)^{-1}QD\beta\}]\,, \end{aligned} \tag{77}$$

where q in (76) is given in (10), and $\bar{N} = (X'X)^{-1}X'(I*M)X = I - N$, by using

$$(I*M) = I - (I*\bar{M})\,. \tag{78}$$

We recall equation (10) for the definitions of $\bar{M}$ and N.

Substituting (74) to (78) in (73), we obtain the expression (14) for the WMSE, to order σ^4, of $\hat{\beta}$ after a little algebraic manipulation.

From (69), we can also obtain the MtxMSE of $\hat{\beta}$ as

$$\text{MtxMSE}(\hat{\beta}) = \sigma^2(E\zeta_1\zeta_1') + \sigma^3(E\zeta_1\zeta_2' + E\zeta_2\zeta_1') + \sigma^4(E\zeta_2\zeta_2' + E\zeta_1\zeta_3' + E\zeta_3\zeta_1') \tag{79}$$

where, using (60) to (62), we have:

$$E(\zeta_1\zeta_1') = (X'X)^{-1}\,, \quad E(\zeta_1\zeta_2') = -\frac{c\nu_1}{\beta'B\beta}(X'X)^{-1}X'(I*M)\iota\beta'D' \tag{80}$$

$$E(\zeta_2\zeta_2') = c^2 q\frac{D\beta\beta'D'}{(\beta'B\beta)^2} \tag{81}$$

$$E(\zeta_1\zeta_3') = -\frac{c}{\beta'B\beta}\Big[(n+\nu_2\bar{N})(X'X)^{-1}D' - \frac{2}{\beta'B\beta}\{(n+\nu_2\bar{N})(X'X)^{-1}B\beta\beta'D'\}\Big]\,. \tag{82}$$

Note that $(E\zeta_2\zeta_1') = (E\zeta_1\zeta_2')'$ and $E(\zeta_3\zeta_1') = (E\zeta_1\zeta_3')'$.

Explanatory Note 13.2

Proof of an Algebraic Inequality

Let us write from equation (10)

$$\phi = \frac{TrN(X'X)^{-1}QD}{Tr(X'X)^{-1}QD} = \frac{TrNL}{TrL}\,,$$

where $N = (X'X)^{-1}X'(I*\bar{M})X$; $\bar{M} = X(X'X)^{-1}X'$, and $L = (X'X)^{-1}QD$. We need to show that

$$0 \le \eta_p \le \phi \le \eta_1 \le 1\,, \tag{83}$$

where η_p and η_1 are the minimum and maximum eigenvalues, respectively, of the matrix N.

First, let us note that N is at least positive semidefinite (see Rao [1973, p. 77, problem 32]). Further, verify that L is positive definite. Thus, $TrNL/TrL$ is a nonnegative quantity, and using Anderson and Gupta [1963, p. 524]

$$0 \le \eta_p \le \frac{TrNL}{TrL} = \frac{\text{Sum of eigenvalues of } NL}{\text{Sum of eigenvalues of } L} \le \eta_1. \tag{84}$$

Now let P be a nonsingular matrix such that $P'X'XP = I$ or $X'X = (PP')^{-1}$. Thus, $\bar{M} = X(X'X)^{-1}X' = XPP'X' = ZZ'$ where $Z=XP$ and $N = (X'X)^{-1}X'(I*\bar{M})X = PP'X'(I*ZZ')X = PZ'(I*ZZ')XPP^{-1} = PZ'(I*ZZ')ZP^{-1}$. Since P is nonsingular, eigenvalues of $P^{-1}NP$ = eigenvalues of $Z'(I*ZZ')Z$ = eigenvalues of N. But $Z'Z = I$ and ZZ' is idempotent. Therefore, using Poincare separation theorem (Rao [1973, p. 64]) eigenvalues of $Z'(I*ZZ')Z$ = eigenvalues of N are positive and ≤ 1. Using this in (84), the inequality in (83) follows.

EXERCISES

13.1 Consider the linear regression $y = X\beta + u$ where the elements of u are independently and symmetrically distributed with finite first four moments. Derive approximate MSE, to order σ^4, of the following estimators

(a) James and Stein $[1 - \frac{cs}{b'X'Xb}]b$

(b) Ridge-type $[I + \frac{cs}{b'X'Xb}(X'X)^{-1}]^{-1}b$

(c) Lindley-type $[I + \frac{cs}{b'Jb}(X'X)^{-1}J]^{-1}b$ where $s = (y-Xb)'(y-Xb)$, b is the OLS estimator and $J = I - \iota(\iota'\iota)^{-1}\iota$; where ι is a column vector of unit elements.

Derive the conditions under which the above estimators will be better than the OLS estimator b. Compare these conditions with the one for normally distributed disturbances. Are the above estimators minimax?

13.2 Suppose the matrix

$$X = \begin{bmatrix} 1/\sqrt{3} & 1/\sqrt{6} \\ 1/\sqrt{3} & -2/\sqrt{6} \\ 1/\sqrt{3} & 1/\sqrt{6} \end{bmatrix}$$

is such that $X'X = I$ and $(XX')^2 = XX'$. Show that the eigenvalues of $X'(I*XX')X$ are ½ and 1, where "$*$" represents Hadamard product. Also show that $0 \leq (I*XX') \leq I$ for any idempotent matrix XX'.

13.3 Write a general expression for α-Trimmed Mean in (39), Winsorized Mean in (40) for arbitrary T and α.

13.4 Plot the ψ curve for Bisquare estimator and compare it with SINE curve in Figure 13-6. Draw corresponding figures for ρ functions.

13.5 For the example based on Klein's data in Chapter 1, compute a robust regression estimator. For the iterative reweighted Huber-M estimate with $K_H = 1.345$ followed by two iterations of Bisquare with $K_B = 6.2$ verify that the regression coefficients are given by (10.2582, -0.37, 1.28), respectively, including the intercept. Note that robust methods give a low weight for 3rd and 9th observation. Do you think that nonnormality of errors is the serious problem here?

REFERENCES

The references contain a few worthwile entries not mentioned in the text to save space.

Abdelmalek, N.N., "An Efficient Method for the Discrete Linear L1 Approximation Problem," Mathematics of Computation, 29(1975), 844-850.

Abrahamse, A.P.J. and J. Koerts, "A Comparison of the Power of the Durbin-Watson Test and the BLUS Test," Journal of the American Statistical Association, 64(1969), 938-948.

Abrahamse, A.P.J. and J. Koerts, "New Estimates of Disturbances in Regression Analysis," Journal of the American Statistical Association, 66(1971), 71-74.

Abrahamse, A.P.J. and A.S. Louter, "On a New Test for Autocorrelation in Least Squares Regression," Biometrika, 58(1971), 53-60.

Adelman, I., M. Grier, and C.T. Morris, "Instruments and Goals in Economic Development," American Economic Review, 59(1969), 409-419.

Aigner, D.J. and G.G. Judge, "Some Applications of Pre-test and Stein Rule Estimators to Economic Data," Econometrica, 45(1977), 1279-1288.

Aitken, A.C., "Studies in Practical Mathematics, IV. On Linear Approximation by Least Squares," Proceedings of the Royal Society of Edinburgh, A, 62(1945), 138-146.

Akaike, H., "Information Theory and the Extension of the Maximum Likelihood Principle," in B.N. Petrov and F. Csaki (eds.) Second International Symposium on Information Theory(Akailseoniai-Kindo, Budapest, 1973), 267-281.

Alam, Khursheed, "A Family of Admissible Estimators of the Mean of a Multivariate Normal Distribution," Annals of Statistics, 1(1973), 517-525.

Alam, K. and J.B. Hawkes, "Estimation of Regression Coefficients," Scandinavian Journal of Statistics, 5(1978), 169-172

Alam, K. and J.B. Hawkes, "Minimax Property of Stein's Estimators," Communications in Statistics, Theor. and Meth., A8(1979), 581-590.

Allen, D.M., "The Relationship Between Variable Selection and Data Augmentation and a Method for Prediction," Technometrics 16(Feb.1974), 125-127.

Almon, S., "The Distributed Lag Between Capital Appropriations and Net Expenditures," Econometrica, 33(1965), 178-196.

Almon, S., "Lags Between Investment Decisions and Their Causes," Review of Economics and Statistics, 49(1968), 193-206.

Amemiya, Takeshi, "On the Use of Principal Components of Independent Variables in Two Stage Least Squares Estimation," International Economic Review, 7(1966), 283-303.

Amemiya, Takeshi, "A Note on a Heteroscedastic Model," Journal of Econometrics, 6(1977), 365-370.

Anderson, R.L. and E.L. Battiste, "The Use of Prior Information in Linear Regression Analysis," Communications in Statistics, 4 (no. 6, 1975), 497-517.

Anderson, T.W. and S. DasGupta, "Some Inequalities on Characteristic Roots of Matrices," Biometrika, 50(1963), 522-524.

Anderson, T.W., An Introduction to Multivariate Statistical Analysis (New York: John Wiley and Sons, Inc., 1958).

Anderson, T.W., The Statistical Analysis of Time Series (New York: John Wiley and Sons, Inc., 1971).

Anderson, T.W. and T. Sawa, "Evaluation of the Distribution Function of the Two-stage Least Squares Estimator," Econometrica, 47(Jan.1979), 163-182.

Andrews, D.F., P.J. Bickel, F.R. Hampel, P.J. Huber, W.H. Rogers,and J.W.Tukey, Robust Estimates of Location (Princeton, N.J.: University Press, 1972).

Andrews, D.F., "A Robust Method for Multiple Linear Regression," Technometrics, 16(1974), 523-531.

Avery, R.B., "Error Components and Seemingly Unrelated Regressions," Econometrica, 45(Jan.1977), 199-209.

Bacon, R.W. and J.A. Hausman, "The Relationship Between Ridge Regression and the Minimum Mean Squared Error Estimator of Chipman," Oxford Bulletin of Economics and Statistics, 36(Feb.1974), 115-124.

Balestra, P. and N. Nerlove, "Pooling Cross-Section and Time Series Data in the Estimation of a Dynamic Model: The Demand for Natural Gas," Econometrica, 34(July 1966), 585-612.

Banerjee, K.S. and R.N. Carr, "A Comment on Ridge Regression. Biased Estimation for Non Orthogonal Problems," Technometrics, 13(Nov.1971), 895-898.

Baranchik, A.J., "Multiple Regression and Estimation of the Mean of a Multivariate Normal Distribution," Technical Report no. 51, (Stanford:Department of Statistics, Stanford University, 1964).

Baranchik, A.J., "A Family of Minimax Estimators of the Mean of a Multivariate Normal Distribution," Annals of Mathematical Statistics, 41(Apr.1970), 642-645.

Baranchik, A.J., "Inadmissibility of Maximum Likelihood Estimators in Some Multiple Regression Problems with Three or More Independent Variables," The Annals of Statistics, 1(1973), 312-321.

Bartlett, M.S., "The Standard Errors of Discriminant Function Coefficients," Journal of the Royal Statistical Society, Supple. 6 (1939), 169-173.

Basar, T. and M. Mintz, "On a Minimax Estimate for the Mean of a Random Vector Under a Generalized Quadratic Loss Function," Annals of Statistics, 1(1973), 127-134.

Bayes, T., "Essays Towards Solving a Problem in the Doctrine of Chances," Biometrika, 45(1958), 293-315 (Reproduction of the 1763 paper).

Beach, C.M. and J.G. MacKinnon, "A Maximum Likelihood Procedure for Regression with Autocorrelated Errors," Econometrica, 46(Jan.1978), 51-58.

Beaton, A.E., D.B. Rubin and J.L. Barone, "The Acceptability of Regression Solutions: Another Look at Computational Accuracy," Journal of the American Statistical Association, 71(Mar.1976), 158-168.

Belsley, D.A. and V.C. Klema, "Detecting and Assessing the Problems Caused by Multicollinearity: A Use of the Singular Value Decomposition," Working paper No.66 (Cambridge, Mass.: National Bureau of Economic Research, Dec.1974).

Berger, J.O., "Minimax Estimation of Location Vectors for a Wide Class of Densities," Annals of Statistics, 3(1975), 1318-1328.

Berger, J.O., "A Robust Generalized Bayes Estimator of a Multivariate Normal Mean," Mimeograph No. 480 (Lafayette, Indiana: Department of Statistics, Purdue University, Dec.1976m).

Berger, J.O., "Admissible Minimax Estimation of a Multivariate Normal Mean with Arbitrary Quadratic Loss," Annals of Statistics, 4(1976), 223-226.

Berger, J.O., "Tail Minimaxity in Location Vector Problems and Its Applications," Annals of Statistics, 4(1976a), 33-50.

Berger, J.O., "Minimax Estimation of a Multivariate Normal Mean with Arbitrary Quadratic Loss," Journal of Multivariate Analysis, 6(1976b), 1-9.

Berger, J.O. and M.E. Bock, "Eliminating Singularities of Stein-Type Estimators of Location Vectors," Journal of the American Statistical Association, B 38 no.2(1976), 166-170.

Berger, J.O. and M.E. Bock, "Combining Independent Normal Mean Estimation Problems with Unknown Variances," Annals of Statistics, 4(Apr.1976), 642-648.

Berk, Robert H., "A Special Structure and Equivalent Estimation," The Annals of Mathematical Statistics, 38(Oct.1967), 1436-1445.

Bettman, J.R., "Perceived Price and Product Perceptual Variables," Journal of Marketing Research, 10 (Feb.1973), 100-102.

Bhattacharya, P.K., "Estimating the Mean of a Multivariate Normal Population with General Quadratic Loss Function," Annals of Mathematical Statistics, 37(Dec.1966), 1818-1825.

Bibby, J., "Minimum Means Square Error Estimation, Ridge Regression, and Some Unanswered Questions," in J. Gani, K. Sarkadi and I. Vinze (eds.), Progress in Statistics, Colloquia Mathematica Societatis Janos Bolyai, European Meeting of Statisticians, Budapest (Hungary) 1972 (Amsterdam: North Holland Publishing Co., 1974), 107-121.

Bibby, J. and H. Toutenburg, Prediction and Improved Estimation in Linear Models (New York: John Wiley and Sons, Inc., 1977).

Billinghurst, R.A., "A Demand and Cost Model for Local Telephone Usage," an M.Sc.(Industrial Engineering)dissertation(Cleveland, Ohio: Cleveland State University, May 1976).

Blattberg, R.C. and T. Sargent, "Regression with Non-Gaussian Stable Disturbances: Some Sampling Results," Econometrica, 39(1971), 501-510.

Blight, B.J.N., "Some General Results on Reduced Mean Squared Error Estimation," The American Statistician, 25(June 1971), 24-25.

Bock, M.E., "A Comparison of the Risk Functions for Preliminary Test and Positive Post Estimators," (Research Paper, University of Illinois, Urbana, 1972).

Bock, M.E., "Minimax Estimators of the Mean of a Multivariate Normal Distribution," Annals of Statistics, 3(1975), 209-218.

Bock, M.E., G.G. Judge and T.A. Yancey, "Some Comments on Estimation in Regression After Preliminary Tests of Significance," Journal of Econometrics, 1(1973), 191-200.

Bock, M.E., T.A. Yancey, and G.G. Judge, "Statistical Consequences of Preliminary Test Estimators in Regression," Journal of the American Statistical Association, 68(1973), 109-116.

Boot, J.C.G. and G.M. Dewit, "Investment Demand: An Empirical Contribution to the Aggregation Problem," International Economic Review, 1(1960), 27-28.

Box, G.E.P., "Use and Abuse of Regression," Technometrics, 8(Nov.1966), 625-629.

Box, G.E.P. and G.M. Jenkins, Time Series Analysis Forecasting and Control (San Francisco: Holden Day, rev.ed.,1976, 1st ed., 1970)

Breusch, T.S., "Testing for Autocorrelation in Dynamic Linear Models," Australian Economic Papers, 17(1978), 334-355.

Breusch, T.S. and A.R. Pagan, "A Simple Test for Heteroscedasticity and Random Coefficient Variation," Econometrica, 47(1979), 1287-1294.

Brillinger, D.R., "Fourier Analysis of Stationary Processes," Proceedings of the IEEE, 62(Dec.1974), 1628-1643.

Brook R.J. and T. Moore, "On the Expected Length of the Least Squares Coefficient Vector," Journal of Econometrics, 12(1980), 245-246.

Brown, L., "On the Admissibility of Invariant Estimators of One or More Location Parameters," Annals of Mathematical Statistics, 37(1966), 1087-1136.

Brown, L.D., "Admissible Estimators, Recurrent Diffusions, and Insoluble Boundary Value Problems," Annals of Mathematical Statistics, 42(1971), 855-903.

Brown, P.J. and Payne, C., "Election Night Forecasting (with discussion)," Journal of the Royal Statistical Society, Ser. A, 138(1975), 463-498.

Brown, P.J., "Centering and Scaling in Ridge Regression," Technometrics, 19 (Feb.1977), 35-36.

Brown, P.J. and J.V. Zidek, "Adaptive Multivariate Ridge Regression," The Annals of Statistics, 8(1980), 64-74.

Brown, W.G.,and B.R. Beattie, "Improving Estimates of Economic Parameters by use of Ridge Regression with Production Function Applications," American Journal of Agricultural Economics, 57(Feb.1975), 21-32.

Bunke O., "Improved Inference in Linear Models with Additional Information," Mathematische Operationsforshung und Statistik, 6(1975), 817-829.

Businger, P.A. and G.H. Golub, "Linear Least Squares Solutions by Householder Transformation," Numerische Math., 7(1965), 269-276.

Cacoullos, T., Discriminant Analysis and Applications (New York: Academic Press, 1973)

Campbell, F. and G. Smith, A Critical Analysis of Ridge Regression. Cowles Foundation Discussion Paper No. 402 (New Haven, Conn.: Yale Station,1975).

Carter, R.A.L. and A. Ullah, "The Finite Sample Properties of OLS and IV Estimators in Two Distributed Lag Models," Research Report, University of Western Ontario(forthcoming in Sankhya, Ser. D, 1981).

Casella, George, "Minimax Ridge Regression," unpublished Ph.D. dissertation(Lafayette, Indiana: Purdue University, 1977).

Casella, George, "Minimax Ridge Regression Estimation," The Annals of Statistics, 8(1980), 1036-1056.

Causey, Beverley D., "A Frequentist Analysis of a Class of Ridge Regression Estimators," Journal of the American Statistical Association, 75(Sept.1980), 736-738.

Chambers, J.M., Computational Methods for Data Analysis (New York: John Wiley and Sons, Inc., 1977).

Charette, M., "The Exact Finite Sample Properties of the Mixed Estimator," unpublished Ph.D. thesis, 1978, University of Western Ontario.

Cheng, D.C. and H.J. Iglarsh, "Principal Components Estimates in Regression Analysis," Review of Economics and Statistics 58(Aug.1976), 229-234.

Chipman, J.S., "On Least Squares with Insufficient Observations," Journal of the American Statistical Association, 59(Dec.1964), 1078-1111.

Chipman, J.S., "Efficiency of Least Squares Estimation of Linear Trend When Residuals Are Autocorrelated," Econometrica, 47(1979), 115-128.

Cleveland, W.S., "Estimation of Parameters in Distributed Lag Econometric Models," in S.E. Fienberg and A. Zellner (eds.), Studies in Bayesian Econometrics and Statistics(Amsterdam: North Holland Publishing Co., 1974).

Cochrane, D. and G.H. Orcutt, "Application of Least Squares Regression to Relationships Containing Autocorrelated Error Terms," Journal of the American Statistical Association, 44(Mar.1949), 32-61.

Coleman, David, Paul Holland, Neil Kaden, Virginia Klema and Stephen C. Peters, "A System of Subroutines for Iteratively Reweighted Least Squares Computations," ACM Transactions on Mathematical Software, 6(Sept.1980), 327-336.

Coniffe, D. and J. Stone, "A Critical View of Ridge Regression," The Statistician, 22(June 1973), 181-187.

Constantine, A.G., "Some Noncentral Distribution Problems in Multivariate Analysis," Annals of Mathematical Statistics, 34(1963), 1270-1285.

Cook, R.D., "Detection of Influential Observations in Linear Regression," Technometrics, 19(Feb.1977), 15-18.

Cooper, W.W. and A.P. Schinnar, "A Model for Demographic Mobility Analysis Under Patterns of Efficient Employment," Economics of Planning, 13(2) (1977), 139-173.

Corradi, C., "A Note on the Computation of Maximum Likelihood Estimates in Linear Regression Models with Autocorrelated Errors," Journal of Econometrics, 11(1979), 303-317.

CRC Standard Mathematical Tables, 13 th Edition (Cleveland, Ohio: The Chemical Rubber Co., 1964) (Random No.s, 250-251).

Crocker, D.C., A Letter, American Statistician, 25(June 1971), 55.

Daniel, C. and F. Wood, Fitting Equations to Data (New York: John Wiley and Sons, Inc., 1971)

Deegan, J.,Jr., "The Process of Political Development," Sociological Methods and Research, 3(May 1975), 384-415.

De Finetti, B., Theory of Probability (New York: John Wiley and Sons, Inc., 1974).

Dempster, A.P., "Alternatives to Least Squares in Multiple Regression" in Kabe and Gupta (eds.), Multivariate Statistical Inference (New York: North Holland Publishing Co., 1973).

Dempster, A.P., M. Schatzoff and N. Wermuth, "A Simulation Study of Alternatives to Ordinary Least Squares," Journal of the American Statistical Association, 72(Mar.1977), 77-91.

Dent, W.T. and S. Cassing, "On Durbin's and Sims' Residuals in Autocorrelation Tests," Econometrica, 46(1978), 1489-1492.

Dent, W.T. and G.P.H. Styan, "Uncorrelated Residuals from Linear Models," Journal of Econometrics, 7(1978), 211-225.

Devlin, S.J., R. Gnanadesikan and J.R. Kettenring, "Robust Estimation and Outlier Detection with Correlation Coefficients," Biometrika, 62(1975), 531-545.

Dhrymes, P.J., Econometrics (New York: Springer-Verlag, 1974).

Dhrymes, P.J., Introductory Econometrics (New York: Springer Verlag, 1978).

DiPillo, P.J., "Application of Bias to Discriminant Analysis," Communications in Statistics, Ser. A, 5(no. 9, 1976), 843-854.

Draper, N.R. and R.C. Van Nostrand, "Ridge Regression and James-Stein Estimation: Review and Comments," Technometrics, 21(1979), 451-466.

Dreze, J., "Bayesian Regression Analysis Using Poly-t Densities," Journal of Econometrics, 6(1977), 329-354.

Dubbelman, C., A.S. Louter and A.P.J. Abrahamse, "On Typical Characteristics of Economic Time Series and the Relative Qualities of Five Autocorrelation Tests," Journal of Econometrics, 8(1978), 295-306.

Durbin, J., "Tests for Serial Correlation in Regression Analysis Based on the Periodogram of Least-Squares Residuals," Biometrika, 56(1969), 1-15.

Durbin, J., "Testing for Serial Correlation in Least Squares Regression When Some of the Regressors are Lagged Dependent Variables," Econometrica, 38(May 1970), 410-421.

Durbin, J., "An Alternative to the Bounds Test for Testing Serial Correlation in Least Squares Regression," Econometrica, 38(1970b), 422-429.

Durbin, J. and G.S. Watson, "Testing for Serial Correlation in Least Squares Regression, I," Biometrika, 37(1950), 409-428.

Durbin, J. and G.S. Watson, "Testing for Serial Correlation in Least Squares Regression, II," Biometrika, 38(1951), 159-178.

Durbin, J. and G.S. Watson, "Testing for Serial Correlation in Least Squares Regression, III," Biometrika, 58(1971), 1-42.

Dwivedi, T.D. and V.K. Srivastava, "Optimality of Least Squares in the Seemingly Unrelated Regression Equation Model," Journal of Econometrics, 7(1978a), 391-395.

Dwivedi, T.D. and V.K. Srivastava, "On the Minimum Mean Squared Error Estimators in a Regression Model," Communications in Statistics, Ser. A, 7(1978b), 487-494.

Dwivedi, T.D., V.K. Srivastava and R.L.Hall, "Finite Sample Properties of Ridge Estimators," Technometrics, 22(May 1980), 205-212.

Efron, B., "Biased Versus Unbiased Estimation," Advances in Mathematics, 16(1975), 259-277.

Efron, B. and C. Morris, "Limiting the Risk of Bayes and Empirical Bayes Estimators — Part II: The Empirical Bayes Case," Journal of the American Statistical Association, 67(Mar.1972), 130-139.

Efron, B. and C. Morris, "Stein's Estimation Rule and Its Competitors — An Empirical Bayes Approach," Journal of the American Statistical Association, 68(Mar.1973), 117-130.

Efron, B. and C. Morris, "Combining Possibly Related Estimation Problems (with Discussion)," Journal of the Royal Statistical Society, B, 35(1973), 379-421.

Efron, B. and C. Morris, "Multivariate Empirical Bayes and Estimation of Covariance Matrices," Annals of Statistics, 4(1976), 22-32.

Efron, B. and C. Morris, "Empirical Bayes on Vector Observations — An Extension of Stein's Method," Biometrika, 59(Aug.1972), 335-347.

Efron, B. and C. Morris, "Data Analysis Using Stein's Estimator and Its Generalizations," Journal of the American Statistical Association, 70(June 1975), 311-319.

Efron, B. and C. Morris, "Families of Minimax Estimators of the Mean of a Multivariate Normal Distribution," The Annals of Statistics, 4(1) (Jan.1976), 11-21.

Eskew, H.L., "Ridge Regression with Non-zero Priors: Some Monte Carlo Results," National Technical Information Service no. AD-769-991 (Springfield, Va.: U.S. Dept. of Commerce 1973).

Fair, R.C., "On the Robust Estimation of Econometric Models," Annals of Economic and Social Measurement, 3(1974), 667-677.

Farebrother, R.W., "The Minimum Mean Square Error Linear Estimator and Ridge Regression," Technometrics, 17(Feb.1975), 127-128.

Farebrother, R.W., "Further Results on the Mean Square Error of Ridge Regression," Journal of the Royal Statistical Society (B) 38(1976), 248-250.

Farebrother, R.W., "A Class of Shrinkage Estimators," Journal of the Royal Statistical Society, B, 40(1978), 47-49.

Farebrother, R.W., "Partitioned Ridge Regression," Technometrics, 20(May 1978b), 121-122.

Farebrother, R.W., "The Restricted Least Squares Estimator and Ridge Regression," unpublished report(Manchester, England: Dept. of Econometrics, University of Manchester, July 1980).

Farebrother, R.W., "Algorithm AS153 Pan's Procedure for the Tail Probabilities of the Durbin-Watson Statistic," Applied Statistics, 29(1980), 224-227.

Farebrother, R.W., "The Durbin-Watson Test for Serial Correlation When there is no Intercept in the Regression," Econometrica, 45(1980) 1553-1563, and Erratum, January 1981.

Farrar, D.E. and R.R. Glauber, "Multicollinearity in Regression Analysis: The Problem Revisited," Review of Economics and Statistics, 49(Feb.1967), 92-107.

Feldstein, M.S., "Multicollinearity and the Mean Squared Error of Alternative Estimators," Econometrica, 41(4) (Mar.1973), 337-346.

Fisher, F.M., "Identifiability Criteria in Non-linear Systems," Econometrica, 33(1965a), 574-590.

Fisher, F.M., "Near Identifiability and the Variances of the Disturbance Terms," Econometrica, 33(1965b), 409-419.

Fisher, F.M., The Identification Problem in Econometrics (New York: McGraw-Hill Book Co., 1966). Fisher, Irving, "A Note on a Short-cut Method for Calculating Distributed Lags," Bulletin de l'Institut International de Statistique,29(troisieme livraison,1937), 323-328.

Fisher, R.A., "The Use of Multiple Measurements in Taxonomic Problems," Annals of Eugenics, 7(1936), 179-188. [Reprinted in Contributions to Mathematical Statistics," in W.A. Shewhart (ed.), John Wiley and Sons, Inc., New York, 1950]

Frisch, R., Statistical Confluence Analysis by Means of Complete Regression Systems (Oslo: Universitets Okonomiske Institut, 1934).

Fomby, T.B., "MSE Evaluation of Shiller's Smoothness Priors," International Economic Review, 20(1979), 203-215.

Fomby, T.B. and S.R. Johnson, "MSE Evalutaion of Ridge Estimators Based on Stochastic Prior Information," Communications in Statistics, Theory and Methods, A6(1977), 1245-1258.

Fujikoshi, Y. and L.G. Veitch, "Estimation of Dimensionality in Canonical Correlation Analysis," Biometrika, 66(1979), 345-351.

Fuller, W.A. and J.N.K. Rao, "Estimation for a Linear Regression Model with Unknown Diagonal," Annals of Statistics, 6(1978), 1149-1158.

Galilei, Galileo, original Italian reference published in 1632 See Leon Harter's Survey (1974) for details.

Gapinski, J.H. and H.P. Tuckman, "Travel Demand Functions for Florida Bound Tourists," Transportation Research, 10(Aug.1976), 267-274.

Gauss, C.F. (1806). A German reference. For details see H. Leon Harter's Survey (1974,p.169).

Gauss, C.F. (1823). German reference reprinted in Werke English translation by H. Trotter, Statistical Techniques Research Group, Princeton University, 1957, 83-126.

Gewirtz, A., H. Sitomer, and A.W. Tucker, "Constructive Linear Algebra" (Englewood Cliffs, N.J.: Prentice Hall, Inc., 1974).

Gibbons, Diane Galarneau, "A Simulation Study of Some Ridge Estimators," Journal of the American Statistical Association, 76(Mar.1981), 131-139.

Giles, D.E.A. and A.C. Rayner, "The Mean Squared Errors of the Maximum Likelihood and Natural Conjugate Bayes Regression Estimators," Journal of Econometrics, 2(1979), 319-334.

Giles, D.E.A., "Bayesian Applications in Econometrics," unpublished Ph.D. thesis, University of Canterbury, Christchurch, New Zealand, 1975.

Glahn, H.R., "Canonical Correlation and its Relationship to Discriminant Analysis and Multiple Regression," Journal of Atmospheric Sciences, 25(1968), 23-31.

Glejser, H., "A New Test for Heteroscedasticity," Journal of the American Statistical Association, 64(1969), 316-323.

Gnanadesikan, Ram, Methods for Statistical Data Analysis of Multivariate Observations (New York: John Wiley and Sons, Inc., 1977)

Godfrey, L.G., "Testing Against General Autoregressive and Moving Average Error Models when the Regressors Include Lagged Dependent Variables," Econometrica, 46(Nov.1978a), 1293-1302.

Godfrey, L.G., "Testing for Higher Order Serial Correlation in Regression Equations when the Regressors Include Lagged Dependent Variables," Econometrica, 46(1978b), 1303-1310.

Godfrey, L.G., "Testing for Multiplicative Heteroscedasticity," Journal of Econometrics, 8(1978c), 227-236.

Goldberger, A.S. Impact Multipliers and Dynamic Properties of the Klein-Goldberger Model (Amsterdam, Holland: North Holland Publishing Co., 1959).

Goldfeld, S.M. and R.E. Quandt, "Some Tests for Homoscedasticity," Journal of the American Statistical Association, 60(1965), 539-547.

Goldfeld, S.M. and R.E. Quandt, Nonlinear Methods in Econometrics (Amsterdam: North-Holland, 1972)

Goldstein, M. and A.F.M. Smith, "Ridge Type Estimators for Regression Analysis," Journal of the Royal Statistical Society, B, 36(Dec.1974), 284-291.

Golub, G., "Matrix Decompositions and Statistical Calculations," in R.C. Milton and J.A. Nedler (eds.), Statistical Computation (New York: Academic Press, 1969)

Golub, G.H. and G.P.H. Styan, "Numerical Computations for Univariate Linear Models," Journal of Statistical Computation and Simulation, 2(1973), 253-274.

Golub, G., Michael Heath and Grace Wahba, "Generalized Cross Validation as a Method for Choosing a Good Ridge Parameter," Technometrics, 21(May 1979), 215-223.

Goodnight, J. and T.D.Wallace, "Operational Techniques and Tables for Making Weak MSE Tests for Restrictions in Regressions," Econometrica, 40 (July 1972), 699-709.

Gorman, J.W., "Fitting Equations to Mixture Data With Restraints on Compositions," Journal of Quality Technology, 2, No. 4 (Oct.1970), 186-194.

Graupe, Daniel, Identification of Systems (New York: Van Nostrand Reinhold Company,1972)

Guilkey, D.K. and J.L. Murphy, "Directed Ridge Regression Techniques in Cases of Multicollinearity," Journal of the American Statistical Association, 70, No. 352(Dec.1975), 769-775.

Haitovsky, Y. and Y. Wax, "Generalized Ridge Regression, Least Squares with Stochastic Prior Information and Bayesian Estimators," presented at NBER-NSF seminar on Bayesian Inference in Econometrics, Ann Arbor, Michigan(Apr.1974).

Hald, A., "Statistical Theory with Engineering Applications," (New York: John Wiley and Sons, Inc., 1952).

Hampel, F., "The Influence Curve and Its Role in Robust Estimation," Journal of the American Statistical Association, 69(1974), 383-393.

Hannan, E.J. and R.D. Terrell, "Testing for Serial Correlation After Least Squares Regression," Econometrica, 36(1968), 133-150.

Harrison, M.J. and B.P. McCabe, "A Test for Heteroscedasticity Based on Ordinary Least Squares Residuals," Journal of the American Statistical Association, 74(1979), 494-499.

Harter, H.L. (1974, 1975 and 1976), "The Method of Least Squares and Some Alternatives," International Statistical Review (1974) Part I, 42, 147-174, (Apr.1974) Part II, 42, 235-264 (1975) Part III, 43, 1-44, (1975) Part IV, 43, 125-190 (1975) Part V, 43, 269-278, (1976) Part VI, 44, 113-159.

Hartley, H.O. and K.S.E. Jayatillake, "Estimation for Linear Models with Unequal Variances," Journal of the American Statistical Association, 68(1973), 189-192.

Harvey, A.C., "Estimation of Parameters in a Heteroscedastic Regression Model," European Meeting of the Econometric Society (1974), Grenoble, France.

Harvey, A.C., "Estimating Regression Models with Multiplicative Heteroscedasticity," Econometrica, 44(1976), 461-465.

Harvey, A.C., The Econometric Analysis of Time Series (Oxford, U.K.: Philip Allan Publishers Limited, 1981).

Harvey, A.C. and G.D.A. Phillips, "A Comparison of the Power of Some Tests for Heteroscedasticity in the General Linear Model," Journal of Econometrics, 2(1974), 307-316.

Hawkins, D.M., "Relations Between Ridge Regression and Eigenanalysis of the Augmented Correlation Matrix," Technometrics, 17(Nov.1975), 477-480.

Hemmerle, W.J., "An Explicit Solution for Generalized Ridge Regression," Technometrics, 17 No. 3(Aug.1975), 309-314.

Hemmerle, W.J. and T.F. Brantle, "Explicit and Constrained Generalized Ridge Estimation," Technometrics, 20(May 1978), 109-120.

Hendry, David, "Monte Carlo Experimentation in Econometrics," CORE, DP 8035 (Louvain-la-neuve, Belgium: CORE, Nov.1980).

Henshaw, R.C., "Testing Single-Equation Least Squares Regression Models for Autocorrelated Disturbances," Econometrica, 34(1966), 646-660.

Hildreth, C. and J. Houck, "Some Estimates for a Linear Model With Random Coefficients," Journal of the American Statistical Association (1968), 584-595.

Hilliard, G.W., "A Simulation Study of Approximations to the Minimum Mean Square Error Linear Estimator," an M.A.(Economics) dissertation, University of Manchester, Manchester, England, April 1975.

Hocking, P.R., "The Analysis and Selection of Variables in Linear Regression," Biometrics, 32(Mar.1976), 1-49.

Hocking, R.R., F.M. Speed and M.J. Lynn, "A Class of Biased Estimators in Linear Regression," Technometrics, 18(1976), 425-438.

Hoerl, A.E., "Optimum Solution of Many Variables Equations," Chemical Engineering Progress, 55(1959) 11-21.

Hoerl, A.E., "Application of Ridge Analysis to Regression Problems," Chemical Engineering Progress, 60(1962), 54-59.

Hoerl, A.E. and R.W. Kennard, "Ridge Regression: Biased Estimation of Nonorthogonal Problems," Technometrics 12(Feb.1970), 55-67.

Hoerl, A.E. and R.W. Kennard, "Ridge Regression: Applications to Nonorthogonal Problems," Technometrics 12(Feb.1970), 69-82.

Hoerl, A.E. and R.W. Kennard, "A Note on a Power Generalization of Ridge Regression," Technometrics, 17(2) (May 1975), 269.

Hoerl A.E. and R.W. Kennard, "Ridge Regression Iterative Estimation of the Biasing Parameter," Communications in Statistics, Ser. A, 5(No. 1, 1976), 77-78.

Hoerl, A.E., R.W. Kennard, and K.F. Baldwin, "Ridge Regression: Some Simulations," Communications in Statistics, 4(1975), 105-123.

Hogg, R.V., "Statistical Robustness: One View of Its Use in Applications Today," The American Statistician, 79(Aug.1979), 108-115.

Holland, P.W., "Weighted Ridge Regression: Combining Ridge and Robust Regression Methods," Working Paper No. 11 (Cambridge, Mass.: National Bureau of Economic Research, Sept.1973).

Hoque, A., "Small Sample Properties of Simple Dynamic Models," unpublished thesis, University of Western Ontario (1980).

Horn, S.D. and R.A. Horn, "Comparison of Estimators of Heteroscedastic Variances in Linear Models," Journal of the Americas Statistical Association, 70(1975), 872-879.

Horn, S.D., Horn, R.A., and D.B. Duncan, "Estimating Heteroscedastic Variances in Linear Models," Journal of the American Statistical Association, 70(1975), 380-385.

Hotelling, Harold, "Relations Between Two Sets of Variables," Biometrika, 28(1936), 321-377.

Houthakker, H.S. and L.D.Taylor. Consumer Demand in the United States, 1929-1970 (Cambridge, Mass.: Harvard University Press, 1966).

Huber, Peter J., "Robust Estimation of a Location Parameter," Annals of Mathematical Statistics, 35(Mar.1964), 73-101.

Huber, Peter J., "Robust Statistics: A Review," Annals of Mathematical Statistics, 43(1972), 1041-1067.

IBM, System/360 Scientific Subrouting Package (360A-CM-03X) H20-0166-5 (White Plains, New York: IBM Technical Publications Dept., 1968).

Imhof, P.J., "Computing the Distribution of Quadratic Forms in Normal Variables," Biometrika, 48(Dec.1961), 419-426.

Jaeckel, Louis A., "Robust Estimates of Location: Symmetry and Asymmetric Contamination," Annals of Mathematical Statistics, 42(1971), 1020-1034.

James, W. and C. Stein. "Estimation with Quadratic Loss," Proceedings Fourth Berkeley Symposium in Mathematical Statistics and Probability, 1(1961), 361-379.

Jeffreys, H., Scientific Inference (Cambridge, U.K., Cambridge University Press, 2nd ed., 1957).

Jeffreys, H., Theory of Probability (Oxford: Clarendon, 1961 and 1966).

Johnson, N.L., "On the Comparison of Estimators,"Biometrika,37 (Dec.1950), 281-287.

Johnston, J., Econometric Methods (New York: McGraw-Hill Book Co., 2nd ed., 1972).

Joshi, V.M., "Joint Admissibility of the Sample Means as Estimators of the Means of Finite Populations," The Annals of Statistics, 7(1979), 995-1002.

Judge, G.G. and M.E. Bock, "A Comparison of Traditional and Stein-Rule Estimators Under Weighted Squared Error Loss," International Economic Review, 17(1976), 234-240

Judge, G.G. and M.E. Bock, The Statistical Implications of Pretest and Stein-Rule Estimators in Econometrics (New York: North Holland Publishing Co., 1978).

Judge, G.G., M.E. Bock and T.A. Yancey, "Post Data Model Evaluation," Review of Economics and Statistics, 56(1974), 245-253.

Judge, G.G., W.E. Griffiths, R.C. Hill and T.C. Lee, The Theory and Practice of Econometrics (New York: John Wiley and Sons, Inc., 1980).

Judge, G.G. and T. Takayama, "Inequality Restrictions in Regression Analysis," Journal of the American Statistical Association, 61 (Mar.1966), 166-181.

Judge, G.G., T.A. Yancey and M.E. Bock, "Properties of Estimators After Preliminary Tests of Significance When Stochastic Restrictions Are Used in Regression," Journal of Econometrics 1(1973), 29-48.

Jureckova, Jana, "Asymptotic Relations of M-Estimates and R-Estimates in Linear Regression Model," Annals of Statistics, 5(1977), 464-472.

Kadane, J.B., "Testing Overidentifying Restrictions When the Disturbances Are Small," Journal of the American Statistical Association, 65(1970), 182-85.

Kadane, J.B., "Comparison of K-Class Estimators When the Disturbances are Small," Econometrica, 39(1971), 723-738.

Kadiyala, K.R. and James R. Nunns, "Estimation of a Simultaneous System of Equations When the Sample Is Undersized," unpublished report(Lafayette, Indiana: Purdue University, 1976).

Kadiyala, Krishna, "Operational Ridge Regression Estimators Under the Prediction Goal," Communications in Statistics, Ser. A 8(1979), 1377-1391.

Kadiyala, Krishna, "Some Finite Sample Properties of Generalized Ridge Regression Estimators," Canadian Journal of Statistics (1980).

Kakwani, N.C., "The Unbiasedness of Zellner's Seemingly Unrelated Regression Equations Estimators," Journal of the American Statistical Association, 62(1967), 141-142.

Kariya, T., "A Robustness Property of the Tests for Serial Correlation," Annals of Statistics, 5(1977), 1212-1220.

Katz, R.W., "Sensitivity Analysis of Statistical Crop-weather Model," in Weather-Climate Modeling for Real-Time Applications in Agriculture and Forest Meteorology, (Lafayette, Indiana: Purdue University, 1977), 1-2.

Kendall, M.G., A Course in Multivariate Analysis (New York: Hafner, 1957).

Kennard, R.W., Letter to the Editor, Technometrics, 18(4) (Nov.1976), 504-505.

Khazzoom, J.D., "An Indirect Least Squares Estimator for Overidentified Equations," Econometrica, 44(July 1976), 741-750.

King, M.L., "The Durbin-Watson Bounds Test and Regressions without an Intercept," Research Report 10, Monash University, 1980a, Forthcoming Australian Economic Papers.

King, M.L., "Robust Tests for Spherical Symmetry and their Application to Least Squares Regression," The Annals of Statistics, (November, 1980b) 1265-1271.

King, M.L., "Testing for Autocorrelation using Linear Unbiased Regression Residuals with Scalar Covariance Matrices," Research Report, Monash University, 1980c.

King., M.L., "The Durbin-Watson Test for Serial Correlation: Bounds for Regressions with Trend and/or Seasonal Dummy Variables," Forthcoming Econometrica, 1981.

Kiviet, J.F., "Effects of ARMA Errors on Tests for Regression Coefficients: Comments on Vinod's Article; Improved and Additional Results," Journal of the American Statistical Association, 75(June 1980), 353-358.

Klee, A.J., "Biased Estimation as a Solution to the Multicollinearity Problem in Multiple Linear Regression," a Ph.D. thesis (Cincinnati, Ohio: University of Cincinnati, Oct.1973).

Kleffe, J., "Optimal Estimation of Variance Components — A Survey," Sankhya, Ser. B, 39(1977), 211-244.

Klein, L.R., Economic Fluctuations in the United States, 1921-1941, (New York: John Wiley and Sons, Inc., 1950).

Klemm, R.J. and V.A. Sposito, "Closed Form Solutions for Least Squares Estimators when the Parameters Are Subject to Inequality Constraints," 1976 Proceedings of the Business and Economic Statistics Section Boston, Mass., Sept.1976 (Washington, D.C.: American Statistical Assoc.), 392-396.

Kloek, T. and L.B.M. Mennes, "Simultaneous Equations Estimation Based on Principal Components of Predetermined Variables," Econometrica, 28(1960), 45-61.

Kmenta, J., Elements of Econometrics (New York: MacMillan Publishing Co., 1971).

Koenker, Roger and Bassett, Gilbert, Jr., "Regression Quantiles," Econometrica, 46(1978), 33-50.

Koerts, J. and A.P.J. Abrahamse, On the theory and Application of the General Linear Model (Rotterdam: Rotterdam University Press, 1969).

Konus, A.A., "The Single Regression Line," Metron, 25(No.1-4, 1966), 1-8.

Koyck, L.M., Distributed Lags and Investment Analysis (Amsterdam: North-Holland Publishing Co., 1954).

Kramer, G., "On the Durbin-Watson Bounds Test in the Case of Regression Through the Origin," Jahrbucher für National Okonomie und Statistik, 185(1971), 345-358.

Kramer, W., "Finite Sample Efficiency of Ordinary least Squares in the Linear Regression Model with Autocorrelated Errors," Journal of the American Statistical Association, 75(Dec.1980), 1005-1009.

Kshirsagar, A.M., Multivariate Analysis (New York: Marcel Dekker, 1972).

Kumar, T.K., "Multicollinearity in Regression Analysis," Review of Economics and Statistics, 57(Aug.1975), 365-366.

Kvalseth, T.O., "Ridge and Bayes Identification for the Quasilinear Human Controller in Compensatory Tracking," IEEE Transactions on Systems Man and Cybernetics, 6(Oct.1976), 705-708.

Lachenbruch, P.A., Discriminant Analysis (New York: Hafner, 1975).

Lachenbruch, P.A. and M. Goldstein, "Discriminant Analysis," Biometrics, 35(Mar.1979), 69-85.

Lawless, J.F. and P. Wang, "A Simulation Study of Ridge and Other Regression Estimators," Communications in Statistics , Ser. A, 5(No. 4, 1976), 307-323.

Leamer, E.E., "Model-Selection Searches: A Bayesian View," Harvard Institute of Economic Research, Discussion Paper 151, December 1970.

Leamer, E.E., "Multicollinearity: A Bayesian Interpretation," The Review of Economics and Statistics, 55(Nov.1973), 371-380.

Leamer, E.E., "False Models and Post Data Model Construction," Journal of the American Statistical Association, 69(Mar.1974), 122-131.

Leamer, E.E., "A result on Sign of Restricted Least Squares Estimates," Journal of Econometrics, 3(1975), 387-390.

Leamer, E.E., "Regression Selection Strategies and Revealed Priors," Journal of the American Statistical Association, 73(Sept.1978), 580-587.

Leamer, E.E., "Information Criteria for Choice of Regression Models: A Comment," Econometrica, 47(Mar.1979),

Leamer, E.E. and G. Chamberlain, "A Bayesian Interpretation of Pretesting," Journal of the Royal Statistical Society, B 38(1976), 85-94.

Lee, B.M.S., "On the Use of Improved Estimators in Econometrics," Ph.D. thesis, Australian National University, 1979.

Lee, B.M.S. and P.K. Trivedi, "Instrumental Variable Estimation of Structural Equations in Undersized Samples," unpublished memorandum, Australian National University,1979.

Legendre, Adrian Marie(1805). See Leon Harter's Survey (1974) for details.

Lehmann, E.L., Testing Statistical Hypotheses (New York: John Wiley, 1959).

Lempers, F.B., Posterior Probabilities of Alternative Linear Models (Rotterdam: Rotterdam University Press, 1971).

L'Esperance, W.L. and D. Taylor, "The Power of Four Tests of Autocorrelation in the Linear Regression Model," Journal of Econometrics, 3(1975), 1-21.

Levenberg, K., "A Method for the Solution of Certain Non-linear Problems in Least Squares," Quarterly of Applied Mathematics, 2(1944), 164-168.

Lewis, E.B., "An Investigation of the Probability Distribution of the Ridge Regression Estimator for Linear Models," an M.Sc. (Operations Research) thesis, Naval Post-graduate School, Monterey, California, Mar.1976.

Liew, C.K., "Inequality Constrained Least Squares Estimation," Journal of the American Statistical Association,71(1976), 746-751.

Lin, Karl and Jan Kmenta, "Some New Results on Ridge Regression Estimation," presented at the Fourth World Congress of the Econometric Society, Sept. 2, 1980 (East Lansing, Michigan: University of Michigan).

Lin, P. and H. Tsai, "Generalized Bayes Minimax Estimators of the Multivariate Normal Mean With Unknown Covariance Matrix," Annals of Statistics , 1(1973), 142-145.

Lindley, D.V., "Discussion of Professor Stein's Paper," Journal of the Royal Statistical Society, Ser. B, 24(1962), 285-287.

Lindley, D.V., "The Choice of Variables in Multiple Regression," Journal of the Royal Statistical Society, Ser. B, 31(1968), 31-66.

Lindley, D.V., Bayesian Statistics,A Review (Philadelphia, Pa.: Society for Industrial and Applied Mathematics, 1971).

Lindley, D.V. and A.F.M. Smith, "Bayes Estimates for the Linear Model," Journal of the Royal Statistical Society, Ser. B, 34(Aug.1972), 1-41.

Lindley, D.V., "The Estimation of Many Parameters," in V.P. Godambe and D.A. Sprott (eds.), Symposium on Foundations of Statistical Inference(Toronto, Canada: Holt, Rinehart and Winston, 1971).

Longley, J.W., "An Appraisal of Least Squares Programs for the Electronic Computer from the Point of View of the User," Journal of the American Statistical Association, 62(Sept.1967), 819-841.

Lovell, M.C. and E. Prescott, "Multiple Regression with Inequality Constraints: Pretesting Bias, Hypothesis Testing and Efficiency," Journal of the American Statistical Association, 65(June 1970), 913-925.

Lowerre, J.M., "On the Mean Square Error of Parameter Estimates for Some Biased Estimators," Technometrics, 16(3) (Aug.1974), 461-464.

Maasoumi, E., "A Study of Improved Method of Estimating Reduced Form Coefficients Based Upon 3SLS," unpublished Ph.D. thesis, (London, UK: London School of Economics, 1977).

Maasoumi, E., "A Modified Stein-like Estimator for the Reduced Form Coefficients of Simultaneous Equations," Econometrica,46(May 1978), 695-703.

Maasoumi, E., "A Ridge-Like Method for Simultaneous Estimation of Simultaneous Equations," Journal of Econometrics, 12(1980), 161-176.

Maddala, G.S., "The Use of Variance Components Models in Pooling Cross Section and Time Series Data," Econometrica, 39(Mar.1971), 341-358.

Maddala, G.S., "Ridge Estimators of Distributed Lag Models," Discussion Paper Number 69, (Cambridge: National Bureau of Economic Research, Oct.1974).

Maddala, G.S., Econometrics (New York: McGraw-Hill Book Co., 1977).

Maeshiro, A., "Autoregressive Transformation, Trended Independent Variables and Autocorrelated Disturbance Terms," Review of Economics and Statistics, 58(1976), 497-500.

Mahalanobis, P.C., "On the Generalized Distance in Statistics," Proceedings of the National Institute of Sciences, India, 12(1936), 49-55.

Makani, S.M., "A Paradox in Admissibility," The Annals of Statistics, 5(1977), 544-546.

Malinvaud, E., Statistical Methods of Econometrics translated by A. Silvey (Chicago: Rand McNally Co., 1966).

Mallela, P., "Necessary and Sufficient Conditions for MINQU Estimation of Heteroscedastic Variances in Linear Model," Journal of the American Statistical Association, 67(1972), 486-487.

Mallows, C.L. "Some Comments on Cp ," Technometrics, 15(Nov.1973), 661-675.

Mallows, C.L., "Robust Methods — Some Examples of their Use," The American Statistician, 33(Nov.1979), 179-184.

Maloney, M.T. and M.E. Ireland, "Fiscal Versus Monetary Policy, An Application of Transfer Functions," Journal of Econometrics, 13(1980) 253-266.

Mantell, L.H., "An Econometric Study of Returns to Scale in the Bell System," Staff Research Paper, Office of Telecommunications Policy, Executive Office of the President, Washington,D.C., Feb.1974.

Mantell E.H., "Factors Affecting Labor Productivity in Post Offices," Journal of the American Statistical Association, 69(June 1974), 303-309.

Margolis, Marvin S., "Perpendicular Projections and Elementary Statistics," The American Statistician, 33(Aug.1979), 131-135.

Mariano, R.S., "The Existence of Moments of the Ordinary Least Squares and Two Stage Least Squares Estimators," Econometrica, 40(1972), 643-652.

Markoff, A.A. (1900), Calculus of Probabilities (in Russian), St. Petersburg For details see H. Leon Harter's (1974,p.26) Survey.

Marquardt, D.W., "An Algorithm for Least Squares Estimation of Nonlinear Parameters," SIAM 11(1963), 431-441.

Marquardt, D.W., "Generalized Inverses, Ridge Regression, Biased Linear Estimation and Nonlinear Estimation," Technometrics 12(Aug.1970), 591-612.

Marquardt, D.W. and R.D. Snee, "Ridge Regression in Practice," The American Statistician, 29(Feb.1975), 3-20.

Mason, R.L., R.F. Gunst and J.T. Webster, "Regression Analysis and Problems of Multicollinearity," Communications in Statistics, 4 (No. 3, 1975), 277-292.

Massy, W.F., "Principal Components Regression in Exploratory Statistical Research," Journal of the American Statistical Association, 60(Mar.1965), 234-256.

Mayer, L.S. and T.A. Willke, "On Biased Estimation in Linear Models," Technometrics, 15(Aug.1973), 497-508.

Meyer, J.R. and R.R. Glauber, Investment Decisions, Economic Forecasting and Public Policy (Boston, Mass.: Harvard Business School, 1964).

McCabe, G.P.,Jr., "Evaluation of Regression Coefficient Estimates Using Alpha-Acceptability," Technometrics, 20(May 1978), 131-139.

McCallum, B.T., "Artificial Orthogonalization in Regression Analysis," Review of Economics and Statistics, 52(Feb.1970), 110-113.

McDonald, G.C. and D.I. Galarneau, "A Monte Carlo Evaluation of Some Ridge-type Estimators," Journal of the American Statistical Association, 70(June 1975), 407-16.

McDonald, G.C. and R.C. Schwing, "Instabilities of Regression Estimates Relating Air Pollution to Mortality," Technometrics, 15(Aug.1973), 463-481.

McElroy, M.B., "Goodness of Fit for Seemingly Unrelated Regressions: Glahn's R(y.x)squared and Hooper's R bar Squared," Journal of Econometrics, 9(1977), 381-387.

McKay, "Variable Selection in Multivariate Regression: An Application of Simultaneous Best Procedures," Journal of the Royal Statistical Society, B 39(1977), 371-380.

Meeter, D.A., "On a Theorem Used in Non-linear Least Squares," SIAM Journal of Applied Mathematics, 14(Sept.1966), 1176-1179.

Mehta, J.S. and P.A.V.B. Swamy, "The Existence of Moments of Some Simple Bayes Estimators of Coefficients in a Simultaneous Equation Model," Journal of Econometrics, 7(1978), 1-13.

Mitchell, Bridger M. and Franklin M. Fisher, "The Choice of Instrumental Variables in the Estimation of Economy-wide Econometric Models: Some Further Thoughts," International Economic Review, 11(June 1970), 226-236.

Moriarty, B.M., "Causal Inference and the Problem of Nonorthogonal Variables," Geographical Analysis, 5(1973), 55-61.

Morrison, D.F., Multivariate Statistical Methods. (New York: McGraw-Hill Book Co.,1967).

Mosteller, F. and J.W. Tukey, Data Analysis and Regression (Reading, Mass: Addison-Wesley, 1977).

Moulaert, F., "Ridge Regression: A Geometrical Revisitation," Regional Science Research Paper No. 15(Lourain, Belgium: Katholieke Universiteit, Centrum voor Economische Studien, 1976).

Mullet, G.M. and T.W. Murray, "A New Method for Examining Rounding Error in Least-Squares Regression Computer Programs," Journal of the American Statistical Association, 66(Sept.1971), 496-498.

Mullet, G.M., "A Graphical Illustration of Simple (Total) and Partial Regression," American Statistician, 26, No. 5(Dec.1972), 25-27.

Mullet, G.M., "Why Regression Coefficients Have the Wrong Sign," Journal of Quality Technology, 8, 3(July 1976), 121-126.

Mundlak, Y., "Empirical Production Functions Free of Management Bias," Journal of Farm Economics, 43(1961), 44-56.

Mundlak, Y., "On the Pooling of Time Series and Cross Section Data," Econometrica, 46(Jan.1978), 69-85.

Myoken, H., "Optimal Estimators of Generalized Ridge Regression, a Monte Carlo Evaluation and Application" in Gordesh.,J. and P.Naeve (eds.), Proceedings in Computational Statistics (COMPSTAT)(Physica- Verlag, Wien,1976), 162-169.

Myoken H. and Y. Uchida, "The Generalized Ridge Estimator and Improved Adjustments for Regression Parameters," Metrika, 24(1977), 113-124.

Nagar, A.L., "The Bias and Moment Matrix of the General k-Class Estimators of the Parameters in Simultaneous Equations," Econometrica, 27(1959), 575-595.

Nagar, A.L. and A. Ullah, "On Nagar's Approximations to Moments of the Two Stage Least Squares Estimation," Indian Economic Review, 5(1970).

Nagar, A.L. and A. Ullah, "Note on Approximatc Skewnesss and Kurtosis of the Two Stage Least Squares Estimator," Indian Economic Review (1973).

Narula, S.C., "Predictive Mean Square Error and Stochastic Regressor Variables," Applied Statistics, 23(1976), 11-17.

Neeleman, D., Multicollinearity in Linear Economic Models (The Netherlands: Tibury University Press, 1973).

Nerlove, M., "Further Evidence on the Estimation of Dynamic Economic Relations from a Time Series of Cross Sections," Econometrica, 39(1971a), 359-382.

Nerlove, M., "A Note on Errors Components Models," Econometrica, 39(Mar.1971b), 383-96.

Neudecker, H., "Abrahamse and Koerts' New Estimator of Disturbances in Regression Analysis," Journal of Econometrics, 5(1977), 129-133.

Newhouse, J.P. and S.D. Oman, "An Evaluation of Ridge Estimators," Report No. R-716-PR (Santa Monica, California: Rand Corp., 1971).

Nicholls, D.F., A.R. Pagan, and R.D. Terrell, "The Estimation and Use of Models with Moving Average Disturbance Terms: A Survey," International Economic Review, 16(1975), 113-134.

Novick, M.R., P.H. Jackson, D.T. Thayer, and N.S. Cole, "Estimating Multiple Regression in M Groups: A Cross Validation Study," British Journal of Mathematical and Statistical Psychology (1972), 33-50.

Obenchain, R.L., "Ridge Analysis Following a Preliminary Test of the Shrunken Hypothesis," Technometrics, 17(Nov.1975), 431-441.

Obenchain, R.L., "Methods of Ridge Regression," Proceedings of the Ninth International Biometric Conference, 1(Sept.1976), 37-57.

Obenchain, R.L., "Classical F Tests and Confidence Regions for Ridge Regression," Technometrics, 19(1977), 429-440.

Obenchain, R.L., "Good and Optimal Ridge Estimators," The Annals of Statistics, 6(5) (1978), 1111-1121.

Obenchain, R.L., "Data Analytic Displays for Ridge Regression," unpublished memorandum, Bell Laboratories, Dec.1980.

Obenchain, R.L., "Maximum Likelihood Ridge Regression and the Pattern Shrinkage Pattern Hypothesis," unpublished memorandum, Bell Laboratories, Jan.1981.

Obenchain, R.L. and H.D. Vinod, "Estimates of Partial Derivatives from Ridge Regression on Ill-Conditioned Data," presented at NBER-NSF seminar on Bayesian Inference in Econometrics, Ann Arbor, Michigan (April 1974).

O'Hagan, J. and B. McCabe, "Tests for the Severity of Multicollinearity in Regression Analysis: A Comment," Review of Economics and Statistics, 57(Aug.1975), 368-370.

Oman, S.D., "A Confidence Bound Approach to Choosing the Biasing Parameter in Ridge Regression," Journal of the American Statistical Association, 76(June 1981), 452-461.

Parikh, A. and P.K. Trivedi, "Estimation of Returns to Inputs in Indian Agriculture," Australian National University unpublished discussion paper,1980.

Pagan, A., "Rational and Polynomial Lags: The Finite Connection," Journal of Econometrics, 8(1978), 247-254.

Park, R.E. and B.M. Mitchell, "Estimating the Autocorrelated Error Model with Trended Data," Journal of Econometrics, 13(1980), 185-201.

Phillips, P.C.B., "Edgeworth and Saddlepoint Approximations in a First Order Autoregression," Biometrika, 65(1978), 91-98.

Phillips, P.C.B., "Exact Distribution of Instrumental Variables Estimator in an Equation Containing n+1 Endogenous Variables," Econometrica, 48(May 1980), 861-878.

Pitman, E.J.G., "The Closest Estimates of Statistical Parameters," Proceedings of the Cambridge Philosophical Society,33 (Apr.1937), 212-222.

Raj, Baldev, "Monte Carlo Study of Simultaneous Equations Estimators Under Normal and Non-normal Errors," Journal of the American Statistical Association, 75(Mar.1980), 221-229.

Raj, B. and A. Ullah, Econometrics: A Varying Coefficients Approach (London, UK: Croom-Helm Publishers, 1981).

Ramsey, J.B., "Classical Model Selection Through Specification Error Tests," in Paul Zarembka (ed.), Frontiers in Econometrics, (New York: Academic Press, 1974), 13-47.

Rao, C.R., "Estimation of Heteroscedastic Variances in Linear Models," Journal of the American Statistical Association, 65(1970), 161-172.

Rao, C.R., "Linear Statistical Inference and Its Applications (New York: John Wiley and Sons, Inc., 1st ed.,1965, 2nd ed. 1973)

Rao, C.R., "Simultaneous Estimation of Parameters in Different Linear Models and Applications to Biometric Problems," Biometrika, 31(June 1975), 545-554.

Rao, C.R., "Estimation of Parameters in a Linear Model," Annals of Statistics, 4(6) (Dec.1976), 1023-1037.

Rao, J.N.K., "On the Estimation of Heteroscedastic Variances," Biometrics, 29(Mar., 1973), 11-24.

Rao, P.S.R.S., "Theory of the MINQUE--A Review" Sankhya, Ser. B, 39(1977), 201-210.

Revankar, N.S., "Some Finite Sample Results in the Context of Two Seemingly Unrelated Regression Equations," Journal of the American Statistical Association, 69(1974), 187-190.

Revankar, N.S., "Use of Restricted Residuals in SUR Systems: Some Finite Sample Results," Journal of the American Statistical Association, 71(1976), 183-188.

Richardson, D.H., "The Exact Distribution of a Structural Coefficient Estimator," Journal of the American Statistical Association, 63(1968), 1214-1226.

Riley, James, "Solving Systems of Linear Equations with a Positive Definite, Symmetric but Possibly Ill-conditioned Matrix," Mathematical Tables and Other Aids to Computation,9(1955), 96-101.

Rolph, J.E., "Choosing Shrinkage Estimates for Regression Problems," Communications in Statistics, Ser. A, 5(No. 9, 1976), 789-802.

Ronning, G., "Regression on Principal Components and Ridge Regression," Diskussionsbeiträge des Fachbereichs Statistik der Universität (Konstanz; Germany, Mar.1976).

Ruppert, David and Raymond J. Carroll, "Trimmed Least Squares Estimation in the Linear Model," Journal of the American Statistical Association, 75(Dec.1980), 828-838.

Sargan, J.D., "Gram Charlier Approximations Applied to t Ratios of k-class Estimators," Econometrica, 43 (Mar.1975), 327-346.

Sargan, J.D., "Econometric Estimators and the Edgeworth Approximation," Econometrica, 45(1976) 421-428.

Sastry, M.V.R., "Some Limits in the Theory of Multicollinearity," American Statistician, 24(Feb.1970), 39-40.

Sathe, S.T. and H.D. Vinod, "Bounds on the Variance of Regression Coefficients Due to Heteroscedastic or Autoregressive Errors," Econometrica, 42(Mar.1974), 333-340.

Savin, N.E. and K.J. White, "Estimation and Testing for Functional Form and Autocorrelation: A Simultaneous Approach," Journal of Econometrics, 8(Aug.1978), 1-12.

Sawa, T., "The Exact Finite Sampling Distribution of Ordinary Least Squares and Two Stage Least Squares Estimators," Journal of the American Statistical Association, 64(1969), 923-936.

Sawa, T., "Almost Unbiased Estimator in Simultaneous Equations Systems," International Economic Review, 14(1972), 97-106.

Sawa, T. and T. Hiromatsu, "Minimax Regret Significance Points for a Preliminary Test in Regression Analysis," Econometrica, 41(1973), 1093-1101.

Schmidt, P., Econometrics (New York: Marcel Dekker, Inc., 1976).

Schmidt, P., "A Note on the Estimation of Seemingly Unrelated Regression Systems," Journal of Econometrics, 7(1978), 259-261.

Schmidt, P., "Notes on Inequality Constraints," unpublished paper, (East Lansing, Mich.: Mich. State University, 1978b).

Schonfeld, P., "A Note on Least Squares Estimation and the BLUE in a Generalized Linear Regression Model," Journal of Econometrics, 3(May 1975), 189-197.

Sclove, S.L., "Improved Estimators for Coefficients in Linear Regression," Journal of the American Statistical Association, 63(1968), 596-606.

Sclove, S.L., C. Morris and R. Radhakrishnan, "Non-optimality of Preliminary Test Estimates for the Multinormal Mean," Annals of Mathematical Statistics, 43(Oct.1972), 1481-1490.

Shiller, R.J., "A distributed Lag Estimator Derived from Smoothness Priors," Econometrica, 41(1973), 429-449.

Shinozaki, N., "Estimation with Quadratic Loss Function," doctoral thesis, Keio University, Japan, 1974.

Shorrock, R.W. and J.V. Zidek, "An Improved Estimator of the Generalized Variance," Annals of Statistics, 4(1976), 629-638.

Sihota, S.S. and K.S. Banerjee, "Biased Estimation in Weighting Designs," Sankhya, 36(b) (Part 1, 1974), 55-66.

Silvey, S.D., "Multicollinearity and Imprecise Estimation," Journal of the Royal Statistical Society, Sec. B, 31 (Apr.1969), 539-552.

Sims, C.A., "A Note on Exact Tests for Serial Correlation," Journal of the American Statistical Association, 70(1975), 162-165.

Singh, B. and A. Ullah, "Estimation of Seemingly Unrelated Regressions with Random Coefficients," Journal of the American Statistical Association, 69(1974), 191-195.

Smidt, R.K. and L.L. McDonald, "Ridge Discriminant Analysis," National Technical Information Service no. AD-A028-728, (Springfield, Va.:U.S.Dept. of Commerce, 1976).

Smith, A.F.M., "A General Bayesian Linear Model," Journal of the Royal Statistical Society, Ser. B, 35(1973), 67-75.

Smith, D.E., "A Ridge Analysis Caveat," National Technical Information Service no. AD-A015-324, (Springfield, Va.: U.S.Dept. of Commerce, 1975).

Smith, G. and F. Campbell, "A Critique of Some Ridge Regression Methods," Journal of the American Statistical Association, with discussion by Thisted, Marquardt, Van Nostrand, Lindley, Obenchain, Peele and Ryan, Vinod and Gunst, 75(Mar.1980), 74-103.

Snee, R.D., "Some Aspects of Nonorthogonal Data Analysis, Part I. Developing Predicition Equations," Journal of Quality Technology, 5(No. 2, 1973), 67-79.

Srinivasan, T.N., "Approximations to Finite Sample Moments of Estimators Whose Exact Sampling Distributions Are Unknown," Econometrica, 38(1970), 533-541.

Srivastava, V.K., "The Efficiency of Estimating Seemingly Unrelated Regression Equations," Annals of the Institute of Statistical Mathematics, 22(1970), 483-493.

Srivastava, V.K., "The Efficiency of an Improved Method of Estimating SUR Equations," Journal of Econometrics, 1(1973), 341-350.

Srivastava, V.K. and T.D.Dwivedi, "Estimation of Seemingly Unrelated Equations, A Brief Survey," Journal of Econometrics, 10(Apr.1979), 15-32.

Srivastava, V.K., R. Tewaria and K.N. Pandey, "Properties of Approximate Minimum Risk k-class Estimators when Disturbances Are Small," Sankhya, Ser. C, 39(1978), 110-127.

Srivastava, V.K. and S. Upadhyaya, "Properties of Stein-like Estimators in Regression Model when Disturbances Are Small," Journal of Statistical Research (1977), 5-21.

Srivastava V.K. and A. Ullah, "On Lindley-like Mean Correction in the Improved Estimation of Linear Regression Models," Economics Letters (1981) to appear.

Stein, C., "A Necessary and Sufficient Condition for Admissibility," Annals of Mathematical Statistics, 36(1955), 518-522.

Stein, C., "Multiple Regression," Chapter 37 in I. Olkin (ed.), Contributions to Probability and Statistics: Essays in Honor of Harold Hotelling (Stanford, California: Stanford University Press, 1960), 424-443.

Stein, C., "Confidence Sets for the Mean of a Multivariate Normal Distribution," Journal of the Royal Statistical Society, Ser. B, 24(Dec.1962), 265-296.

Stein, C., "Inadmissibility of the Usual Estimator for the Mean of a Multivariate Normal Distribution," Proceedings of the Third Berkeley Symposium on Mathematical Statistics and Probability, vol.1 (Berkeley: University of California Press, 1956), 197-206.

Stein, C., "Confidence Sets for the Mean of a Multivariate Normal Distribution," Journal of Royal Statistical Society, B 24(1962), 265-96.

Stein, C., "An Approach to the Recovery of Inter-block Information on Balanced In-complete Block Designs," in F.N.David (ed.), Research Papers in Statistics (New York: John Wiley and Sons, Inc., 1966), 351-366.

Stein, C., "Estimation of the Mean of a Multivariate Normal Distribution," Proceedings of the Prague Symposium on Asymptotic Statistics, Sept.1973.

Stein, C.M., "Estimation of the Parameters of a Multivariate Normal Distribution, I: Estimation of Means," Technical Report No. 63 (Stanford, California: Dept. of Statistics, Stanford University, Nov.1974).

Stein, C., B. Efron and C. Morris, "Improving the Usual Estimator of a normal Covariance Matrix," Technical Report no. 37, Dept. of Statistics, Stanford University, 1972.

Stewart, G.W., "On the Continuity of the Generalized Inverse," SIAM Journal of Applied Mathematics, 17(Jan.1969), 33-45.

Stewart, G.W., Introduction to Matrix Computations (New York: Academic Press, 1973).

Strawderman, W.E., "Proper Bayes Minimax Estimators of the Multivariate Normal Mean," Annals of Mathematical Statistics, 42(1971), 385-388.

Strawderman, W.E., "On the Existence of Proper Bayes Minimax Estimators of the Mean of a Multivariate Normal Distribution," Proceedings of the Sixth Berkeley Symposium on Mathematical Statistics and Probability, 1(1972), 51-55.

Strawderman, W.E., "Minimax Estimators of Location Parameters for Certain Spherically Symetric Distributions," Journal of Multivariate Analysis, 4(1974), 255-264.

Strawderman, W.E., "Minimax Adaptive Generalized Ridge Regression Estimators," Journal of the American Statistical Association 73(1978), 623-627.

Swamy, P.A.V.B., "Efficient Inference in a Random Coefficient Regression Model," Econometrica, 38(1970), 311-323.

Swamy, P.A.V.B., Statistical Inference in Random Coefficient Regression Models (New York: Springer-Verlag, 1971).

Swamy, P.A.V.B., "Linear Models With Random Coefficients," in Paul Zarembka (ed.), Frontiers in Econometrics, (New York: Academic Press, 1973).

Swamy, P.A.V.B., "Criteria, Constraints, and Multicollinearity in Random Coefficient Regression Models," Annals of Economic and Social Measurement, 2(Oct.1973).

Swamy, P.A.V.B., "A Comparison of Estimators for Undersized Samples," Journal of Econometrics, 12(Oct.1980), 161-181.

Swamy, P.A.V.B. and J. Holmes, "The Use of Undersized Samples in the Estimation of Simultaneous Equation Systems," Econometrica, 39(1971), 455-459.

Swamy, P.A.V.B. and J.S. Mehta, "On Theil's Mixed Regression Estimator," Journal of the American Statistical Association, 64(1969), 273-276.

Swamy, P.A.V.B. and J.S. Mehta, "Minimum Average Risk Estimators for Coefficients in Linear Models," Communications in Statistics, Ser. A, 5(9) (1976), 803-818.

Swamy, P.A.V.B. and J.S. Mehta, "A Note on Minimum Average Risk Estimators for Coefficients in Linear Models," Communications in Statistics, A6(1977).

Swamy, P.A.V.B. and J.S. Mehta, "Ridge Regression Estimation of the Rotterdam Model," Special Studies Paper, 136(Washington, D.C: Federal Reserve Board, Dec.1979a)

Swamy, P.A.V.B. and J.S. Mehta, "Estimation of Common Coefficients in Two Regression Equations," Journal of Econometrics, 10(1979b), 1-14.

Swamy, P.A.V.B. and P.N. Rappoport, "Relative Efficiences of Some Simple Bayes Estimators of Coefficients in Dynamic Models — I," Journal of Econometrics, 3(Sept.1975), 273-296.

Swamy, P.A.V.B. and P.A. Tinsley, "Linear Prediction and Estimation Methods for Regression Models With Stationary Stochastic Coefficients," Special Studies Paper no.78 (Washington, D.C.: Federal Reserve Board).

Swindel, B.F., "Instability of Regression Coefficients Illustrated," The American Statistician, 28(May 1974), 63-65.

Swindel, B.F. and D.D. Chapman, "Good Ridge Estimators," Abstracts Booklet, 1973 Joint Statistical Meetings in New York City, Dec. 1973, page 126.

Szroeter, J., "A Class of Parametric Tests for Heteroscedasticity in Linear Econometric Models," Econometrica, 46(1978), 1311-1328.

Taylor, Lester D., "Estimation by Minimizing the Sum of Absolute Errors," in P. Zarembka (ed.), Frontiers in Econometrics (New York: Academic Press, 1974).

Taylor, W.E., "Small Sample Properties of a Class of Two-Stage Aitken Estimators," Econometrica, 45(1977), 497-508.

Taylor, W.E., "The Heteroscedastic Linear Model: Exact Finite Sample Results," Econometrica, 46(1978), 663-675.

Teräsvirta, T., "A Note on Bias in the Almon Distributed Lag Estimator," Econometrica, 44(1976), 1317-1322.

Theil, H., Econometric Forecasts and Policy (Amsterdam: North-Holland Publishing Co., 1961).

Theil, H., "On the Use of Incomplete Prior Information in Regression Analysis," Journal of the American Statistical Association, 58(June 1963), 401-414.

Theil, H., Principles of Econometrics (New York: John Wiley and Sons, Inc., 1971).

Theil, H. and A.S. Goldberger, "On Pure and Mixed Statistical Estimation in Economics," International Economic Review, 2(Jan.1961), 65-78.

Theil, H. and K. Laitinen, "Singular Moment Matrices in Applied Econometrics," in P.R.Krishnaiah (ed.), Multivariate Analysis, V. (New York: North Holland Publishing Co.,1980), 629-649.

Theil, H. and A.L. Nagar, "Testing the Independence of Regression Disturbances," Journal of the American Statistical Association, 56(1961), 793-806.

Theobald, C.M., "Generalizations of Mean Square Error Applied to Ridge Regression," Journal of the Royal Statistical Society, Series B, 36(Aug.1974), 103-106.

Timm, N.H., Multivariate Analysis with Applications in Education and Psychology (Belmont: Brooks/Cole , 1975).

Timm, N.H. and J.E.Carlson, "Part and Bipartial Canonical Correlation Analysis," Psychometrika, 41(June 1976), 159-176.

Thisted, R.A., "Ridge Regression, Minimax Estimation, and Empirical Bayes Methods," Tech. Report No.28(Stanford, California: Division of Biostatistics, Stanford University, Dec.1976).

Thompson, J.R., "Some Shrinkage Techniques for Estimating the Mean," Journal of the American Statistical Association, 63(Mar.1968), 113-122.

Thomson, M., "Inequality Constrained Least Squares and Other Related Estimators: A Comparison of the MSE in a Linear Model with Two Regressors," Econometrics Workshop, paper No. 7907(East Lansing, Mich.: Mich. State University, 1979).

Thomson, M. and P. Schmidt, "A Comparison of the Statistical Properties of Inequality Constrained Least Squares and Other Related Estimators," Econometrics Workshop Paper No. 7813 (East Lansing, Mich. : Mich. State University, 1979).

Toro-Vizcarrondo, C. and T.D. Wallace, "A Test of the Mean Square Error Criterion for Restrictions in Linear Regression," Journal of the American Statistical Association, 63(June 1968), 558-572.

Trivedi, P.K., "Estimation of a Distributed Lag Model Under Quadratic Loss," Econometrica, 46(1978), 1181-1192.

Trivedi, P.K., "Small Sample and Collateral Information: An Application of the Hyperparameter Model," Journal of Econometrics, 12(1980), 301-318.

Trivedi, P.K. and A.R. Pagan, "Polynomial Distributed Lags, A Unified Treatment," Economic Studies Quarterly, 1979

Tukey, J.W., "Instead of Gauss Markov Least Squares, What?," R.P.Gupta (ed.), Applied Statistics in Proceedings of the Conference at Dalhousie University, Halifax, May 1974, (New York: Elsevier Publishing Co., 1975), 351-372.

Ullah, A., "On the Sampling Distribution of Improved Estimators for Coefficients in Linear Regression," Journal of Econometrics, 2(1974), 143-150.

Ullah, A., "Lindley and Smith Type Improved Estimators of Regression Coefficients," Center for Operations Research and Econometrics, Research Report (Louvain-la-Neuve, Belgium: C.O.R.E., 1980a)

Ullah, A., "MSE Properties of the Bayes Estimator for the Linear Model under Exchangeable Prior: A Note," Center for Operations Research and Econometrics, Research Report (Louvain-la-Neuve, Belgium: C.O.R.E., 1980b)

Ullah, A., "The Exact Large -Sample and Small-Disturbance Conditions of Dominance of Biased Estimators in Linear Models," Economics Letters, 6(1980), 339-344.

Ullah, A. and B. Singh, "Note on Zellner's SURE Estimator when the Disturbances are Heteroscedastic," Research Report, Southern Methodist University, 1973.

Ullah, A., V.K. Srivastava and R. Chandar, "Properties of Improved Estimators in Linear Regression when Disturbances are Non-normal," Center for Operations Research and Econometrics, Research Report (Louvain-la-Neuve, Belgium: C.O.R.E., 1980)

Ullah, A. and V.K. Srivastava, "Improved Estimators of Structural Coefficients in the Simultaneous Equations Model," unpublished working paper, 1980.

Ullah A. and A.L. Nagar, "The Exact Mean of the Two-stage Least Squares Estimator of the Structural Parameter in a Equation Having Three Endogenous Variables," Econometrica, 42(July 1974), 749-758.

Ullah, A. and M. Rafiquzzaman "On the Sampling Distribution of Zellner's SUR Estimator," Indian Economic Review, 1977.

Ullah, A. and B. Raj, "A Distributed Lag Estimatoor Derived From Shiller's Smoothness Priors," Economics Letters, 2(1979), 219-223.

Ullah, A. and S. Ullah, "Double k-Class Estimators of Coefficients in Linear Regression," Econometrica, 46(May 1978) and Erratum (Mar.1981).

Ullah, A., H.D. Vinod and K. Kadiyala, "A Family of Improved Shrinkage Factors for the Ordinary Ridge Estimator," European Econometric Society Meetings, Athens, Greece, 1979, in E.G. Charatsis (ed.) Selected Papers on Contemporary Econometric Problems (Amsterdam: North Holland Publishing Co., 1981).

Ullah, A. and H.D. Vinod, "MSE Properties of Ridge Estimators," unpublished, 1979.

Vinod, H.D., "Econometrics of Joint Production," Econometrica, 36(1968), 322-336.

Vinod, H.D., "Econometrics of Joint Production — A Reply," Econometrica, 37(1969), 739-740.

Vinod, H.D., "Non-homogeneous Production Functions and Applications to Telecommunications," Bell Journal of Economics and Management Science, 3(Autumn 1972), 531-543.

Vinod, H.D., "Generalization of the Durbin-Watson Statistic for Higher Order Autoregressive Process," Communications in Statistics, 2(1973), 115-144.

Vinod, H.D., "Ridge Estimation of a Trans-log Production Function," 1974 Business and Economic Statistics Section Proceedings, St. louis, Missouri, Aug.1974 (Washington, D.C.: American Statistical Association), 596-601.

Vinod, H.D., "Canonical Ridge and Econometrics of Joint Production," Journal of Econometrics, 4(June 1976e), 147-166.

Vinod, H.D., "Bell System Scale Economies and Estimation of Joint Production Functions," Federal Communications Commission, (FCC) Docket 20003, Bell Exhibit 59, a part of Fifth Supplemental Response by A.T. and T. Company, New York, Aug.1976f.

Vinod, H.D., "Effects of ARMA Errors on the Significance Tests for Regression Coefficients," Journal of the American Statistical Association, 71(Dec.1976i), 929-933.

Vinod, H.D., "Application of New Ridge Regression Methods to a Study of Bell System Scale Economies," Journal of the American Statistical Association, 71(Dec.1976j), 835-841.

Vinod, H.D., "Bounds on the Bias in Ridge Regression Coefficients," 1976 Business and Economic Statistics Section Proceedings, Boston,Mass.,Sept.1976p (Washington, D.C.:American Statistical Association), 628-634.

Vinod, H.D., "Simulation and Extension of a Minimum MSE Estimator in Comparison with Stein's," Technometrics, 18(Nov.1976t), 491-496.

Vinod, H.D., Letter to the Editor Technometrics, 18(Nov.1976u), 504.

Vinod, H.D., "Estimating the Largest Acceptable k and a Confidence Interval for Ridge Regression Parameters," presented at the Econometric Society European Meeting, Vienna, Sept.1977.

Vinod H.D., "Equivariance of Ridge Estimators Through Standardization, A Note," Communications in Statistics, A7(12) (1978c), 1159-1167.

Vinod, H.D.," A Ridge Estimator Whose MSE Dominates OLS, "International Economic Review, 19(Oct.1978i), 727-737.

Vinod, H.D., "A Survey of Ridge Regression and Related Techniques for Improvements over Ordinary Least Squares," Review of Economics and Statistics, 60(Feb.1978r), 121-131.

Vinod, H.D., Letter to the Editor, Technometrics, 21(Feb.1979), 138.

Vinod, H.D., "Improved Stein-Rule Estimator for Regression Problems," Journal of Econometrics, 12(1980), 143-150.

Vinod, H.D., "New Confidence Intervals for Ridge Regression Parameters," Bell Laboratories Economics Discussion Paper No. 172 (Murray Hill,N.J.: Bell Laboratories, 1980b).

Vinod, H.D., "Enduring Regression Estimator," Bell Laboratories Economics Discussion Paper No.188 (Murray Hill, N.J.: Bell Laboratories, 1981).

Vinod, H.D., "Maximum Entropy Measurement Error Estimates of Singular Covariance Matrices," Bell Laboratories Economics Discussion Paper No. 220 (Murray Hill, N.J.: Bell Labs, 1981b).

Vinod, H.D. and B. Raj, "Bell System Scale Economies from Randomly Varying Parameter Model," 1978 Business and Economic Statistics Section Proceedings (Washington D.C. : American Statistical Association), 596-599.

Vinod H.D., A. Ullah and K. Kadiyala, "Evaluation of the Mean Squared Error of Certain Generalized Ridge Estimators Using Confluent Hypergeometric Functions," Bell Laboratories

Economic Discussion Paper 137 (Murarray Hill, N.J.: Bell Labs, Jan.1979), also in Sankhya, Ser. C, 42(1981).

Visco, I., "On Obtaining the Ridge Sign of a Coefficient Estimate by Omitting a Variable from the Regression," Journal of Econometrics, 6(1978), 115-118.

Von Neumann, J., "Distribution of the Ratio of the Mean-Square Successive Difference to the Variance," Annals of Mathematical Statistics, 12(1941), 367-395.

Wahba, G., "A Survey of Some Smoothing Problems and the Method of Generalized Cross-Validation for Solving Them," in Paruchuri R. Krishnaiah (ed.) Applications of Statistics (New York: North Holland Publishing Co.,1977), 507-523.

Wallace, T.D., "Weaker Criteria and Tests for Linear Restrictions in Regression," Econometrica, 40(July 1972), 689-698.

Wallace, T.D. and A.Hussain, "The Use of Error Components Models in Combining Cross Section With Time Series Data," Econometrica, 37(Jan.1969), 55-72.

Wallace, T.D. and C.E. Toro-Vizcarrondo, "Tables for the Mean Square Error Test for Exact Linear Restrictions in Regression," Journal of the American Statistical Association, 64(1969), 1649-1663.

Wallis, Kenneth F., "Testing for Fourth Order Autocorrelation in Quarterly Regression Equations," Econometrica, 40(July 1972), 617-636.

Wampler, R.H., "A Report on the Accuracy of Some Widely Used Least Squares Computer Programs," Journal of the American Statistical Association, 65(June 1970), 549-565.

Watson, G.S., "Serial Correlation in Regression Analysis I," Biometrika, 42(Dec.1955), 327-341.

Watson, D.E. and K.J. White, "Forecasting the Demand for Money Under Changing Term Structure of Interest Rates: An Application of Ridge Regression," Southern Economic Journal, 43(Oct.1976), 1096-1105.

Watson, G.S. and E.J. Hannan, "Serial Correlation in Regression Analysis II," Biometrika, 43(Dec.1956), 436-448.

Waugh, F.W., "Regression Between Sets of Variates," Econometrica, 10(1942), 290-310.

Webster, J.T., R.F. Gunst and R.L. Mason, "Latent Root Regression Analysis," Technometrics, 16(Nov.1974), 513-522

White, H., "A Heteroscedasticity-Consistent Covariance Matrix Estimator and a Direct Test for Heteroscedasticity," Econometrica, 48(1980), 817-838.

Wichern, D.W. and G.A. Churchill, "A Comparison of Ridge Estimators," Technometrics, 20(Aug.1978), 301-312.

Wichers, C.R., "The Detection of Multicollinearity: A Comment," Review of Economics and Statistics, 57(Aug.1975), 366-368.

Wilkinson, J.H., "The Algebraic Eigenvalue Problem," (New York: Oxford University Press, Clarendon, 1965).

Wilkinson, R.K. and C.A. Archer, "Measuring the Determinants of Relative House Prices," Environment and Planning, 5(1973), 357-367.

Wonnacott, R.J. and T.H. Wonnacott, Econometrics, (New York: John Wiley and Sons, Inc., 2nd ed.,1978, 1st ed.,1970).

Yancey, T.A. and G.G. Judge, "A Monte Carlo Comparison of Traditional and Stein-Rule Estimates Under Squared Error Loss," Journal of Econometrics 4 (Aug.1976), 285-294.

Zellner, A., "An Efficient Method of Estimating Seemingly Unrelated Regressions and Tests for Aggregation Bias," Journal of the American Statistical Association, 57(June 1962), 348-368.

Zellner, A., An Introduction to Bayesian Inference in Econometrics (New York: John Wiley and Sons, Inc., 1971).

Zellner, A., "Estimation of Functions of Population Means and Regression Coefficients Including Structural Coefficients: As Minimum Expected Loss (MELO) Approach," Journal of Econometrics, 8(1978), 127-158.

Zellner, A. and Soo-Bin Park, "Minimum Expected Loss (MELO) Estimators for Functions of Parameters and Structural Coefficients of Econometric Models," Journal of the American Statistical Association, 74(1979), 185-193.

Zellner, A. and W. Vandaele, "Bayes-Stein Estimators for k Means, Regression and Simultaneous Equation Models," in S.E. Fienberg and A. Zellner (eds.), Studies in Bayesian Econometrics and Statistics (Amsterdam: North Holland Publishing Co., 1974).

INDEX

Acceptable Range for k, 176-178
Asymptotic expansion, 215
Asymptotic properties, 36
  unbiasedness, 37
  variance, 37, 245
  efficiency, 38-39
Autocorrelation:
  coefficients, 91,99
  testing for, 107
Autoregressive moving average
    errors, 90
  testing for, 109-111

Bayesian estimator, 58
Bayesian regression:
  under conjugate priors, 77-80
  under diffuse priors, 80-81
  estimator, 81, 274
  Lindley's model, 84, 232, 255
Bayes theorem, 55
Bisquare, 319, 324
BLUS residuals, 92
Bock, 148-149

Canonical correlation, 293-300
Canonical reduction, 5
Condition number, 41, 121
Confidence intervals, 193-194,
    220-221
Consistency, 37
Correlation matrix, 10, 19
  robust, 326
Cochrane-Orcutt estimator, 106
C_p, 197-198

Data standardization, 179

Declining deltas, 192
Diagonally weighted least squares,
    323
Direction cosines, 19
Discriminant analysis, 300-303
Distance:
  absolute, 152
  Euclidean, 32, 214
  Mahalanobis, 43, 214, 302-303
  Prokhorov, 44
  Strawderman, 214
  Cook, 44
Distributed lag, 226-240
Double f-class, 208
Double h-class, 213-214, 222-225
Durbin-Watson test, 90
  without tables, 107

Empirical Bayes, 197
Error components model, 250, 254
Efficiency:
  definition, 30
  properties, 38-39
Eigenvectors, 18
Estimability, 123
Estimator:
  better, 46
  admissible, 46
  minimax, 48
Exchangeability, 187-188
  between equations, 255
Existence theorem, 175

Farebrother generalization, 218
Fixed effect estimator, 250, 254
Full information estimators
    (FIML), 285-289

Gauss-Markov theorem, 31
Generalized least squares, 93
under autocorrelation, 105
under heteroscedasticity, 112, 114
under SUR, 244, 252
under FIML, 287
GIVE, 281
GRE, 172, 276-277

Hadamard product, 113
Hat matrix, 42, 114
Hemmerle's iterative ridge, 208
Heteroscedasticity, 111
estimation, 113-115
testing for, 116-117
HKB, 195, 213, 317
Horizontal m scale, 181-184, 190
Hyperparameters, 84, 255

Identificability restrictions, 269
Improved estimators, 305-308
Indirect least squares, 272
Inequality constraints, 186
Inequality restricted estimator, 87
Influence function, 43-45, 326
Information matrix, 39, 320, 322
Instrumental variable, 280
ISRM, 182, 195, 200
Iteratively reweighted least squares, 314

Jointly dependent, 263

k-class, 275
Kurtosis, 306, 308, 312

L-Estimates, 319
Likelihood ratio test, 63-64, 67-68, 94-96
Likelihood surface, 14, 17
Lindley's estimator, 138, 145, 316
LW, 195

MADM, 319
Maximum entropy, 285
Maximum likelihood estimation, 17
Maximum likelihood estimator, 4
Maximum likelihood under autocorrelation, 106
Mean squared error (MSE):
definition, 30
predictive, 33
weighted, 33
matrix, 33
MELO, 274, 276
M-Estimates, 318
Minimaxity, 313-317, 322
Minimum norm estimator, 274
MINQUE estimator, 114
Mixed estimator, 230
Multicollinearity, 120-122, 150
effects of, 12, 124-129
remedies, 130, 192-194
Multi-regression, 241
MWMSE, 146

Noncentrality parameter, 215
Nonnormal errors, 305, 308-331
Normalizing condition, 263

Outlier detection, 45
ORE, 169, 277

Parameterization, 5-6
PCSA, 192
Perturbation, 127-129
Pooling (see Temporal cross section model)
Predetermined, 263
Preliminary test, 76-77
Principal components, 16, 283, 292-293
Priors:
conjugate and diffuse (*see* Bayesian regression)
smoothness, 227
stochastic smoothness, 229
Probability limit, 37, 281, 288
Projection matrix, 23

R^2, 4, 7-8
Random coefficients, 251-254

Rank deficiency, 181
Recentering, 25
R-Estimates, 320
Regression quantiles, 325
Restricted least squares, 61-62, 186, 219, 229
Ridge estimators, 172, 173, 210, 213, 237-238, 307, 315
  properties, 175-176, 211-225
Ridge regression, 15, 169
Ridge trace, 178-183, 195, 201-203
Risk function, 32
Robustness, 42
Robust estimators, 317-325
Rotation of axes, 23

Seemingly unrelated regressions, 242
  Stein-rule estimator, 246
  ridge estimators, 242-243
  generalized least squares, 244
Shiller-type estimator, 230, 235
Shrinkage factors, 49-55, 172, 207
  acceptable range, 49-54, 176-178
  operational form, 208-209
  selection of, 54-55
Simultaneous equations, 262
  identification, 267, 269-271
  estimation, 271
  ridge and Stein-type estimators, 276-280, 288-289
Singular value decomposition, 5, 21, 41, 172
Skewness, 306-307
Small sigma asymptotics, 149, 158, 162-167, 215-216, 309-310, 326-330
Squared loss function, 32
SSCBC, 193
Stability properties, 40-41, 127-129
Stable region, 181
Standardization, 8-10, 25
Stein-rule estimators, 136, 144, 146-148, 153-156, 307, 314
  sampling distribution, 151
Stochastic restriction, 70
Studentized residuals, 42
Sufficient statistic, 40
Sum of Squares (SSR, SSE), 3
SUR, 242

Temporal cross section model, 248
Three stage least squares (3SLS), 286
Trimmed mean, 317
Two stage least squares (2SLS), 273

Unbiasedness, 30
Uncorrelation components, 171
Undersized sample, 283-285

Variance inflation factors, 124-125, 193

Winsorized mean, 318
WMSE, 33

Zero restrictions, 267